Gauge Theory

Contents

CONTENTS

xi

Chapter 1

Gauge theory

For a more accessible and less technical introduction to this topic, see Introduction to gauge theory.

In physics, a **gauge theory** is a type of field theory in which the Lagrangian is invariant under a continuous group of local transformations.

The term *gauge* refers to redundant degrees of freedom in the Lagrangian. The transformations between possible gauges, called *gauge transformations*, form a Lie group—referred to as the *symmetry group* or the *gauge group* of the theory. Associated with any Lie group is the Lie algebra of group generators. For each group generator there necessarily arises a corresponding vector field called the *gauge field*. Gauge fields are included in the Lagrangian to ensure its invariance under the local group transformations (called *gauge invariance*). When such a theory is quantized, the quanta of the gauge fields are called *gauge bosons*. If the symmetry group is non-commutative, the gauge theory is referred to as *non-abelian*, the usual example being the Yang–Mills theory.

Many powerful theories in physics are described by Lagrangians that are invariant under some symmetry transformation groups. When they are invariant under a transformation identically performed at *every* point in the space in which the physical processes occur, they are said to have a global symmetry. The requirement of local symmetry, the cornerstone of gauge theories, is a stricter constraint. In fact, a global symmetry is just a local symmetry whose group's parameters are fixed in space-time.

Gauge theories are important as the successful field theories explaining the dynamics of elementary particles. Quantum electrodynamics is an abelian gauge theory with the symmetry group $U(1)$ and has one gauge field, the electromagnetic four-potential, with the photon being the gauge boson. The Standard Model is a non-abelian gauge theory with the symmetry group $U(1) \times SU(2) \times SU(3)$ and has a total of twelve gauge bosons: the photon, three weak bosons and eight gluons.

Gauge theories are also important in explaining gravitation in the theory of general relativity. Its case is somewhat unique in that the gauge field is a tensor, the Lanczos tensor. Theories of quantum gravity, beginning with gauge gravitation theory, also postulate the existence of a gauge boson known as the graviton. Gauge symmetries can be viewed as analogues of the principle of general covariance of general relativity in which the coordinate system can be chosen freely under arbitrary diffeomorphisms of spacetime. Both gauge invariance and diffeomorphism invariance reflect a redundancy in the description of the system. An alternative theory of gravitation, gauge theory gravity, replaces the principle of general covariance with a true gauge principle with new gauge fields.

Historically, these ideas were first stated in the context of classical electromagnetism and later in general relativity. However, the modern importance of gauge symmetries appeared first in the relativistic quantum mechanics of electrons – quantum electrodynamics, elaborated on below. Today, gauge theories are useful in condensed matter, nuclear and high energy physics among other subfields.

1.1   History and importance

The earliest field theory having a gauge symmetry was Maxwell's formulation of electrodynamics in 1864. The importance of this symmetry remained unnoticed in the earliest formulations. Similarly unnoticed, Hilbert had derived the Einstein field equations by postulating the invariance of the action under a general coordinate transformation. Later Hermann Weyl, in an attempt to unify general relativity and electromagnetism, conjectured that *Eichinvarianz* or invariance under the change of scale (or "gauge") might also be a local symmetry of general relativity. After the development of quantum mechanics, Weyl, Vladimir Fock and Fritz London modified gauge by replacing the scale factor with a complex quantity and turned the scale transformation into a change of phase, which is a U(1) gauge symmetry. This explained the electromagnetic field effect on the wave function of a charged quantum mechanical particle. This was the first widely recognised gauge theory, popularised by Pauli in the 1940s.[1]

In 1954, attempting to resolve some of the great confusion in elementary particle physics, Chen Ning Yang and Robert Mills introduced **non-abelian gauge theories** as models to understand the strong interaction holding together nucleons in atomic nuclei. (Ronald Shaw, working under Abdus Salam, independently introduced the same notion in his doctoral thesis.) Generalizing the gauge invariance of electromagnetism, they attempted to construct a theory based on the action of the (non-abelian) SU(2) symmetry group on the isospin doublet of protons and neutrons. This is similar to the action of the U(1) group on the spinor fields of quantum electrodynamics. In particle physics the emphasis was on using **quantized gauge theories**.

This idea later found application in the quantum field theory of the weak force, and its unification with electromagnetism in the electroweak theory. Gauge theories became even more attractive when it was realized that non-abelian gauge theories reproduced a feature called asymptotic freedom. Asymptotic freedom was believed to be an important characteristic of strong interactions. This motivated searching for a strong force gauge theory. This theory, now known as quantum chromodynamics, is a gauge theory with the action of the SU(3) group on the color triplet of quarks. The Standard Model unifies the description of electromagnetism, weak interactions and strong interactions in the language of gauge theory.

In the 1970s, Sir Michael Atiyah began studying the mathematics of solutions to the classical Yang–Mills equations. In 1983, Atiyah's student Simon Donaldson built on this work to show that the differentiable classification of smooth 4-manifolds is very different from their classification up to homeomorphism. Michael Freedman used Donaldson's work to exhibit exotic $\mathbf{R}^4$s, that is, exotic differentiable structures on Euclidean 4-dimensional space. This led to an increasing interest in gauge theory for its own sake, independent of its successes in fundamental physics. In 1994, Edward Witten and Nathan Seiberg invented gauge-theoretic techniques based on supersymmetry that enabled the calculation of certain topological invariants (the Seiberg–Witten invariants). These contributions to mathematics from gauge theory have led to a renewed interest in this area.

The importance of gauge theories in physics is exemplified in the tremendous success of the mathematical formalism in providing a unified framework to describe the quantum field theories of electromagnetism, the weak force and the strong force. This theory, known as the Standard Model, accurately describes experimental predictions regarding three of the four fundamental forces of nature, and is a gauge theory with the gauge group SU(3) × SU(2) × U(1). Modern theories like string theory, as well as general relativity, are, in one way or another, gauge theories.

See Pickering[2] for more about the history of gauge and quantum field theories.

1.2   Description

1.2.1   Global and local symmetries

In physics, the mathematical description of any physical situation usually contains excess degrees of freedom; the same physical situation is equally well described by many equivalent mathematical configurations. For instance, in Newtonian dynamics, if two configurations are related by a Galilean transformation (an inertial change of reference frame) they represent the same physical situation. These transformations form a group of "symmetries" of the theory, and a physical situation corresponds not to an individual mathematical configuration but to a class of configurations related to one another by this symmetry group.

This idea can be generalized to include local as well as global symmetries, analogous to much more abstract "changes of coordinates" in a situation where there is no preferred "inertial" coordinate system that covers the entire physical system. A gauge theory is a mathematical model that has symmetries of this kind, together with a set of techniques for making physical predictions consistent with the symmetries of the model.

1.2.2 Example of global symmetry

When a quantity occurring in the mathematical configuration is not just a number but has some geometrical significance, such as a velocity or an axis of rotation, its representation as numbers arranged in a vector or matrix is also changed by a coordinate transformation. For instance, if one description of a pattern of fluid flow states that the fluid velocity in the neighborhood of (x=1, y=0) is 1 m/s in the positive x direction, then a description of the same situation in which the coordinate system has been rotated clockwise by 90 degrees states that the fluid velocity in the neighborhood of (x=0, y=1) is 1 m/s in the positive y direction. The coordinate transformation has affected both the coordinate system used to identify the *location* of the measurement and the basis in which its *value* is expressed. As long as this transformation is performed globally (affecting the coordinate basis in the same way at every point), the effect on values that represent the *rate of change* of some quantity along some path in space and time as it passes through point P is the same as the effect on values that are truly local to P.

1.2.3 Use of fiber bundles to describe local symmetries

In order to adequately describe physical situations in more complex theories, it is often necessary to introduce a "coordinate basis" for some of the objects of the theory that do not have this simple relationship to the coordinates used to label points in space and time. (In mathematical terms, the theory involves a fiber bundle in which the fiber at each point of the base space consists of possible coordinate bases for use when describing the values of objects at that point.) In order to spell out a mathematical configuration, one must choose a particular coordinate basis at each point (a *local section* of the fiber bundle) and express the values of the objects of the theory (usually "fields" in the physicist's sense) using this basis. Two such mathematical configurations are equivalent (describe the same physical situation) if they are related by a transformation of this abstract coordinate basis (a change of local section, or *gauge transformation*).

In most gauge theories, the set of possible transformations of the abstract gauge basis at an individual point in space and time is a finite-dimensional Lie group. The simplest such group is U(1), which appears in the modern formulation of quantum electrodynamics (QED) via its use of complex numbers. QED is generally regarded as the first, and simplest, physical gauge theory. The set of possible gauge transformations of the entire configuration of a given gauge theory also forms a group, the *gauge group* of the theory. An element of the gauge group can be parameterized by a smoothly varying function from the points of spacetime to the (finite-dimensional) Lie group, such that the value of the function and its derivatives at each point represents the action of the gauge transformation on the fiber over that point.

A gauge transformation with constant parameter at every point in space and time is analogous to a rigid rotation of the geometric coordinate system; it represents a global symmetry of the gauge representation. As in the case of a rigid rotation, this gauge transformation affects expressions that represent the rate of change along a path of some gauge-dependent quantity in the same way as those that represent a truly local quantity. A gauge transformation whose parameter is *not* a constant function is referred to as a local symmetry; its effect on expressions that involve a derivative is qualitatively different from that on expressions that don't. (This is analogous to a non-inertial change of reference frame, which can produce a Coriolis effect.)

1.2.4 Gauge fields

The "gauge covariant" version of a gauge theory accounts for this effect by introducing a gauge field (in mathematical language, an Ehresmann connection) and formulating all rates of change in terms of the covariant derivative with respect to this connection. The gauge field becomes an essential part of the description of a mathematical configuration. A configuration in which the gauge field can be eliminated by a gauge transformation has the property that its field strength (in mathematical language, its curvature) is zero everywhere; a gauge theory is *not* limited to these configurations. In other words, the distinguishing characteristic of a gauge theory is that the gauge field does not merely compensate for a poor choice of coordinate system; there is generally no gauge transformation that makes the gauge field vanish.

When analyzing the dynamics of a gauge theory, the gauge field must be treated as a dynamical variable, similarly to other objects in the description of a physical situation. In addition to its interaction with other objects via the covariant derivative, the gauge field typically contributes energy in the form of a "self-energy" term. One can obtain the equations for the gauge theory by:

- starting from a naïve ansatz without the gauge field (in which the derivatives appear in a "bare" form);

- listing those global symmetries of the theory that can be characterized by a continuous parameter (generally an abstract equivalent of a rotation angle);

- computing the correction terms that result from allowing the symmetry parameter to vary from place to place; and

- reinterpreting these correction terms as couplings to one or more gauge fields, and giving these fields appropriate self-energy terms and dynamical behavior.

This is the sense in which a gauge theory "extends" a global symmetry to a local symmetry, and closely resembles the historical development of the gauge theory of gravity known as general relativity.

1.2.5   Physical experiments

Gauge theories are used to model the results of physical experiments, essentially by:

- limiting the universe of possible configurations to those consistent with the information used to set up the experiment, and then

- computing the probability distribution of the possible outcomes that the experiment is designed to measure.

The mathematical descriptions of the "setup information" and the "possible measurement outcomes" (loosely speaking, the "boundary conditions" of the experiment) are generally not expressible without reference to a particular coordinate system, including a choice of gauge. (If nothing else, one assumes that the experiment has been adequately isolated from "external" influence, which is itself a gauge-dependent statement.) Mishandling gauge dependence in boundary conditions is a frequent source of anomalies in gauge theory calculations, and gauge theories can be broadly classified by their approaches to anomaly avoidance.

1.2.6   Continuum theories

The two gauge theories mentioned above (continuum electrodynamics and general relativity) are examples of continuum field theories. The techniques of calculation in a continuum theory implicitly assume that:

- given a completely fixed choice of gauge, the boundary conditions of an individual configuration can in principle be completely described;

- given a completely fixed gauge and a complete set of boundary conditions, the principle of least action determines a unique mathematical configuration (and therefore a unique physical situation) consistent with these bounds;

- the likelihood of possible measurement outcomes can be determined by:

  - establishing a probability distribution over all physical situations determined by boundary conditions that are consistent with the setup information,

  - establishing a probability distribution of measurement outcomes for each possible physical situation, and

- convolving these two probability distributions to get a distribution of possible measurement outcomes consistent with the setup information; and

- fixing the gauge introduces no anomalies in the calculation, due either to gauge dependence in describing partial information about boundary conditions or to incompleteness of the theory.

These assumptions are close enough to be valid across a wide range of energy scales and experimental conditions, to allow these theories to make accurate predictions about almost all of the phenomena encountered in daily life, from light, heat, and electricity to eclipses and spaceflight. They fail only at the smallest and largest scales (due to omissions in the theories themselves) and when the mathematical techniques themselves break down (most notably in the case of turbulence and other chaotic phenomena).

1.2.7 Quantum field theories

Other than these classical continuum field theories, the most widely known gauge theories are quantum field theories, including quantum electrodynamics and the Standard Model of elementary particle physics. The starting point of a quantum field theory is much like that of its continuum analog: a gauge-covariant action integral that characterizes "allowable" physical situations according to the principle of least action. However, continuum and quantum theories differ significantly in how they handle the excess degrees of freedom represented by gauge transformations. Continuum theories, and most pedagogical treatments of the simplest quantum field theories, use a gauge fixing prescription to reduce the orbit of mathematical configurations that represent a given physical situation to a smaller orbit related by a smaller gauge group (the global symmetry group, or perhaps even the trivial group).

More sophisticated quantum field theories, in particular those that involve a non-abelian gauge group, break the gauge symmetry within the techniques of perturbation theory by introducing additional fields (the Faddeev–Popov ghosts) and counterterms motivated by anomaly cancellation, in an approach known as BRST quantization. While these concerns are in one sense highly technical, they are also closely related to the nature of measurement, the limits on knowledge of a physical situation, and the interactions between incompletely specified experimental conditions and incompletely understood physical theory . The mathematical techniques that have been developed in order to make gauge theories tractable have found many other applications, from solid-state physics and crystallography to low-dimensional topology.

1.3 Classical gauge theory

1.3.1 Classical electromagnetism

Historically, the first example of gauge symmetry discovered was classical electromagnetism. In electrostatics, one can either discuss the electric field, **E**, or its corresponding electric potential, *V*. Knowledge of one makes it possible to find the other, except that potentials differing by a constant, $V \to V + C$, correspond to the same electric field. This is because the electric field relates to *changes* in the potential from one point in space to another, and the constant C would cancel out when subtracting to find the change in potential. In terms of vector calculus, the electric field is the gradient of the potential, $\mathbf{E} = -\nabla V$. Generalizing from static electricity to electromagnetism, we have a second potential, the vector potential **A**, with

$$\mathbf{E} = -\nabla V - \frac{\partial \mathbf{A}}{\partial t}$$
$$\mathbf{B} = \nabla \times \mathbf{A}$$

The general gauge transformations now become not just $V \to V + C$ but

$$\mathbf{A} \to \mathbf{A} + \nabla f$$
$$V \to V - \frac{\partial f}{\partial t}$$

where f is any function that depends on position and time. The fields remain the same under the gauge transformation, and therefore Maxwell's equations are still satisfied. That is, Maxwell's equations have a gauge symmetry.

1.3.2   An example: Scalar O(n) gauge theory

The remainder of this section requires some familiarity with classical or quantum field theory, and the use of Lagrangians.

Definitions in this section: gauge group, gauge field, interaction Lagrangian, gauge boson.

The following illustrates how local gauge invariance can be "motivated" heuristically starting from global symmetry properties, and how it leads to an interaction between originally non-interacting fields.

Consider a set of n non-interacting real scalar fields, with equal masses m. This system is described by an action that is the sum of the (usual) action for each scalar field φ_i

$$\mathcal{S} = \int \mathrm{d}^4 x \sum_{i=1}^{n} \left[\frac{1}{2} \partial_\mu \varphi_i \partial^\mu \varphi_i - \frac{1}{2} m^2 \varphi_i^2 \right]$$

The Lagrangian (density) can be compactly written as

$$\mathcal{L} = \frac{1}{2} (\partial_\mu \Phi)^T \partial^\mu \Phi - \frac{1}{2} m^2 \Phi^T \Phi$$

by introducing a vector of fields

$$\Phi = (\varphi_1, \varphi_2, \dots, \varphi_n)^T$$

The term ∂_μ is Einstein notation for the partial derivative of Φ in each of the four dimensions. It is now transparent that the Lagrangian is invariant under the transformation

$$\Phi \mapsto \Phi' = G\Phi$$

whenever G is a *constant* matrix belonging to the n-by-n orthogonal group O(n). This is seen to preserve the Lagrangian, since the derivative of Φ transforms identically to Φ and both quantities appear inside dot products in the Lagrangian (orthogonal transformations preserve the dot product).

$$(\partial_\mu \Phi) \mapsto (\partial_\mu \Phi)' = G \partial_\mu \Phi$$

This characterizes the *global* symmetry of this particular Lagrangian, and the symmetry group is often called the **gauge group**; the mathematical term is **structure group**, especially in the theory of G-structures. Incidentally, Noether's theorem implies that invariance under this group of transformations leads to the conservation of the *currents*

$$J_\mu^a = i \partial_\mu \Phi^T T^a \Phi$$

where the T^a matrices are generators of the SO(n) group. There is one conserved current for every generator.

Now, demanding that this Lagrangian should have *local* O(n)-invariance requires that the G matrices (which were earlier constant) should be allowed to become functions of the space-time coordinates x.

Unfortunately, the *G* matrices do not "pass through" the derivatives, when $G = G(x)$,

$$\partial_\mu(G\Phi) \neq G(\partial_\mu\Phi)$$

The failure of the derivative to commute with "G" introduces an additional term (in keeping with the product rule), which spoils the invariance of the Lagrangian. In order to rectify this we define a new derivative operator such that the derivative of Φ again transforms identically with Φ

$$(D_\mu\Phi)' = GD_\mu\Phi$$

This new "derivative" is called a (gauge) covariant derivative and takes the form

$$D_\mu = \partial_\mu + igA_\mu$$

Where *g* is called the coupling constant; a quantity defining the strength of an interaction. After a simple calculation we can see that the **gauge field** $A(x)$ must transform as follows

$$A'_\mu = GA_\mu G^{-1} + \frac{i}{g}(\partial_\mu G)G^{-1}$$

The gauge field is an element of the Lie algebra, and can therefore be expanded as

$$A_\mu = \sum_a A^a_\mu T^a$$

There are therefore as many gauge fields as there are generators of the Lie algebra.

Finally, we now have a *locally gauge invariant* Lagrangian

$$\mathcal{L}_{\text{loc}} = \frac{1}{2}(D_\mu\Phi)^T D^\mu\Phi - \frac{1}{2}m^2\Phi^T\Phi$$

Pauli uses the term *gauge transformation of the first type* to mean the transformation of Φ, while the compensating transformation in *A* is called a *gauge transformation of the second type*.

The difference between this Lagrangian and the original *globally gauge-invariant* Lagrangian is seen to be the **interaction Lagrangian**

$$\mathcal{L}_{\text{int}} = i\frac{g}{2}\Phi^T A^T_\mu \partial^\mu\Phi + i\frac{g}{2}(\partial_\mu\Phi)^T A^\mu\Phi - \frac{g^2}{2}(A_\mu\Phi)^T A^\mu\Phi$$

This term introduces interactions between the *n* scalar fields just as a consequence of the demand for local gauge invariance. However, to make this interaction physical and not completely arbitrary, the mediator $A(x)$ needs to propagate in space. That is dealt with in the next section by adding yet another term, $\mathcal{L}_{\text{gf}}$, to the Lagrangian. In the quantized version of the obtained classical field theory, the quanta of the gauge field $A(x)$ are called gauge bosons. The interpretation of the interaction Lagrangian in quantum field theory is of scalar bosons interacting by the exchange of these gauge bosons.

1.3.3 The Yang–Mills Lagrangian for the gauge field

Main article: Yang–Mills theory

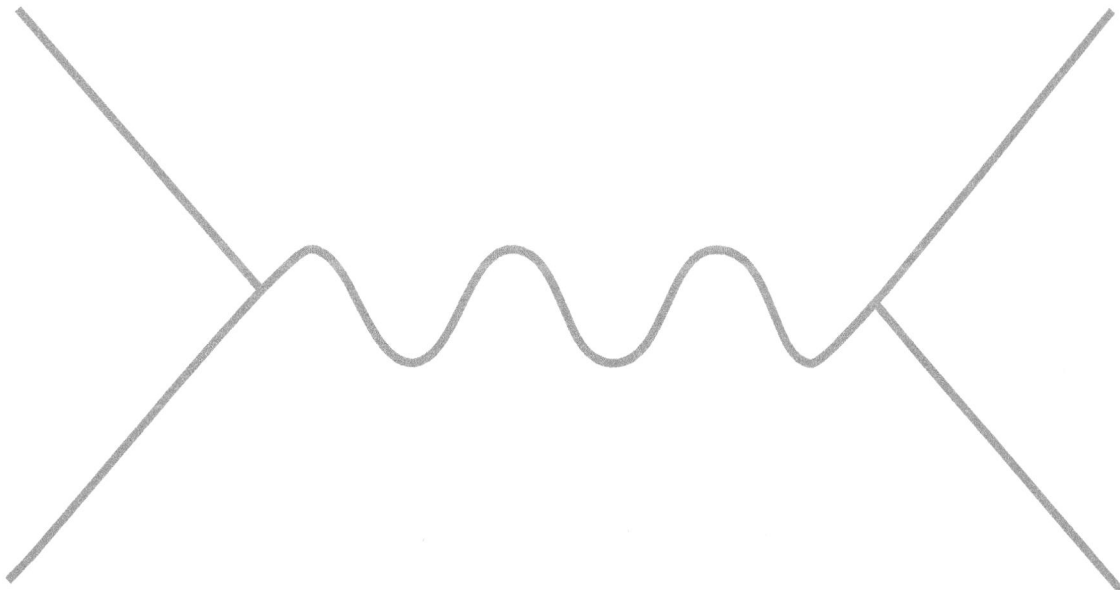

Feynman diagram of scalar bosons interacting via a gauge boson

The picture of a classical gauge theory developed in the previous section is almost complete, except for the fact that to define the covariant derivatives D, one needs to know the value of the gauge field $A(x)$ at all space-time points. Instead of manually specifying the values of this field, it can be given as the solution to a field equation. Further requiring that the Lagrangian that generates this field equation is locally gauge invariant as well, one possible form for the gauge field Lagrangian is (conventionally) written as

$$\mathcal{L}_{\text{gf}} = -\frac{1}{2}\,\text{Tr}(F^{\mu\nu}F_{\mu\nu})$$

with

$$F_{\mu\nu} = \frac{1}{ig}[D_\mu, D_\nu]$$

and the trace being taken over the vector space of the fields. This is called the **Yang–Mills action**. Other gauge invariant actions also exist (e.g., nonlinear electrodynamics, Born–Infeld action, Chern–Simons model, theta term, etc.).

Note that in this Lagrangian term there is no field whose transformation counterweighs the one of A. Invariance of this term under gauge transformations is a particular case of *a priori* classical (geometrical) symmetry. This symmetry must be restricted in order to perform quantization, the procedure being denominated gauge fixing, but even after restriction, gauge transformations may be possible.[3]

The complete Lagrangian for the gauge theory is now

$$\mathcal{L} = \mathcal{L}_{\text{loc}} + \mathcal{L}_{\text{gf}} = \mathcal{L}_{\text{global}} + \mathcal{L}_{\text{int}} + \mathcal{L}_{\text{gf}}$$

1.3.4   An example: Electrodynamics

As a simple application of the formalism developed in the previous sections, consider the case of electrodynamics, with only the electron field. The bare-bones action that generates the electron field's Dirac equation is

$$\mathcal{S} = \int \bar{\psi}(i\hbar c\, \gamma^\mu \partial_\mu - mc^2)\psi \, \mathrm{d}^4 x$$

The global symmetry for this system is

$$\psi \mapsto e^{i\theta}\psi$$

The gauge group here is U(1), just rotations of the phase angle of the field, with the particular rotation determined by the constant θ.

"Localising" this symmetry implies the replacement of θ by $\theta(x)$. An appropriate covariant derivative is then

$$D_\mu = \partial_\mu - i\frac{e}{\hbar}A_\mu$$

Identifying the "charge" e (not to be confused with the mathematical constant e in the symmetry description) with the usual electric charge (this is the origin of the usage of the term in gauge theories), and the gauge field $A(x)$ with the four-vector potential of electromagnetic field results in an interaction Lagrangian

$$\mathcal{L}_{\text{int}} = \frac{e}{\hbar}\bar{\psi}(x)\gamma^\mu\psi(x)A_\mu(x) = J^\mu(x)A_\mu(x)$$

where $J^\mu(x)$ is the usual four vector electric current density. The gauge principle is therefore seen to naturally introduce the so-called minimal coupling of the electromagnetic field to the electron field.

Adding a Lagrangian for the gauge field $A_\mu(x)$ in terms of the field strength tensor exactly as in electrodynamics, one obtains the Lagrangian used as the starting point in quantum electrodynamics.

$$\mathcal{L}_{\text{QED}} = \bar{\psi}(i\hbar c\,\gamma^\mu D_\mu - mc^2)\psi - \frac{1}{4\mu_0}F_{\mu\nu}F^{\mu\nu}$$

See also: Dirac equation, Maxwell's equations, Quantum electrodynamics

1.4 Mathematical formalism

Gauge theories are usually discussed in the language of differential geometry. Mathematically, a *gauge* is just a choice of a (local) section of some principal bundle. A **gauge transformation** is just a transformation between two such sections.

Although gauge theory is dominated by the study of connections (primarily because it's mainly studied by high-energy physicists), the idea of a connection is not central to gauge theory in general. In fact, a result in general gauge theory shows that affine representations (i.e., affine modules) of the gauge transformations can be classified as sections of a jet bundle satisfying certain properties. There are representations that transform covariantly pointwise (called by physicists gauge transformations of the first kind), representations that transform as a connection form (called by physicists gauge transformations of the second kind, an affine representation)—and other more general representations, such as the B field in BF theory. There are more general nonlinear representations (realizations), but these are extremely complicated. Still, nonlinear sigma models transform nonlinearly, so there are applications.

If there is a principal bundle P whose base space is space or spacetime and structure group is a Lie group, then the sections of P form a principal homogeneous space of the group of gauge transformations.

Connections (gauge connection) define this principal bundle, yielding a covariant derivative ∇ in each associated vector bundle. If a local frame is chosen (a local basis of sections), then this covariant derivative is represented by the connection form A, a Lie algebra-valued 1-form, which is called the **gauge potential** in physics. This is evidently not an intrinsic but a frame-dependent quantity. The curvature form F, a Lie algebra-valued 2-form that is an intrinsic quantity, is constructed from a connection form by

$$\mathbf{F} = d\mathbf{A} + \mathbf{A} \wedge \mathbf{A}$$

where d stands for the exterior derivative and $\wedge$ stands for the wedge product. ($\mathbf{A}$ is an element of the vector space spanned by the generators T^a, and so the components of $\mathbf{A}$ do not commute with one another. Hence the wedge product $\mathbf{A} \wedge \mathbf{A}$ does not vanish.)

Infinitesimal gauge transformations form a Lie algebra, which is characterized by a smooth Lie-algebra-valued scalar, ε. Under such an infinitesimal gauge transformation,

$$\delta_\varepsilon \mathbf{A} = [\varepsilon, \mathbf{A}] - d\varepsilon$$

where $[\cdot, \cdot]$ is the Lie bracket.

One nice thing is that if $\delta_\varepsilon X = \varepsilon X$, then $\delta_\varepsilon DX = \varepsilon DX$ where D is the covariant derivative

$$DX \overset{\text{def}}{=} dX + \mathbf{A}X$$

Also, $\delta_\varepsilon \mathbf{F} = \varepsilon \mathbf{F}$, which means $\mathbf{F}$ transforms covariantly.

Not all gauge transformations can be generated by infinitesimal gauge transformations in general. An example is when the base manifold is a compact manifold without boundary such that the homotopy class of mappings from that manifold to the Lie group is nontrivial. See instanton for an example.

The *Yang–Mills action* is now given by

$$\frac{1}{4g^2} \int \mathrm{Tr}[*F \wedge F]$$

where * stands for the Hodge dual and the integral is defined as in differential geometry.

A quantity which is **gauge-invariant** (i.e., invariant under gauge transformations) is the Wilson loop, which is defined over any closed path, γ, as follows:

$$\chi^{(\rho)} \left(\mathcal{P} \left\{ e^{\int_\gamma A} \right\} \right)$$

where χ is the character of a complex representation ρ and $\mathcal{P}$ represents the path-ordered operator.

1.5   Quantization of gauge theories

Main article: Quantum gauge theory

Gauge theories may be quantized by specialization of methods which are applicable to any quantum field theory. However, because of the subtleties imposed by the gauge constraints (see section on Mathematical formalism, above) there are many technical problems to be solved which do not arise in other field theories. At the same time, the richer structure of gauge theories allows simplification of some computations: for example Ward identities connect different renormalization constants.

1.5.1 Methods and aims

The first gauge theory quantized was quantum electrodynamics (QED). The first methods developed for this involved gauge fixing and then applying canonical quantization. The Gupta–Bleuler method was also developed to handle this problem. Non-abelian gauge theories are now handled by a variety of means. Methods for quantization are covered in the article on quantization.

The main point to quantization is to be able to compute quantum amplitudes for various processes allowed by the theory. Technically, they reduce to the computations of certain correlation functions in the vacuum state. This involves a renormalization of the theory.

When the running coupling of the theory is small enough, then all required quantities may be computed in perturbation theory. Quantization schemes intended to simplify such computations (such as canonical quantization) may be called **perturbative quantization schemes**. At present some of these methods lead to the most precise experimental tests of gauge theories.

However, in most gauge theories, there are many interesting questions which are non-perturbative. Quantization schemes suited to these problems (such as lattice gauge theory) may be called **non-perturbative quantization schemes**. Precise computations in such schemes often require supercomputing, and are therefore less well-developed currently than other schemes.

1.5.2 Anomalies

Some of the symmetries of the classical theory are then seen not to hold in the quantum theory; a phenomenon called an **anomaly**. Among the most well known are:

- The scale anomaly, which gives rise to a *running coupling constant*. In QED this gives rise to the phenomenon of the Landau pole. In Quantum Chromodynamics (QCD) this leads to asymptotic freedom.

- The chiral anomaly in either chiral or vector field theories with fermions. This has close connection with topology through the notion of instantons. In QCD this anomaly causes the decay of a pion to two photons.

- The gauge anomaly, which must cancel in any consistent physical theory. In the electroweak theory this cancellation requires an equal number of quarks and leptons.

1.6 Pure gauge

A pure gauge is the set of field configurations obtained by a gauge transformation on the null-field configuration, i.e., a gauge-transform of zero. So it is a particular "gauge orbit" in the field configuration's space.

Thus, in the abelian case, where $A_\mu(x) \to A'_\mu(x) = A_\mu(x) + \partial_\mu f(x)$, the pure gauge is just the set of field configurations $A'_\mu(x) = \partial_\mu f(x)$ for all $f(x)$.

1.7 See also

1.8 References

[1] Wolfgang Pauli (1941) "Relativistic Field Theories of Elementary Particles," *Rev. Mod. Phys.* **13**: 203–32.

[2] Pickering, A. (1984). *Constructing Quarks*. University of Chicago Press. ISBN 0-226-66799-5.

[3] Sakurai, *Advanced Quantum Mechanics*, sect 1–4

1.9   Bibliography

General readers

- Schumm, Bruce (2004) *Deep Down Things*. Johns Hopkins University Press. Esp. chpt. 8. A serious attempt by a physicist to explain gauge theory and the Standard Model with little formal mathematics.

Texts

- Bromley, D.A. (2000). *Gauge Theory of Weak Interactions*. Springer. ISBN 3-540-67672-4.

- Cheng, T.-P.; Li, L.-F. (1983). *Gauge Theory of Elementary Particle Physics*. Oxford University Press. ISBN 0-19-851961-3.

- Frampton, P. (2008). *Gauge Field Theories* (3rd ed.). Wiley-VCH.

- Kane, G.L. (1987). *Modern Elementary Particle Physics*. Perseus Books. ISBN 0-201-11749-5.

Articles

- Becchi, C. (1997). "Introduction to Gauge Theories". p. 5211. arXiv:hep-ph/9705211. Bibcode:1997hep.ph....5211B.

- Gross, D. (1992). "Gauge theory – Past, Present and Future" (PDF). Retrieved 2009-04-23.

- Jackson, J.D. (2002). "From Lorenz to Coulomb and other explicit gauge transformations". *Am.J.Phys* **70** (9): 917–928. arXiv:physics/0204034. Bibcode:2002AmJPh..70..917J. doi:10.1119/1.1491265.

- Svetlichny, George (1999). "Preparation for Gauge Theory". p. 2027. arXiv:math-ph/9902027. Bibcode:1999math.ph...2027S.

1.10   External links

- Hazewinkel, Michiel, ed. (2001), "Gauge transformation", *Encyclopedia of Mathematics*, Springer, ISBN 978-1-55608-010-4

- Yang–Mills equations on DispersiveWiki

- Gauge theories on Scholarpedia

Chapter 2

Introduction to gauge theory

This article is a non-technical introduction to the subject. For the main encyclopedia article, see Gauge theory.

A **gauge theory** is a type of theory in physics. Modern physical theories, such as the theory of electromagnetism, describe the nature of reality in terms of fields, e.g., the electromagnetic field, the gravitational field, and fields for the electron and all other elementary particles. A general feature of these field theories is that the fundamental fields cannot be directly measured; however, there are observable quantities that can be measured experimentally, such as charges, energies, and velocities. In field theories, different configurations of the unobservable fields can result in identical observable quantities. A transformation from one such field configuration to another is called a **gauge transformation**;[1][2] the lack of change in the measurable quantities, despite the field being transformed, is a property called **gauge invariance**. Since any kind of invariance under a field transformation is considered a symmetry, gauge invariance is sometimes called **gauge symmetry**. Generally, any theory that has the property of gauge invariance is considered a gauge theory.

For example, in electromagnetism the electric and magnetic fields, $\mathbf{E}$ and $\mathbf{B}$, are observable, while the potentials V ("voltage") and $\mathbf{A}$ (the vector potential) are not.[3] Under a gauge transformation in which a constant is added to V, no observable change occurs in $\mathbf{E}$ or $\mathbf{B}$.

With the advent of quantum mechanics in the 1920s, and with successive advances in quantum field theory, the importance of gauge transformations has steadily grown. Gauge theories constrain the laws of physics, because all the changes induced by a gauge transformation have to cancel each other out when written in terms of observable quantities. Over the course of the 20th century, physicists gradually realized that all forces (fundamental interactions) arise from the constraints imposed by *local* gauge symmetries, in which case the transformations vary from point to point in space and time. Perturbative quantum field theory (usually employed for scattering theory) describes forces in terms of force-mediating particles called gauge bosons. The nature of these particles is determined by the nature of the gauge transformations. The culmination of these efforts is the Standard Model, a quantum field theory explaining all of the fundamental interactions except gravity.

2.1   History and importance

The earliest field theory having a gauge symmetry was Maxwell's formulation of electrodynamics in 1864–65 ("A Dynamical Theory of the Electromagnetic Field"). The importance of this symmetry remained unnoticed in the earliest formulations. Similarly unnoticed, Hilbert had derived Einstein's equations of general relativity by postulating a symmetry under any change of coordinates. Later Hermann Weyl, in an attempt to unify general relativity and electromagnetism, conjectured (incorrectly, as it turned out) that invariance under the change of scale or "gauge" (a term inspired by the various track gauges of railroads) might also be a local symmetry of general relativity. Although Weyl's choice of the gauge was incorrect, the name "gauge" stuck to the approach. After the development of quantum mechanics, Weyl, Fock and London modified their gauge choice by replacing the scale factor with a change of wave phase, and applying it successfully to electromagnetism. Gauge symmetry was generalized mathematically in 1954 by Chen Ning Yang and Robert Mills in an attempt to describe the strong nuclear forces. This idea, dubbed Yang–Mills theory, later found application in the quantum field theory of the weak force, and its unification with electromagnetism in the electroweak theory.

The importance of gauge theories for physics stems from their tremendous success in providing a unified framework to describe the quantum-mechanical behavior of electromagnetism, the weak force and the strong force. This gauge theory, known as the Standard Model, accurately describes experimental predictions regarding three of the four fundamental forces of nature.

2.2   In classical physics

2.2.1   Electromagnetism

Main article: Gauge fixing

Historically, the first example of gauge symmetry to be discovered was classical electromagnetism. A static electric field can be described in terms of an electric potential (voltage) that is defined at every point in space, and in practical work it is conventional to take the Earth as a physical reference that defines the zero level of the potential, or ground. But only *differences* in potential are physically measurable, which is the reason that a voltmeter must have two probes, and can only report the voltage difference between them. Thus one could choose to define all voltage differences relative to some other standard, rather than the Earth, resulting in the addition of a constant offset.[4] If the potential V is a solution to Maxwell's equations then, after this gauge transformation, the new potential $V \to V + C$ is also a solution to Maxwell's equations and no experiment can distinguish between these two solutions. In other words the laws of physics governing electricity and magnetism (that is, Maxwell equations) are invariant under gauge transformation.[5] That is, Maxwell's equations have a gauge symmetry.

Generalizing from static electricity to electromagnetism, we have a second potential, the magnetic vector potential **A**, which can also undergo gauge transformations. These transformations may be local. That is, rather than adding a constant onto V, one can add a function that takes on different values at different points in space and time. If **A** is also changed in certain corresponding ways, then the same **E** and **B** fields result. The detailed mathematical relationship between the fields **E** and **B** and the potentials V and **A** is given in the article Gauge fixing, along with the precise statement of the nature of the gauge transformation. The relevant point here is that the fields remain the same under the gauge transformation, and therefore Maxwell's equations are still satisfied.

Gauge symmetry is closely related to charge conservation. Suppose that there existed some process by which one could violate conservation of charge, at least temporarily, by creating a charge q at a certain point in space, 1, moving it to some other point 2, and then destroying it. We might imagine that this process was consistent with conservation of energy. We could posit a rule stating that creating the charge required an input of energy $E_1 = qV_1$ and destroying it released $E_2 = qV_2$, which would seem natural since qV measures the extra energy stored in the electric field because of the existence of a charge at a certain point. (There may also be energy associated, e.g., with the rest mass of the particle, but that is not relevant to the present argument.) Conservation of energy would be satisfied, because the net energy released by creation and destruction of the particle, $qV_2 - qV_1$, would be equal to the work done in moving the particle from 1 to 2, $qV_2 - qV_1$. But although this scenario salvages conservation of energy, it violates gauge symmetry. Gauge symmetry requires that the laws of physics be invariant under the transformation $V \to V + C$, which implies that no experiment should be able to measure the absolute potential, without reference to some external standard such as an electrical ground. But the proposed rules $E_1 = qV_1$ and $E_2 = qV_2$ for the energies of creation and destruction *would* allow an experimenter to determine the absolute potential, simply by checking how much energy input was required in order to create the charge q at a particular point in space. The conclusion is that if gauge symmetry holds, and energy is conserved, then charge must be conserved.[6]

2.2.2   General relativity

As discussed above, the gauge transformations for classical (i.e., non-quantum mechanical) general relativity are arbitrary coordinate transformations.[7] (Technically, the transformations must be invertible, and both the transformation and its inverse must be smooth, in the sense of being differentiable an arbitrary number of times.)

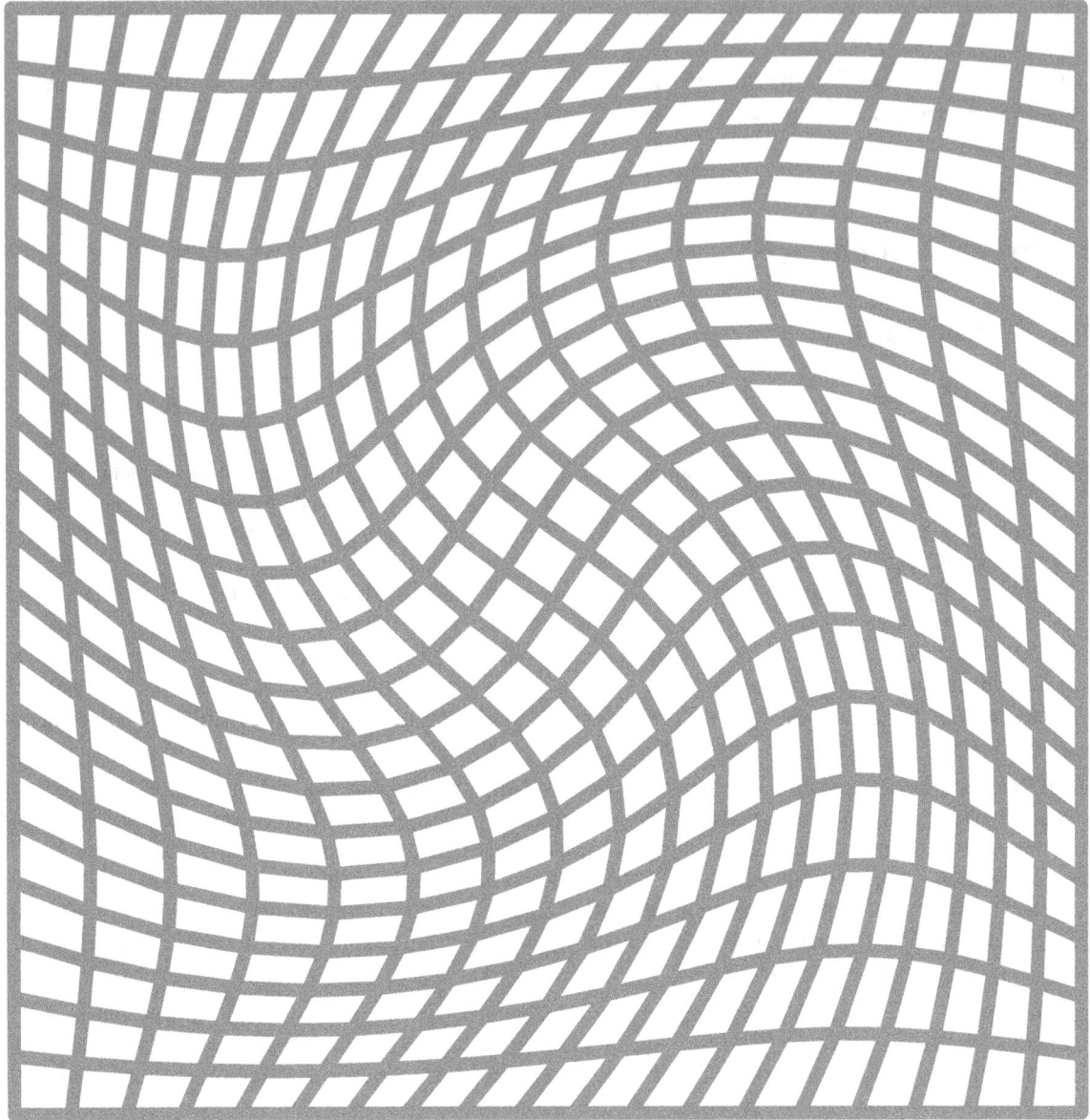

The Cartesian coordinate grid on this square has been distorted by a coordinate transformation, so that there is a nonlinear relationship between the old (x, y) coordinates and the new ones. Einstein's equations of general relativity are still valid in the new coordinate system. Such changes of coordinate system are the gauge transformations of general relativity.

An example of a symmetry in a physical theory: translation invariance

Some global symmetries under changes of coordinate predate both general relativity and the concept of a gauge. For example, translation invariance was introduced in the era of Galileo, who eliminated the Aristotelian concept that various places in space, such as the earth and the heavens, obeyed different physical rules.

Suppose, for example, that one observer examines the properties of a hydrogen atom on Earth, the other—on the Moon (or any other place in the universe), the observer will find that their hydrogen atoms exhibit completely identical properties. Again, if one observer had examined a hydrogen atom today and the other—100 years ago (or any other time in the past or in the future), the two experiments would again produce completely identical results. The invariance of the properties of a hydrogen atom with respect to the time and place where these properties were investigated is called translation invariance.

Recalling our two observers from different ages: the time in their experiments is shifted by 100 years. If the time when the older observer did the experiment was t, the time of the modern experiment is $t+100$ years. Both observers discover the same laws of physics. Because light from hydrogen atoms in distant galaxies may reach the earth after having traveled across space for billions of years, in effect one can do such observations covering periods of time almost all the way back to the Big Bang, and they show that the laws of physics have always been the same.

In other words, if in the theory we change the time t to $t+100$ years (or indeed any other time shift) the theoretical predictions do not change.[8]

Another example of a symmetry: the invariance of Einstein's field equation under arbitrary coordinate transformations

In Einstein's general relativity, coordinates like x, y, z, and t are not only "relative" in the global sense of translations like $t \to t + C$, rotations, etc., but become completely arbitrary, so that for example one can define an entirely new timelike coordinate according to some arbitrary rule such as $t \to t + t^3/t_0^2$, where t_0 has units of time, and yet Einstein's equations will have the same form.[7][9]

Invariance of the form of an equation under an arbitrary coordinate transformation is customarily referred to as general covariance and equations with this property are referred to as written in the covariant form. General covariance is a special case of gauge invariance.

Maxwell's equations can also be expressed in a generally covariant form, which is as invariant under general coordinate transformation as Einstein's field equation.

2.3   In quantum mechanics

2.3.1   Quantum electrodynamics

Until the advent of quantum mechanics, the only well known example of gauge symmetry was in electromagnetism, and the general significance of the concept was not fully understood. For example, it was not clear whether it was the fields **E** and **B** or the potentials V and **A** that were the fundamental quantities; if the former, then the gauge transformations could be considered as nothing more than a mathematical trick.

2.3.2   Aharonov–Bohm experiment

Main article: Aharonov–Bohm effect

In quantum mechanics a particle, such as an electron, is also described as a wave. For example, if the double-slit experiment is performed with electrons, then a wave-like interference pattern is observed. The electron has the highest probability of being detected at locations where the parts of the wave passing through the two slits are in phase with one another, resulting in constructive interference. The frequency of the electron *wave* is related to the kinetic energy of an individual electron *particle* via the quantum-mechanical relation $E = hf$. If there are no electric or magnetic fields present in this experiment, then the electron's energy is constant, and, for example, there will be a high probability of detecting the electron along the central axis of the experiment, where by symmetry the two parts of the wave are in phase.

But now suppose that the electrons in the experiment are subject to electric or magnetic fields. For example, if an electric field was imposed on one side of the axis but not on the other, the results of the experiment would be affected. The part of the electron wave passing through that side oscillates at a different rate, since its energy has had $-eV$ added to it, where $-e$ is the charge of the electron and V the electrical potential. The results of the experiment will be different, because phase relationships between the two parts of the electron wave have changed, and therefore the locations of constructive and destructive interference will be shifted to one side or the other. It is the electric potential that occurs here, not the electric field, and this is a manifestation of the fact that it is the potentials and not the fields that are of fundamental significance in quantum mechanics.

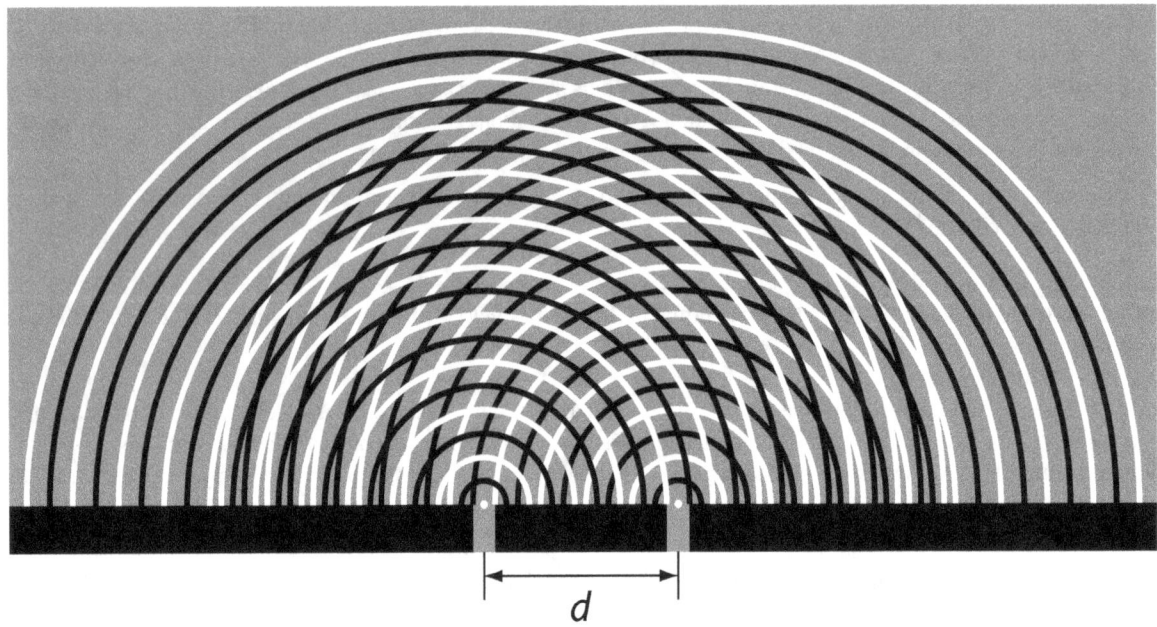

Double-slit diffraction and interference pattern

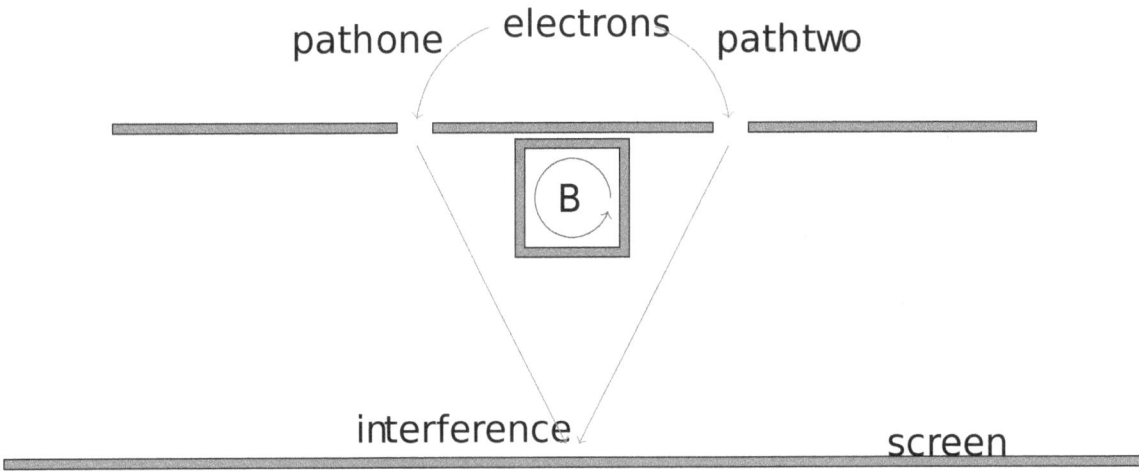

*Schematic of double-slit experiment in which Aharonov–Bohm effect can be observed: electrons pass through two slits, interfering at an observation screen, with the interference pattern shifted when a magnetic field **B** is turned on in the cylindrical solenoid, marked in blue on the diagram.*

Explanation with potentials

It is even possible to have cases in which an experiment's results differ when the potentials are changed, even if no charged particle is ever exposed to a different field. One such example is the Aharonov–Bohm effect, shown in the figure.[10] In this example, turning on the solenoid only causes a magnetic field **B** to exist within the solenoid. But the solenoid has been positioned so that the electron cannot possibly pass through its interior. If one believed that the fields were the fundamental quantities, then one would expect that the results of the experiment would be unchanged. In reality, the results are different, because turning on the solenoid changed the vector potential **A** in the region that the electrons do pass through. Now that it has been established that it is the potentials V and **A** that are fundamental, and not the fields **E** and **B**, we can see that the gauge transformations, which change V and **A**, have real physical significance, rather than being merely mathematical artifacts.

Gauge invariance: the results of the experiments are independent of the choice of the gauge for the potentials

Note that in these experiments, the only quantity that affects the result is the *difference* in phase between the two parts of the electron wave. Suppose we imagine the two parts of the electron wave as tiny clocks, each with a single hand that sweeps around in a circle, keeping track of its own phase. Although this cartoon ignores some technical details, it retains the physical phenomena that are important here.[11] If both clocks are sped up by the same amount, the phase relationship between them is unchanged, and the results of experiments are the same. Not only that, but it is not even necessary to change the speed of each clock by a *fixed* amount. We could change the angle of the hand on each clock by a *varying* amount θ, where θ could depend on both the position in space and on time. This would have no effect on the result of the experiment, since the final observation of the location of the electron occurs at a single place and time, so that the phase shift in each electron's "clock" would be the same, and the two effects would cancel out. This is another example of a gauge transformation: it is local, and it does not change the results of experiments.

2.3.3   Summary

In summary, gauge symmetry attains its full importance in the context of quantum mechanics. In the application of quantum mechanics to electromagnetism, i.e., quantum electrodynamics, gauge symmetry applies to both electromagnetic waves and electron waves. These two gauge symmetries are in fact intimately related. If a gauge transformation θ is applied to the electron waves, for example, then one must also apply a corresponding transformation to the potentials that describe the electromagnetic waves.[12] Gauge symmetry is required in order to make quantum electrodynamics a renormalizable theory, i.e., one in which the calculated predictions of all physically measurable quantities are finite.

2.3.4   Types of gauge symmetries

The description of the electrons in the subsection above as little clocks is in effect a statement of the mathematical rules according to which the phases of electrons are to be added and subtracted: they are to be treated as ordinary numbers, except that in the case where the result of the calculation falls outside the range of $0 \leq \theta < 360°$, we force it to "wrap around" into the allowed range, which covers a circle. Another way of putting this is that a phase angle of, say, 5° is considered to be completely equivalent to an angle of 365°. Experiments have verified this testable statement about the interference patterns formed by electron waves. Except for the "wrap-around" property, the algebraic properties of this mathematical structure are exactly the same as those of the ordinary real numbers.

In mathematical terminology, electron phases form an Abelian group under addition, called the circle group or $U(1)$. "Abelian" means that addition commutes, so that $\theta + \varphi = \varphi + \theta$. Group means that addition associates and has an identity element, namely "0". Also, for every phase there exists an inverse such that the sum of a phase and its inverse is 0. Other examples of abelian groups are the integers under addition, 0, and negation, and the nonzero fractions under product, 1, and reciprocal.

As a way of visualizing the choice of a gauge, consider whether it is possible to tell if a cylinder has been twisted. If the cylinder has no bumps, marks, or scratches on it, we cannot tell. We could, however, draw an arbitrary curve along the cylinder, defined by some function $\theta(x)$, where x measures distance along the axis of the cylinder. Once this arbitrary choice (the choice of gauge) has been made, it becomes possible to detect it if someone later twists the cylinder.

In 1954, Chen Ning Yang and Robert Mills proposed to generalize these ideas to noncommutative groups. A noncommutative gauge group can describe a field that, unlike the electromagnetic field, interacts with itself. For example, general relativity states that gravitational fields have energy, and special relativity concludes that energy is equivalent to mass. Hence a gravitational field induces a further gravitational field. The nuclear forces also have this self-interacting property.

2.3.5   Gauge bosons

Surprisingly, gauge symmetry can give a deeper explanation for the existence of interactions, such as the electrical and nuclear interactions. This arises from a type of gauge symmetry relating to the fact that all particles of a given type are experimentally indistinguishable from one other. Imagine that Alice and Betty are identical twins, labeled at birth by

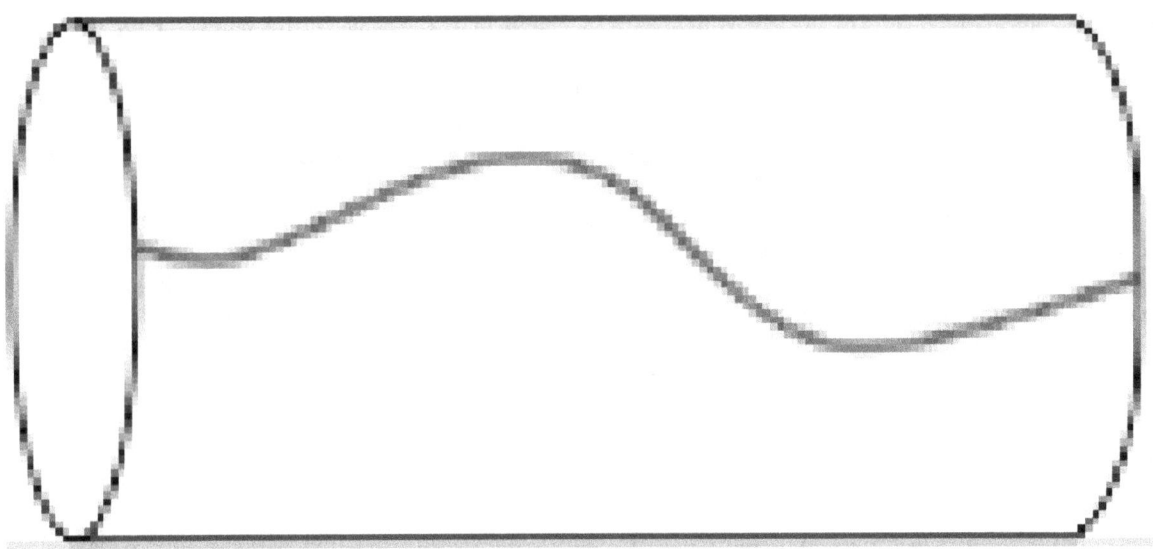

Gauge fixing of a twisted *cylinder.*

bracelets reading A and B. Because the girls are identical, nobody would be able to tell if they had been switched at birth; the labels A and B are arbitrary, and can be interchanged. Such a permanent interchanging of their identities is like a global gauge symmetry. There is also a corresponding local gauge symmetry, which describes the fact that from one moment to the next, Alice and Betty could swap roles while nobody was looking, and nobody would be able to tell. If we observe that Mom's favorite vase is broken, we can only infer that the blame belongs to one twin or the other, but we cannot tell whether the blame is 100% Alice's and 0% Betty's, or vice versa. If Alice and Betty are in fact quantum-mechanical particles rather than people, then they also have wave properties, including the property of superposition, which allows waves to be added, subtracted, and mixed arbitrarily. It follows that we are not even restricted to a complete swaps of identity. For example, if we observe that a certain amount of energy exists in a certain location in space, there is no experiment that can tell us whether that energy is 100% A's and 0% B's, 0% A's and 100% B's, or 20% A's and 80% B's, or some other mixture. The fact that the symmetry is local means that we cannot even count on these proportions to remain fixed as the particles propagate through space. The details of how this is represented mathematically depend on technical issues relating to the spins of the particles, but for our present purposes we consider a spinless particle, for which it turns out that the mixing can be specified by some arbitrary choice of gauge $\theta(x)$, where an angle $\theta = 0°$ represents 100% A and 0% B, $\theta = 90°$ means 0% A and 100% B, and intermediate angles represent mixtures.

According to the principles of quantum mechanics, particles do not actually have trajectories through space. Motion can only be described in terms of waves, and the momentum p of an individual particle is related to its wavelength λ by $p = h/\lambda$. In terms of empirical measurements, the wavelength can only be determined by observing a change in the wave between one point in space and another nearby point (mathematically, by differentiation). A wave with a shorter wavelength oscillates more rapidly, and therefore changes more rapidly between nearby points. Now suppose that we arbitrarily fix a gauge at one point in space, by saying that the energy at that location is 20% A's and 80% B's. We then measure the two waves at some other, nearby point, in order to determine their wavelengths. But there are two entirely different reasons that the waves could have changed. They could have changed because they were oscillating with a certain wavelength, or they could have changed because the gauge function changed from a 20-80 mixture to, say, 21-79. If we ignore the second possibility, the resulting theory doesn't work; strange discrepancies in momentum will show up, violating the principle of conservation of momentum. Something in the theory must be changed.

Again there are technical issues relating to spin, but in several important cases, including electrically charged particles and particles interacting via nuclear forces, the solution to the problem is to impute physical reality to the gauge function $\theta(x)$. We say that if the function θ oscillates, it represents a new type of quantum-mechanical wave, and this new wave has its own momentum $p = h/\lambda$, which turns out to patch up the discrepancies that otherwise would have broken conservation of momentum. In the context of electromagnetism, the particles A and B would be charged particles such as electrons, and the quantum mechanical wave represented by θ would be the electromagnetic field. (Here we ignore the technical issues raised by the fact that electrons actually have spin 1/2, not spin zero. This oversimplification is the reason that the gauge field θ comes out to be a scalar, whereas the electromagnetic field is actually represented by a vector consisting of V and $\mathbf{A}$.) The result is that we have an explanation for the presence of electromagnetic interactions: if we try to construct a gauge-symmetric theory of identical, non-interacting particles, the result is not self-consistent, and can only be repaired by adding electrical and magnetic fields that cause the particles to interact.

Although the function θ(x) describes a wave, the laws of quantum mechanics require that it also have particle properties. In the case of electromagnetism, the particle corresponding to electromagnetic waves is the photon. In general, such particles are called gauge bosons, where the term "boson" refers to a particle with integer spin. In the simplest versions of the theory, gauge bosons are massless, but it is also possible to construct versions in which they have mass, as is the case for the gauge bosons that transmit the nuclear decay forces.

2.4   References

[1] Donald H. Perkins (1982) *Introduction to High-Energy Physics*. Addison-Wesley: 22.

[2] Roger Penrose (2004) *The Road to Reality*, p. 451. For an alternative formulation in terms of symmetries of the Lagrangian density, see p. 489. Also see J. D. Jackson (1975) *Classical Electrodynamics*, 2nd ed. Wiley and Sons: 176.

[3] For an argument that V and $\mathbf{A}$ are more fundamental, see Feynman, Leighton, and Sands, **The Feynman Lectures**, Addison Wesley Longman, 1970, II-15-7,8,12, but this is partly a matter of personal preference.

[4] Edward Purcell (1963) *Electricity and Magnetism*. McGraw-Hill: 38.

[5] J.D. Jackson (1975) *Classical Electrodynamics*, 2nd ed. Wiley and Sons: 176.

[6] Donald H. Perkins (1982) *Introduction to High-Energy Physics*. Addison-Wesley: 92.

[7] Robert M. Wald (1984) *General Relativity*. University of Chicago Press: 260.

[8] Charles Misner, Kip Thorne, and John A. Wheeler (1973) *Gravitation*. W. H. Freeman: 68.

[9] Misner, Thorne, and Wheeler (1973) *Gravitation*. W. H. Freeman: 967.

[10] Feynman, Leighton, and Sands (1970) *The Feynman Lectures on Physics*. Addison Wesley, vol. II, chpt. 15, section 5.

[11] Richard Feynman (1985) *QED: The Strange Theory of Light and Matter*. Princeton University Press.

[12] Donald H. Perkins (1982) *Introduction to High-Energy Physics*. Addison-Wesley: 332.

2.5   Further reading

These books are intended for general readers and employ the barest minimum of mathematics.

- 't Hooft, Gerard: **"Gauge Theories of the Force between Elementary Particles,"** *Scientific American*, **242(6):104-138 (June 1980).**

- "Press Release: The 1999 Nobel Prize in Physics". *Nobelprize.org*. Nobel Media AB 2013. Web. 20 Aug 2013. <http://www.nobelprize.org/nobel_prizes/physics/laureates/1999/press.html>

- Schumm, Bruce (2004) *Deep Down Things*. Johns Hopkins University Press. A serious attempt by a physicist to explain gauge theory and the Standard Model.

- Feynman, Richard (2006) *QED: The Strange Theory of Light and Matter*. Princeton University Press. A nontechnical description of quantum field theory (not specifically about gauge theory).

2.6   Related information

Chapter 3

Quantum gauge theory

See gauge theory for the classical preliminaries.

In quantum physics, in order to quantize a gauge theory, like for example Yang-Mills theory, Chern-Simons or BF model, one method is to perform a gauge fixing. This is done in the BRST and Batalin-Vilkovisky formulation. Another is to factor out the symmetry by dispensing with vector potentials altogether (they're not physically observable anyway) and work directly with Wilson loops, Wilson lines contracted with other charged fields at its endpoints and spin networks.

Older approaches to quantization for Abelian models use the Gupta-Bleuler formalism with a "semi-Hilbert space" with an indefinite sesquilinear form. However, it is much more elegant to just work with the quotient space of vector field configurations by gauge transformations.

An alternative approach using lattice approximations is covered in (Wick rotated) lattice gauge theory.

To establish the existence of the Yang-Mills theory and a mass gap is one of the seven Millennium Prize Problems of the Clay Mathematics Institute.

3.1 References

Chapter 4

Quantum field theory

"Relativistic quantum field theory" redirects here. For other uses, see Relativity.

In theoretical physics, **quantum field theory** (**QFT**) is a theoretical framework for constructing quantum mechanical models of subatomic particles in particle physics and quasiparticles in condensed matter physics. A QFT treats particles as excited states of an underlying physical field, so these are called field quanta.

In QFT, quantum mechanical interactions between particles are described by interaction terms between the corresponding underlying fields.

4.1 Definition

Quantum electrodynamics (QED) has one electron field and one photon field; quantum chromodynamics (QCD) has one field for each type of quark; and, in condensed matter, there is an atomic displacement field that gives rise to phonon particles. Edward Witten describes QFT as "by far" the most difficult theory in modern physics.[1]

4.1.1 Dynamics

See also: Relativistic dynamics

Ordinary quantum mechanical systems have a fixed number of particles, with each particle having a finite number of degrees of freedom. In contrast, the excited states of a QFT can represent any number of particles. This makes quantum field theories especially useful for describing systems where the particle count/number may change over time, a crucial feature of relativistic dynamics.

4.1.2 States

QFT interaction terms are similar in spirit to those between charges with electric and magnetic fields in Maxwell's equations. However, unlike the classical fields of Maxwell's theory, fields in QFT generally exist in quantum superpositions of states and are subject to the laws of quantum mechanics.

Because the fields are continuous quantities over space, there exist excited states with arbitrarily large numbers of particles in them, providing QFT systems with an effectively infinite number of degrees of freedom. Infinite degrees of freedom can easily lead to divergences of calculated quantities (i.e., the quantities become infinite). Techniques such as renormalization of QFT parameters or discretization of spacetime, as in lattice QCD, are often used to avoid such infinities so as to yield physically meaningful results.

4.1.3 Fields and radiation

The gravitational field and the electromagnetic field are the only two fundamental fields in nature that have infinite range and a corresponding classical low-energy limit, which greatly diminishes and hides their "particle-like" excitations. Albert Einstein in 1905, attributed "particle-like" and discrete exchanges of momenta and energy, characteristic of "field quanta", to the electromagnetic field. Originally, his principal motivation was to explain the thermodynamics of radiation. Although the photoelectric effect and Compton scattering strongly suggest the existence of the photon, it might alternately be explained by a mere quantization of emission; more definitive evidence of the quantum nature of radiation is now taken up into modern quantum optics as in the antibunching effect.[2]

4.2 Theories

There is currently no complete quantum theory of the remaining fundamental force, gravity. Many of the proposed theories to describe gravity as a QFT postulate the existence of a graviton particle that mediates the gravitational force. Presumably, the as yet unknown correct quantum field-theoretic treatment of the gravitational field will behave like Einstein's general theory of relativity in the low-energy limit. Quantum field theory of the fundamental forces itself has been postulated to be the low-energy effective field theory limit of a more fundamental theory such as superstring theory.

Most theories in standard particle physics are formulated as **relativistic quantum field theories**, such as QED, QCD, and the Standard Model. QED, the quantum field-theoretic description of the electromagnetic field, approximately reproduces Maxwell's theory of electrodynamics in the low-energy limit, with small non-linear corrections to the Maxwell equations required due to virtual electron–positron pairs.

In the perturbative approach to quantum field theory, the full field interaction terms are approximated as a perturbative expansion in the number of particles involved. Each term in the expansion can be thought of as forces between particles being mediated by other particles. In QED, the electromagnetic force between two electrons is caused by an exchange of photons. Similarly, intermediate vector bosons mediate the weak force and gluons mediate the strong force in QCD. The notion of a force-mediating particle comes from perturbation theory, and does not make sense in the context of non-perturbative approaches to QFT, such as with bound states.

4.3 History

Main article: History of quantum field theory

4.3.1 Foundations

The early development of the field involved Dirac, Fock, Pauli, Heisenberg and Bogolyubov. This phase of development culminated with the construction of the theory of quantum electrodynamics in the 1950s.

4.3.2 Gauge theory

Gauge theory was formulated and quantized, leading to the **unification of forces** embodied in the standard model of particle physics. This effort started in the 1950s with the work of Yang and Mills, was carried on by Martinus Veltman and a host of others during the 1960s and completed by the 1970s through the work of Gerard 't Hooft, Frank Wilczek, David Gross and David Politzer.

4.3.3   Grand synthesis

Parallel developments in the understanding of phase transitions in condensed matter physics led to the study of the renormalization group. This in turn led to the **grand synthesis** of theoretical physics, which unified theories of particle and condensed matter physics through quantum field theory. This involved the work of Michael Fisher and Leo Kadanoff in the 1970s, which led to the seminal reformulation of quantum field theory by Kenneth G. Wilson in 1975.

4.4   Principles

4.4.1   Classical and quantum fields

Main article: Classical field theory

A classical field is a function defined over some region of space and time.[3] Two physical phenomena which are described by classical fields are Newtonian gravitation, described by Newtonian gravitational field $\mathbf{g}(\mathbf{x}, t)$, and classical electromagnetism, described by the electric and magnetic fields $\mathbf{E}(\mathbf{x}, t)$ and $\mathbf{B}(\mathbf{x}, t)$. Because such fields can in principle take on distinct values at each point in space, they are said to have infinite degrees of freedom.[3]

Classical field theory does not, however, account for the quantum-mechanical aspects of such physical phenomena. For instance, it is known from quantum mechanics that certain aspects of electromagnetism involve discrete particles—photons—rather than continuous fields. The business of *quantum* field theory is to write down a field that is, like a classical field, a function defined over space and time, but which also accommodates the observations of quantum mechanics. This is a *quantum field*.

It is not immediately clear *how* to write down such a quantum field, since quantum mechanics has a structure very unlike a field theory. In its most general formulation, quantum mechanics is a theory of abstract operators (observables) acting on an abstract state space (Hilbert space), where the observables represent physically observable quantities and the state space represents the possible states of the system under study.[4] For instance, the fundamental observables associated with the motion of a single quantum mechanical particle are the position and momentum operators $\hat{x}$ and $\hat{p}$. Field theory, in contrast, treats x as a way to index the field rather than as an operator.[5]

There are two common ways of developing a quantum field: the path integral formalism and canonical quantization.[6] The latter of these is pursued in this article.

Lagrangian formalism

Quantum field theory frequently makes use of the Lagrangian formalism from classical field theory. This formalism is analogous to the Lagrangian formalism used in classical mechanics to solve for the motion of a particle under the influence of a field. In classical field theory, one writes down a Lagrangian density, $\mathcal{L}$, involving a field, $\varphi(\mathbf{x},t)$, and possibly its first derivatives ($\partial\varphi/\partial t$ and $\nabla\varphi$), and then applies a field-theoretic form of the Euler–Lagrange equation. Writing coordinates $(t, \mathbf{x}) = (x^0, x^1, x^2, x^3) = x^\mu$, this form of the Euler–Lagrange equation is[3]

$$\frac{\partial}{\partial x^\mu}\left[\frac{\partial\mathcal{L}}{\partial(\partial\phi/\partial x^\mu)}\right] - \frac{\partial\mathcal{L}}{\partial\phi} = 0,$$

where a sum over μ is performed according to the rules of Einstein notation.

By solving this equation, one arrives at the "equations of motion" of the field.[3] For example, if one begins with the Lagrangian density

$$\mathcal{L}(\phi, \nabla\phi) = -\rho(t, \mathbf{x})\,\phi(t, \mathbf{x}) - \frac{1}{8\pi G}|\nabla\phi|^2,$$

and then applies the Euler–Lagrange equation, one obtains the equation of motion

$$4\pi G\rho(t, \mathbf{x}) = \nabla^2\phi.$$

This equation is Newton's law of universal gravitation, expressed in differential form in terms of the gravitational potential $\varphi(t, \mathbf{x})$ and the mass density $\rho(t, \mathbf{x})$. Despite the nomenclature, the "field" under study is the gravitational potential, φ, rather than the gravitational field, $\mathbf{g}$. Similarly, when classical field theory is used to study electromagnetism, the "field" of interest is the electromagnetic four-potential $(V/c, \mathbf{A})$, rather than the electric and magnetic fields $\mathbf{E}$ and $\mathbf{B}$.

Quantum field theory uses this same Lagrangian procedure to determine the equations of motion for quantum fields. These equations of motion are then supplemented by commutation relations derived from the canonical quantization procedure described below, thereby incorporating quantum mechanical effects into the behavior of the field.

4.4.2 Single- and many-particle quantum mechanics

Main articles: Quantum mechanics and First quantization

In quantum mechanics, a particle (such as an electron or proton) is described by a complex wavefunction, $\psi(x, t)$, whose time-evolution is governed by the Schrödinger equation:

$$-\frac{\hbar^2}{2m}\frac{\partial^2}{\partial x^2}\psi(x, t) + V(x)\psi(x, t) = i\hbar\frac{\partial}{\partial t}\psi(x, t).$$

Here m is the particle's mass and $V(x)$ is the applied potential. Physical information about the behavior of the particle is extracted from the wavefunction by constructing expected values for various quantities; for example, the expected value of the particle's position is given by integrating $\psi^*(x)\, x\, \psi(x)$ over all space, and the expected value of the particle's momentum is found by integrating $-i\hbar\psi^*(x)\mathrm{d}\psi/\mathrm{d}x$. The quantity $\psi^*(x)\psi(x)$ is itself in the Copenhagen interpretation of quantum mechanics interpreted as a probability density function. This treatment of quantum mechanics, where a particle's wavefunction evolves against a classical background potential $V(x)$, is sometimes called *first quantization*.

This description of quantum mechanics can be extended to describe the behavior of multiple particles, so long as the number and the type of particles remain fixed. The particles are described by a wavefunction $\psi(x_1, x_2, ..., xN, t)$, which is governed by an extended version of the Schrödinger equation.

Often one is interested in the case where N particles are all of the same type (for example, the 18 electrons orbiting a neutral argon nucleus). As described in the article on identical particles, this implies that the state of the entire system must be either symmetric (bosons) or antisymmetric (fermions) when the coordinates of its constituent particles are exchanged. This is achieved by using a Slater determinant as the wavefunction of a fermionic system (and a Slater permanent for a bosonic system), which is equivalent to an element of the symmetric or antisymmetric subspace of a tensor product.

For example, the general quantum state of a system of N bosons is written as

$$|\phi_1 \cdots \phi_N\rangle = \sqrt{\frac{\prod_j N_j!}{N!}} \sum_{p \in S_N} |\phi_{p(1)}\rangle \otimes \cdots \otimes |\phi_{p(N)}\rangle,$$

where $|\phi_i\rangle$ are the single-particle states, Nj is the number of particles occupying state j, and the sum is taken over all possible permutations p acting on N elements. In general, this is a sum of $N!$ (N factorial) distinct terms. $\sqrt{\frac{\prod_j N_j!}{N!}}$ is a normalizing factor.

There are several shortcomings to the above description of quantum mechanics, which are addressed by quantum field theory. First, it is unclear how to extend quantum mechanics to include the effects of special relativity.[7] Attempted replacements for the Schrödinger equation, such as the Klein–Gordon equation or the Dirac equation, have many unsatisfactory qualities; for instance, they possess energy eigenvalues that extend to $-\infty$, so that there seems to be no easy definition of a ground state. It turns out that such inconsistencies arise from relativistic wavefunctions not having a well-defined probabilistic interpretation in position space, as probability conservation is not a relativistically covariant concept. The second shortcoming, related to the first, is that in quantum mechanics there is no mechanism to describe particle creation and annihilation;[8] this is crucial for describing phenomena such as pair production, which result from the conversion between mass and energy according to the relativistic relation $E = mc^2$.

4.4.3   Second quantization

Main article: Second quantization

In this section, we will describe a method for constructing a quantum field theory called **second quantization**. This basically involves choosing a way to index the quantum mechanical degrees of freedom in the space of multiple identical-particle states. It is based on the Hamiltonian formulation of quantum mechanics.

Several other approaches exist, such as the Feynman path integral,[9] which uses a Lagrangian formulation. For an overview of some of these approaches, see the article on quantization.

Bosons

For simplicity, we will first discuss second quantization for bosons, which form perfectly symmetric quantum states. Let us denote the mutually orthogonal single-particle states which are possible in the system by $|\phi_1\rangle, |\phi_2\rangle, |\phi_3\rangle$, and so on. For example, the 3-particle state with one particle in state $|\phi_1\rangle$ and two in state $|\phi_2\rangle$ is

$$\frac{1}{\sqrt{3}} \left[|\phi_1\rangle|\phi_2\rangle|\phi_2\rangle + |\phi_2\rangle|\phi_1\rangle|\phi_2\rangle + |\phi_2\rangle|\phi_2\rangle|\phi_1\rangle \right].$$

The first step in second quantization is to express such quantum states in terms of **occupation numbers**, by listing the number of particles occupying each of the single-particle states $|\phi_1\rangle, |\phi_2\rangle$, etc. This is simply another way of labelling the states. For instance, the above 3-particle state is denoted as

$$|1, 2, 0, 0, 0, \dots\rangle.$$

An N-particle state belongs to a space of states describing systems of N particles. The next step is to combine the individual N-particle state spaces into an extended state space, known as Fock space, which can describe systems of any number of particles. This is composed of the state space of a system with no particles (the so-called vacuum state, written as $|0\rangle$), plus the state space of a 1-particle system, plus the state space of a 2-particle system, and so forth. States describing a definite number of particles are known as Fock states: a general element of Fock space will be a linear combination of Fock states. There is a one-to-one correspondence between the occupation number representation and valid boson states in the Fock space.

At this point, the quantum mechanical system has become a quantum field in the sense we described above. The field's elementary degrees of freedom are the occupation numbers, and each occupation number is indexed by a number j indicating which of the single-particle states $|\phi_1\rangle, |\phi_2\rangle, \dots, |\phi_j\rangle, \dots$ it refers to:

$$|N_1, N_2, N_3, \dots, N_j, \dots\rangle.$$

The properties of this quantum field can be explored by defining creation and annihilation operators, which add and subtract particles. They are analogous to ladder operators in the quantum harmonic oscillator problem, which added and subtracted energy quanta. However, these operators literally create and annihilate particles of a given quantum state. The bosonic annihilation operator a_2 and creation operator $a_2^{\dagger}$ are easily defined in the occupation number representation as having the following effects:

$$a_2|N_1, N_2, N_3, \dots\rangle = \sqrt{N_2}\,|\,N_1, (N_2 - 1), N_3, \dots\rangle,$$

$$a_2^\dagger|N_1, N_2, N_3, \dots\rangle = \sqrt{N_2 + 1}\,|\,N_1, (N_2 + 1), N_3, \dots\rangle.$$

It can be shown that these are operators in the usual quantum mechanical sense, i.e. linear operators acting on the Fock space. Furthermore, they are indeed Hermitian conjugates, which justifies the way we have written them. They can be shown to obey the commutation relation

$$[a_i, a_j] = 0 \quad , \quad \left[a_i^\dagger, a_j^\dagger\right] = 0 \quad , \quad \left[a_i, a_j^\dagger\right] = \delta_{ij},$$

where δ stands for the Kronecker delta. These are precisely the relations obeyed by the ladder operators for an infinite set of independent quantum harmonic oscillators, one for each single-particle state. Adding or removing bosons from each state is therefore analogous to exciting or de-exciting a quantum of energy in a harmonic oscillator.

Applying an annihilation operator a_k followed by its corresponding creation operator $a_k^\dagger$ returns the number N_k of particles in the kth single-particle eigenstate:

$$a_k^\dagger a_k|\dots, N_k, \dots\rangle = N_k|\dots, N_k, \dots\rangle.$$

The combination of operators $a_k^\dagger a_k$ is known as the number operator for the kth eigenstate.

The Hamiltonian operator of the quantum field (which, through the Schrödinger equation, determines its dynamics) can be written in terms of creation and annihilation operators. For instance, for a field of free (non-interacting) bosons, the total energy of the field is found by summing the energies of the bosons in each energy eigenstate. If the kth single-particle energy eigenstate has energy E_k and there are N_k bosons in this state, then the total energy of these bosons is $E_k N_k$. The energy in the *entire* field is then a sum over k :

$$E_{\text{tot}} = \sum_k E_k N_k$$

This can be turned into the Hamiltonian operator of the field by replacing N_k with the corresponding number operator, $a_k^\dagger a_k$. This yields

$$H = \sum_k E_k\, a_k^\dagger a_k.$$

Fermions

It turns out that a different definition of creation and annihilation must be used for describing fermions. According to the Pauli exclusion principle, fermions cannot share quantum states, so their occupation numbers N_i can only take on the value 0 or 1. The fermionic annihilation operators c and creation operators $c^\dagger$ are defined by their actions on a Fock state thus

$$c_j|N_1, N_2, \dots, N_j = 0, \dots\rangle = 0$$

$$c_j|N_1, N_2, \dots, N_j = 1, \dots\rangle = (-1)^{(N_1 + \dots + N_{j-1})}|N_1, N_2, \dots, N_j = 0, \dots\rangle$$

$$c_j^\dagger|N_1, N_2, \dots, N_j = 0, \dots\rangle = (-1)^{(N_1 + \dots + N_{j-1})}|N_1, N_2, \dots, N_j = 1, \dots\rangle$$

$$c_j^\dagger | N_1, N_2, \ldots, N_j = 1, \ldots \rangle = 0.$$

These obey an anticommutation relation:

$$\{c_i, c_j\} = 0 \quad , \quad \left\{ c_i^\dagger, c_j^\dagger \right\} = 0 \quad , \quad \left\{ c_i, c_j^\dagger \right\} = \delta_{ij}.$$

One may notice from this that applying a fermionic creation operator twice gives zero, so it is impossible for the particles to share single-particle states, in accordance with the exclusion principle.

Field operators

We have previously mentioned that there can be more than one way of indexing the degrees of freedom in a quantum field. Second quantization indexes the field by enumerating the single-particle quantum states. However, as we have discussed, it is more natural to think about a "field", such as the electromagnetic field, as a set of degrees of freedom indexed by position.

To this end, we can define *field operators* that create or destroy a particle at a particular point in space. In particle physics, these operators turn out to be more convenient to work with, because they make it easier to formulate theories that satisfy the demands of relativity.

Single-particle states are usually enumerated in terms of their momenta (as in the particle in a box problem.) We can construct field operators by applying the Fourier transform to the creation and annihilation operators for these states. For example, the bosonic field annihilation operator $\phi(\mathbf{r})$ is

$$\phi(\mathbf{r}) \overset{\text{def}}{=} \sum_j e^{i\mathbf{k}_j \cdot \mathbf{r}} a_j.$$

The bosonic field operators obey the commutation relation

$$[\phi(\mathbf{r}), \phi(\mathbf{r}')] = 0 \quad , \quad \left[\phi^\dagger(\mathbf{r}), \phi^\dagger(\mathbf{r}') \right] = 0 \quad , \quad \left[\phi(\mathbf{r}), \phi^\dagger(\mathbf{r}') \right] = \delta^3(\mathbf{r} - \mathbf{r}')$$

where $\delta(x)$ stands for the Dirac delta function. As before, the fermionic relations are the same, with the commutators replaced by anticommutators.

The field operator is not the same thing as a single-particle wavefunction. The former is an operator acting on the Fock space, and the latter is a quantum-mechanical amplitude for finding a particle in some position. However, they are closely related, and are indeed commonly denoted with the same symbol. If we have a Hamiltonian with a space representation, say

$$H = -\frac{\hbar^2}{2m} \sum_i \nabla_i^2 + \sum_{i<j} U(|\mathbf{r}_i - \mathbf{r}_j|)$$

where the indices i and j run over all particles, then the field theory Hamiltonian (in the non-relativistic limit and for negligible self-interactions) is

$$H = -\frac{\hbar^2}{2m} \int d^3r \; \phi^\dagger(\mathbf{r}) \nabla^2 \phi(\mathbf{r}) + \frac{1}{2} \int d^3r \int d^3r' \; \phi^\dagger(\mathbf{r}) \phi^\dagger(\mathbf{r}') U(|\mathbf{r} - \mathbf{r}'|) \phi(\mathbf{r}') \phi(\mathbf{r}).$$

This looks remarkably like an expression for the expectation value of the energy, with ϕ playing the role of the wavefunction. This relationship between the field operators and wavefunctions makes it very easy to formulate field theories starting from space-projected Hamiltonians.

4.4.4 Dynamics

Once the Hamiltonian operator is obtained as part of the canonical quantization process, the time dependence of the state is described with the Schrödinger equation, just as with other quantum theories. Alternatively, the Heisenberg picture can be used where the time dependence is in the operators rather than in the states.

4.4.5 Implications

Unification of fields and particles

The "second quantization" procedure that we have outlined in the previous section takes a set of single-particle quantum states as a starting point. Sometimes, it is impossible to define such single-particle states, and one must proceed directly to quantum field theory. For example, a quantum theory of the electromagnetic field *must* be a quantum field theory, because it is impossible (for various reasons) to define a wavefunction for a single photon.[10] In such situations, the quantum field theory can be constructed by examining the mechanical properties of the classical field and guessing the corresponding quantum theory. For free (non-interacting) quantum fields, the quantum field theories obtained in this way have the same properties as those obtained using second quantization, such as well-defined creation and annihilation operators obeying commutation or anticommutation relations.

Quantum field theory thus provides a unified framework for describing "field-like" objects (such as the electromagnetic field, whose excitations are photons) and "particle-like" objects (such as electrons, which are treated as excitations of an underlying electron field), so long as one can treat interactions as "perturbations" of free fields. There are still unsolved problems relating to the more general case of interacting fields that may or may not be adequately described by perturbation theory. For more on this topic, see Haag's theorem.

Physical meaning of particle indistinguishability

The second quantization procedure relies crucially on the particles being identical. We would not have been able to construct a quantum field theory from a distinguishable many-particle system, because there would have been no way of separating and indexing the degrees of freedom.

Many physicists prefer to take the converse interpretation, which is that *quantum field theory explains what identical particles are*. In ordinary quantum mechanics, there is not much theoretical motivation for using symmetric (bosonic) or antisymmetric (fermionic) states, and the need for such states is simply regarded as an empirical fact. From the point of view of quantum field theory, particles are identical if and only if they are excitations of the same underlying quantum field. Thus, the question "why are all electrons identical?" arises from mistakenly regarding individual electrons as fundamental objects, when in fact it is only the electron field that is fundamental.

Particle conservation and non-conservation

During second quantization, we started with a Hamiltonian and state space describing a fixed number of particles (N), and ended with a Hamiltonian and state space for an arbitrary number of particles. Of course, in many common situations N is an important and perfectly well-defined quantity, e.g. if we are describing a gas of atoms sealed in a box. From the point of view of quantum field theory, such situations are described by quantum states that are eigenstates of the number operator $\hat{N}$, which measures the total number of particles present. As with any quantum mechanical observable, $\hat{N}$ is conserved if it commutes with the Hamiltonian. In that case, the quantum state is trapped in the N-particle subspace of the total Fock space, and the situation could equally well be described by ordinary N-particle quantum mechanics. (Strictly speaking, this is only true in the noninteracting case or in the low energy density limit of renormalized quantum field theories)

For example, we can see that the free-boson Hamiltonian described above conserves particle number. Whenever the Hamiltonian operates on a state, each particle destroyed by an annihilation operator a_k is immediately put back by the creation operator $a_k^\dagger$.

On the other hand, it is possible, and indeed common, to encounter quantum states that are *not* eigenstates of $\hat{N}$, which do not have well-defined particle numbers. Such states are difficult or impossible to handle using ordinary quantum mechanics, but they can be easily described in quantum field theory as quantum superpositions of states having different values of N. For example, suppose we have a bosonic field whose particles can be created or destroyed by interactions with a fermionic field. The Hamiltonian of the combined system would be given by the Hamiltonians of the free boson and free fermion fields, plus a "potential energy" term such as

$$H_I = \sum_{k,q} V_q (a_q + a_{-q}^\dagger) c_{k+q}^\dagger c_k,$$

where $a_k^\dagger$ and a_k denotes the bosonic creation and annihilation operators, $c_k^\dagger$ and c_k denotes the fermionic creation and annihilation operators, and V_q is a parameter that describes the strength of the interaction. This "interaction term" describes processes in which a fermion in state k either absorbs or emits a boson, thereby being kicked into a different eigenstate $k + q$. (In fact, this type of Hamiltonian is used to describe interaction between conduction electrons and phonons in metals. The interaction between electrons and photons is treated in a similar way, but is a little more complicated because the role of spin must be taken into account.) One thing to notice here is that even if we start out with a fixed number of bosons, we will typically end up with a superposition of states with different numbers of bosons at later times. The number of fermions, however, is conserved in this case.

In condensed matter physics, states with ill-defined particle numbers are particularly important for describing the various superfluids. Many of the defining characteristics of a superfluid arise from the notion that its quantum state is a superposition of states with different particle numbers. In addition, the concept of a coherent state (used to model the laser and the BCS ground state) refers to a state with an ill-defined particle number but a well-defined phase.

4.4.6   Axiomatic approaches

The preceding description of quantum field theory follows the spirit in which most physicists approach the subject. However, it is not mathematically rigorous. Over the past several decades, there have been many attempts to put quantum field theory on a firm mathematical footing by formulating a set of axioms for it. These attempts fall into two broad classes.

The first class of axioms, first proposed during the 1950s, include the Wightman, Osterwalder–Schrader, and Haag–Kastler systems. They attempted to formalize the physicists' notion of an "operator-valued field" within the context of functional analysis, and enjoyed limited success. It was possible to prove that any quantum field theory satisfying these axioms satisfied certain general theorems, such as the spin-statistics theorem and the CPT theorem. Unfortunately, it proved extraordinarily difficult to show that any realistic field theory, including the Standard Model, satisfied these axioms. Most of the theories that could be treated with these analytic axioms were physically trivial, being restricted to low-dimensions and lacking interesting dynamics. The construction of theories satisfying one of these sets of axioms falls in the field of constructive quantum field theory. Important work was done in this area in the 1970s by Segal, Glimm, Jaffe and others.

During the 1980s, a second set of axioms based on geometric ideas was proposed. This line of investigation, which restricts its attention to a particular class of quantum field theories known as topological quantum field theories, is associated most closely with Michael Atiyah and Graeme Segal, and was notably expanded upon by Edward Witten, Richard Borcherds, and Maxim Kontsevich. However, most of the physically relevant quantum field theories, such as the Standard Model, are not topological quantum field theories; the quantum field theory of the fractional quantum Hall effect is a notable exception. The main impact of axiomatic topological quantum field theory has been on mathematics, with important applications in representation theory, algebraic topology, and differential geometry.

Finding the proper axioms for quantum field theory is still an open and difficult problem in mathematics. One of the Millennium Prize Problems—proving the existence of a mass gap in Yang–Mills theory—is linked to this issue.

4.5 Associated phenomena

In the previous part of the article, we described the most general features of quantum field theories. Some of the quantum field theories studied in various fields of theoretical physics involve additional special ideas, such as renormalizability, gauge symmetry, and supersymmetry. These are described in the following sections.

4.5.1 Renormalization

Main article: Renormalization

Early in the history of quantum field theory, it was found that many seemingly innocuous calculations, such as the perturbative shift in the energy of an electron due to the presence of the electromagnetic field, give infinite results. The reason is that the perturbation theory for the shift in an energy involves a sum over all other energy levels, and there are infinitely many levels at short distances that each give a finite contribution which results in a divergent series.

Many of these problems are related to failures in classical electrodynamics that were identified but unsolved in the 19th century, and they basically stem from the fact that many of the supposedly "intrinsic" properties of an electron are tied to the electromagnetic field that it carries around with it. The energy carried by a single electron—its self energy— is not simply the bare value, but also includes the energy contained in its electromagnetic field, its attendant cloud of photons. The energy in a field of a spherical source diverges in both classical and quantum mechanics, but as discovered by Weisskopf with help from Furry, in quantum mechanics the divergence is much milder, going only as the logarithm of the radius of the sphere.

The solution to the problem, presciently suggested by Stueckelberg, independently by Bethe after the crucial experiment by Lamb, implemented at one loop by Schwinger, and systematically extended to all loops by Feynman and Dyson, with converging work by Tomonaga in isolated postwar Japan, comes from recognizing that all the infinities in the interactions of photons and electrons can be isolated into redefining a finite number of quantities in the equations by replacing them with the observed values: specifically the electron's mass and charge: this is called renormalization. The technique of renormalization recognizes that the problem is essentially purely mathematical, that extremely short distances are at fault. In order to define a theory on a continuum, first place a cutoff on the fields, by postulating that quanta cannot have energies above some extremely high value. This has the effect of replacing continuous space by a structure where very short wavelengths do not exist, as on a lattice. Lattices break rotational symmetry, and one of the crucial contributions made by Feynman, Pauli and Villars, and modernized by 't Hooft and Veltman, is a symmetry-preserving cutoff for perturbation theory (this process is called regularization). There is no known symmetrical cutoff outside of perturbation theory, so for rigorous or numerical work people often use an actual lattice.

On a lattice, every quantity is finite but depends on the spacing. When taking the limit of zero spacing, we make sure that the physically observable quantities like the observed electron mass stay fixed, which means that the constants in the Lagrangian defining the theory depend on the spacing. Hopefully, by allowing the constants to vary with the lattice spacing, all the results at long distances become insensitive to the lattice, defining a continuum limit.

The renormalization procedure only works for a certain class of quantum field theories, called **renormalizable quantum field theories**. A theory is **perturbatively renormalizable** when the constants in the Lagrangian only diverge at worst as logarithms of the lattice spacing for very short spacings. The continuum limit is then well defined in perturbation theory, and even if it is not fully well defined non-perturbatively, the problems only show up at distance scales that are exponentially small in the inverse coupling for weak couplings. The Standard Model of particle physics is perturbatively renormalizable, and so are its component theories (quantum electrodynamics/electroweak theory and quantum chromodynamics). Of the three components, quantum electrodynamics is believed to not have a continuum limit, while the asymptotically free $SU(2)$ and $SU(3)$ weak hypercharge and strong color interactions are nonperturbatively well defined.

The renormalization group describes how renormalizable theories emerge as the long distance low-energy effective field theory for any given high-energy theory. Because of this, renormalizable theories are insensitive to the precise nature of the underlying high-energy short-distance phenomena. This is a blessing because it allows physicists to formulate low energy theories without knowing the details of high energy phenomenon. It is also a curse, because once a renormalizable theory like the standard model is found to work, it gives very few clues to higher energy processes. The only way high energy processes can be seen in the standard model is when they allow otherwise forbidden events, or if they predict quantitative relations between the coupling constants.

4.5.2 Haag's theorem

See also: Haag's theorem

From a mathematically rigorous perspective, there exists no interaction picture in a Lorentz-covariant quantum field theory. This implies that the perturbative approach of Feynman diagrams in QFT is not strictly justified, despite producing vastly precise predictions validated by experiment. This is called Haag's theorem, but most particle physicists relying on QFT largely shrug it off.

4.5.3 Gauge freedom

A gauge theory is a theory that admits a symmetry with a local parameter. For example, in every quantum theory the global phase of the wave function is arbitrary and does not represent something physical. Consequently, the theory is invariant under a global change of phases (adding a constant to the phase of all wave functions, everywhere); this is a global symmetry. In quantum electrodynamics, the theory is also invariant under a *local* change of phase, that is – one may shift the phase of all wave functions so that the shift may be different at every point in space-time. This is a *local* symmetry. However, in order for a well-defined derivative operator to exist, one must introduce a new field, the gauge field, which also transforms in order for the local change of variables (the phase in our example) not to affect the derivative. In quantum electrodynamics this gauge field is the electromagnetic field. The change of local gauge of variables is termed gauge transformation. It is worth noting that by Noether's theorem, for every such symmetry there exists an associated conserved current. The aforementioned symmetry of the wavefunction under global phase changes implies the conservation of electric charge.

In quantum field theory the excitations of fields represent particles. The particle associated with excitations of the gauge field is the gauge boson, which is the photon in the case of quantum electrodynamics.

The degrees of freedom in quantum field theory are local fluctuations of the fields. The existence of a gauge symmetry reduces the number of degrees of freedom, simply because some fluctuations of the fields can be transformed to zero by gauge transformations, so they are equivalent to having no fluctuations at all, and they therefore have no physical meaning. Such fluctuations are usually called "non-physical degrees of freedom" or *gauge artifacts*; usually some of them have a negative norm, making them inadequate for a consistent theory. Therefore, if a classical field theory has a gauge symmetry, then its quantized version (i.e. the corresponding quantum field theory) will have this symmetry as well. In other words, a gauge symmetry cannot have a quantum anomaly. If a gauge symmetry is anomalous (i.e. not kept in the quantum theory) then the theory is non-consistent: for example, in quantum electrodynamics, had there been a gauge anomaly, this would require the appearance of photons with longitudinal polarization and polarization in the time direction, the latter having a negative norm, rendering the theory inconsistent; another possibility would be for these photons to appear only in intermediate processes but not in the final products of any interaction, making the theory non-unitary and again inconsistent (see optical theorem).

In general, the gauge transformations of a theory consist of several different transformations, which may not be commutative. These transformations are together described by a mathematical object known as a gauge group. Infinitesimal gauge transformations are the gauge group generators. Therefore the number of gauge bosons is the group dimension (i.e. number of generators forming a basis).

All the fundamental interactions in nature are described by gauge theories. These are:

- Quantum chromodynamics, whose gauge group is $\mathbf{SU}(3)$. The gauge bosons are eight gluons.

- The electroweak theory, whose gauge group is $\mathbf{U}(1) \times \mathbf{SU}(2)$, (a direct product of $\mathbf{U}(1)$ and $\mathbf{SU}(2)$).

- Gravity, whose classical theory is general relativity, admits the equivalence principle, which is a form of gauge symmetry. However, it is explicitly non-renormalizable.

4.5.4 Multivalued gauge transformations

The gauge transformations which leave the theory invariant involve, by definition, only single-valued gauge functions $\Lambda(x_i)$ which satisfy the Schwarz integrability criterion

$$\partial_{x_i x_j}\Lambda = \partial_{x_j x_i}\Lambda.$$

An interesting extension of gauge transformations arises if the gauge functions $\Lambda(x_i)$ are allowed to be multivalued functions which violate the integrability criterion. These are capable of changing the physical field strengths and are therefore not proper symmetry transformations. Nevertheless, the transformed field equations describe correctly the physical laws in the presence of the newly generated field strengths. See the textbook by H. Kleinert cited below for the applications to phenomena in physics.

4.5.5 Supersymmetry

Main article: Supersymmetry

Supersymmetry assumes that every fundamental fermion has a superpartner that is a boson and vice versa. It was introduced in order to solve the so-called Hierarchy Problem, that is, to explain why particles not protected by any symmetry (like the Higgs boson) do not receive radiative corrections to its mass driving it to the larger scales (GUT, Planck...). It was soon realized that supersymmetry has other interesting properties: its gauged version is an extension of general relativity (Supergravity), and it is a key ingredient for the consistency of string theory.

The way supersymmetry protects the hierarchies is the following: since for every particle there is a superpartner with the same mass, any loop in a radiative correction is cancelled by the loop corresponding to its superpartner, rendering the theory UV finite.

Since no superpartners have yet been observed, if supersymmetry exists it must be broken (through a so-called soft term, which breaks supersymmetry without ruining its helpful features). The simplest models of this breaking require that the energy of the superpartners not be too high; in these cases, supersymmetry is expected to be observed by experiments at the Large Hadron Collider. The Higgs particle has been detected at the LHC, and no such superparticles have been discovered.

4.6 See also

- Abraham–Lorentz force

- Basic concepts of quantum mechanics

- Common integrals in quantum field theory

- Constructive quantum field theory

- Einstein–Maxwell–Dirac equations

- Feynman path integral

- Form factor (quantum field theory)

- Fundamental equation of unified field theory

- Green–Kubo relations

- Green's function (many-body theory)

- Invariance mechanics

- List of quantum field theories

- Pauli exclusion principle

- Photon polarization

- Pseudoscalar Field

- Quantum field theory in curved spacetime

- Quantum flavordynamics

- Quantum geometrodynamics

- Quantum hydrodynamics

- Quantum magnetodynamics

- Quantum triviality

- Relation between Schrödinger's equation and the path integral formulation of quantum mechanics

- Relationship between string theory and quantum field theory

- Schwinger–Dyson equation

- Static forces and virtual-particle exchange

- Symmetry in quantum mechanics

- Theoretical and experimental justification for the Schrödinger equation

- Ward–Takahashi identity

- Wheeler–Feynman absorber theory

- Wigner's classification

- Wigner's theorem

4.7 Notes

4.8 References

[1] "Beautiful Minds, Vol. 20: Ed Witten". la Repubblica. 2010. Retrieved 22 June 2012. See here.

[2] J. J. Thorn et al. (2004). Observing the quantum behavior of light in an undergraduate laboratory. . J. J. Thorn, M. S. Neel, V. W. Donato, G. S. Bergreen, R. E. Davies, and M. Beck. American Association of Physics Teachers, 2004.DOI: 10.1119/1.1737397.

[3] David Tong, *Lectures on Quantum Field Theory*, chapter 1.

[4] Srednicki, Mark. *Quantum Field Theory* (1st ed.). p. 19.

[5] Srednicki, Mark. *Quantum Field Theory* (1st ed.). pp. 25–6.

[6] Zee, Anthony. *Quantum Field Theory in a Nutshell* (2nd ed.). p. 61.

[7] David Tong, *Lectures on Quantum Field Theory*, Introduction.

[8] Zee, Anthony. *Quantum Field Theory in a Nutshell* (2nd ed.). p. 3.

[9] Abraham Pais, *Inward Bound: Of Matter and Forces in the Physical World* ISBN 0-19-851997-4. Pais recounts how his astonishment at the rapidity with which Feynman could calculate using his method. Feynman's method is now part of the standard methods for physicists.

[10] Newton, T.D.; Wigner, E.P. (1949). "Localized states for elementary systems". *Reviews of Modern Physics* **21** (3): 400–406. Bibcode:1949RvMP...21..400N. doi:10.1103/RevModPhys.21.400.

4.9   Further reading

General readers

- Feynman, R.P. (2001) [1964]. *The Character of Physical Law*. MIT Press. ISBN 0-262-56003-8.

- Feynman, R.P. (2006) [1985]. *QED: The Strange Theory of Light and Matter*. Princeton University Press. ISBN 0-691-12575-9.

- Gribbin, J. (1998). *Q is for Quantum: Particle Physics from A to Z*. Weidenfeld & Nicolson. ISBN 0-297-81752-3.

- Schumm, Bruce A. (2004) *Deep Down Things*. Johns Hopkins Univ. Press. Chpt. 4.

Introductory texts

- McMahon, D. (2008). *Quantum Field Theory*. McGraw-Hill. ISBN 978-0-07-154382-8.

- Bogoliubov, N.; Shirkov, D. (1982). *Quantum Fields*. Benjamin-Cummings. ISBN 0-8053-0983-7.

- Frampton, P.H. (2000). *Gauge Field Theories. Frontiers in Physics (2nd ed.). Wiley.*

- Greiner, W; Müller, B. (2000). *Gauge Theory of Weak Interactions*. Springer. ISBN 3-540-67672-4.

- Itzykson, C.; Zuber, J.-B. (1980). *Quantum Field Theory*. McGraw-Hill. ISBN 0-07-032071-3.

- Kane, G.L. (1987). *Modern Elementary Particle Physics*. Perseus Books. ISBN 0-201-11749-5.

- Kleinert, H.; Schulte-Frohlinde, Verena (2001). *Critical Properties of φ^4-Theories*. World Scientific. ISBN 981-02-4658-7.

- Kleinert, H. (2008). *Multivalued Fields in Condensed Matter, Electrodynamics, and Gravitation* (PDF). World Scientific. ISBN 978-981-279-170-2.

- Loudon, R (1983). *The Quantum Theory of Light*. Oxford University Press. ISBN 0-19-851155-8.

- Mandl, F.; Shaw, G. (1993). *Quantum Field Theory*. John Wiley & Sons. ISBN 978-0-471-94186-6.

- Peskin, M.; Schroeder, D. (1995). *An Introduction to Quantum Field Theory*. Westview Press. ISBN 0-201-50397-2.

- Ryder, L.H. (1985). *Quantum Field Theory*. Cambridge University Press. ISBN 0-521-33859-X.

- Schwartz, M.D. (2014). *Quantum Field Theory and the Standard Model*. Cambridge University Press. ISBN 978-1107034730.

- Srednicki, Mark (2007) *Quantum Field Theory*. Cambridge Univ. Press.

- Ynduráin, F.J. (1996). *Relativistic Quantum Mechanics and Introduction to Field Theory* (1st ed.). Springer. ISBN 978-3-540-60453-2.

- Zee, A. (2003). *Quantum Field Theory in a Nutshell*. Princeton University Press. ISBN 0-691-01019-6.

Advanced texts

- Brown, Lowell S. (1994). *Quantum Field Theory*. Cambridge University Press. ISBN 978-0-521-46946-3.

- Bogoliubov, N.; Logunov, A.A.; Oksak, A.I.; Todorov, I.T. (1990). *General Principles of Quantum Field Theory*. Kluwer Academic Publishers. ISBN 978-0-7923-0540-8.

- Weinberg, S. (1995). *The Quantum Theory of Fields* **1–3**. Cambridge University Press.

Articles:

- Gerard 't Hooft (2007) "The Conceptual Basis of Quantum Field Theory" in Butterfield, J., and John Earman, eds., *Philosophy of Physics, Part A*. Elsevier: 661–730.

- Frank Wilczek (1999) "Quantum field theory", *Reviews of Modern Physics* 71: S83–S95. Also doi=10.1103/Rev. Mod. Phys. 71.

4.10   External links

- Hazewinkel, Michiel, ed. (2001), "Quantum field theory", *Encyclopedia of Mathematics*, Springer, ISBN 978-1-55608-010-4

- Stanford Encyclopedia of Philosophy: "Quantum Field Theory", by Meinard Kuhlmann.

- Siegel, Warren, 2005. *Fields*. A free text, also available from arXiv:hep-th/9912205.

- Quantum Field Theory by P. J. Mulders

Chapter 5

Field (physics)

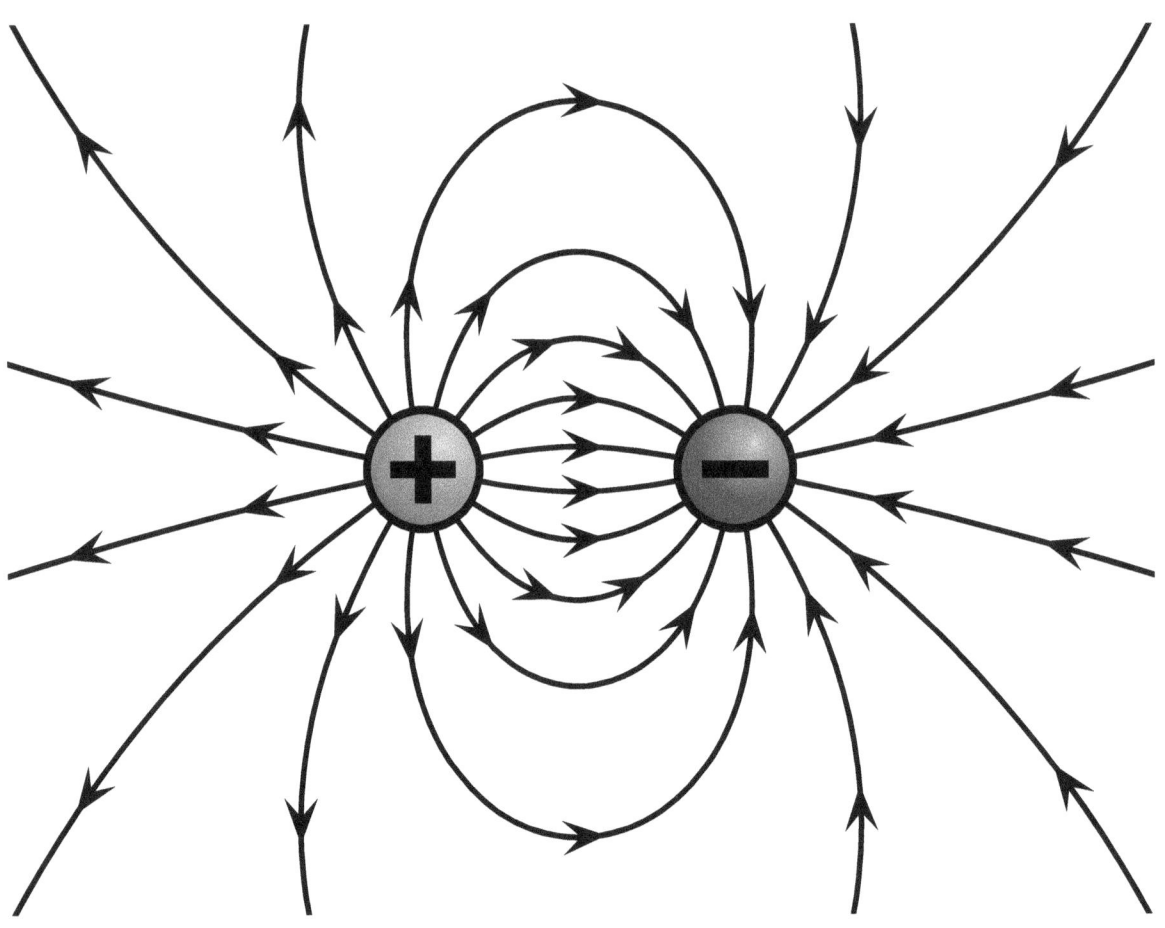

Illustration of the electric field surrounding a positive (red) and a negative (blue) charge.

In physics, a **field** is a physical quantity that has a value for each point in space and time.[1][2][3] For example, on a weather map, the surface wind velocity is described by assigning a vector to each point on a map. Each vector represents the speed and direction of the movement of air at that point. As another example, an electric field can be thought of as a "condition in space"[4] emanating from an electric charge and extending throughout the whole of space. When a test electric charge is placed in this electric field, the particle accelerates due to a force. Physicists have found the notion of a field to be of such practical utility for the analysis of forces that they have come to think of a force as due to a field.[5]

In the modern framework of the quantum theory of fields, even without referring to a test particle, a field occupies space, contains energy, and its presence eliminates a true vacuum.[6] This lead physicists to consider electromagnetic fields to be a physical entity, making the field concept a supporting paradigm of the edifice of modern physics. "The fact that the electromagnetic field can possess momentum and energy makes it very real... a particle makes a field, and a field acts on another particle, and the field has such familiar properties as energy content and momentum, just as particles can have".[7] In practice, the strength of most fields has been found to diminish with distance to the point of being undetectable. For instance the strength of many relevant classical fields, such as the gravitational field in Newton's theory of gravity or the electrostatic field in classical electromagnetism, is inversely proportional to the square of the distance from the source (i.e. they follow the Gauss's law). One consequence is that the Earth's gravitational field quickly becomes undetectable on cosmic scales.

A field can be classified as a scalar field, a vector field, a spinor field or a tensor field according to whether the represented physical quantity is a scalar, a vector, a spinor or a tensor, respectively. A field has a unique tensorial character in every point where it is defined: i.e. a field cannot be a scalar field somewhere and a vector field somewhere else. For example, the Newtonian gravitational field is a vector field: specifying its value at a point in spacetime requires three numbers, the components of the gravitational field vector at that point. Moreover, within each category (scalar, vector, tensor), a field can be either a *classical field* or a *quantum field*, depending on whether it is characterized by numbers or quantum operators respectively. In fact in this theory an equivalent representation of field is a field particle, namely a boson.[8]

5.1   History

To Isaac Newton his law of universal gravitation simply expressed the gravitational force that acted between any pair of massive objects. When looking at the motion of many bodies all interacting with each other, such as the planets in the Solar System, dealing with the force between each pair of bodies separately rapidly becomes computationally inconvenient. In the eighteenth century, a new quantity was devised to simplify the bookkeeping of all these gravitational forces. This quantity, the gravitational field, gave at each point in space the total gravitational force which would be felt by an object with unit mass at that point. This did not change the physics in any way: it did not matter if you calculated all the gravitational forces on an object individually and then added them together, or if you first added all the contributions together as a gravitational field and then applied it to an object.[9]

The development of the independent concept of a field truly began in the nineteenth century with the development of the theory of electromagnetism. In the early stages, André-Marie Ampère and Charles-Augustin de Coulomb could manage with Newton-style laws that expressed the forces between pairs of electric charges or electric currents. However, it became much more natural to take the field approach and express these laws in terms of electric and magnetic fields; in 1849 Michael Faraday became the first to coin the term "field".[9]

The independent nature of the field became more apparent with James Clerk Maxwell's discovery that waves in these fields propagated at a finite speed. Consequently, the forces on charges and currents no longer just depended on the positions and velocities of other charges and currents at the same time, but also on their positions and velocities in the past.[9]

Maxwell, at first, did not adopt the modern concept of a field as fundamental quantity that could independently exist. Instead, he supposed that the electromagnetic field expressed the deformation of some underlying medium—the luminiferous aether—much like the tension in a rubber membrane. If that were the case, the observed velocity of the electromagnetic waves should depend upon the velocity of the observer with respect to the aether. Despite much effort, no experimental evidence of such an effect was ever found; the situation was resolved by the introduction of the special theory of relativity, by Albert Einstein in 1905. This theory changed the way the viewpoints of moving observers should be related to each other in such a way that velocity of electromagnetic waves in Maxwell's theory would be the same for all observers. By doing away with the need for a background medium, this development opened the way for physicists to start thinking about fields as truly independent entities.[9]

In the late 1920s, the new rules of quantum mechanics were first applied to the electromagnetic fields. In 1927, Paul Dirac used quantum fields to successfully explain how the decay of an atom to lower quantum state lead to the spontaneous emission of a photon, the quantum of the electromagnetic field. This was soon followed by the realization (following the work of Pascual Jordan, Eugene Wigner, Werner Heisenberg, and Wolfgang Pauli) that all particles, including electrons and protons, could be understood as the quanta of some quantum field, elevating fields to the status of the most fundamental objects in nature.[9] That said, John Wheeler and Richard Feynman seriously considered Newton's pre-field concept of action at a distance (although they set it aside because of the ongoing utility of the field concept for research in general relativity and quantum electrodynamics).

5.2 Classical fields

Main article: Classical field theory

There are several examples of classical fields. Classical field theories remain useful wherever quantum properties do not arise, and can be active areas of research. Elasticity of materials, fluid dynamics and Maxwell's equations are cases in point.

Some of the simplest physical fields are vector force fields. Historically, the first time that fields were taken seriously was with Faraday's lines of force when describing the electric field. The gravitational field was then similarly described.

5.2.1 Newtonian gravitation

A classical field theory describing gravity is Newtonian gravitation, which describes the gravitational force as a mutual interaction between two masses.

Any body with mass M is associated with a gravitational field $\mathbf{g}$ which describes its influence on other bodies with mass. The gravitational field of M at a point $\mathbf{r}$ in space corresponds to the ratio between force $\mathbf{F}$ that M exerts on a small or negligible test mass m located at $\mathbf{r}$ and the test mass itself:[10]

$$\mathbf{g}(\mathbf{r}) = \frac{\mathbf{F}(\mathbf{r})}{m}.$$

Stipulating that m is much smaller than M ensures that the presence of m has a negligible influence on the behavior of M. According to Newton's law of universal gravitation, $\mathbf{F}(\mathbf{r})$ is given by[10]

$$\mathbf{F}(\mathbf{r}) = -\frac{GMm}{r^2}\hat{\mathbf{r}},$$

where $\hat{\mathbf{r}}$ is a unit vector lying along the line joining M and m and pointing from m to M. Therefore, the gravitational field of $\mathbf{M}$ is[10]

$$\mathbf{g}(\mathbf{r}) = \frac{\mathbf{F}(\mathbf{r})}{m} = -\frac{GM}{r^2}\hat{\mathbf{r}}.$$

The experimental observation that inertial mass and gravitational mass are equal to an unprecedented level of accuracy leads to the identity that gravitational field strength is identical to the acceleration experienced by a particle. This is the starting point of the equivalence principle, which leads to general relativity.

Because the gravitational force $\mathbf{F}$ is conservative, the gravitational field $\mathbf{g}$ can be rewritten in terms of the gradient of a scalar function, the gravitational potential $\Phi(\mathbf{r})$:

$$\mathbf{g}(\mathbf{r}) = -\nabla\Phi(\mathbf{r}).$$

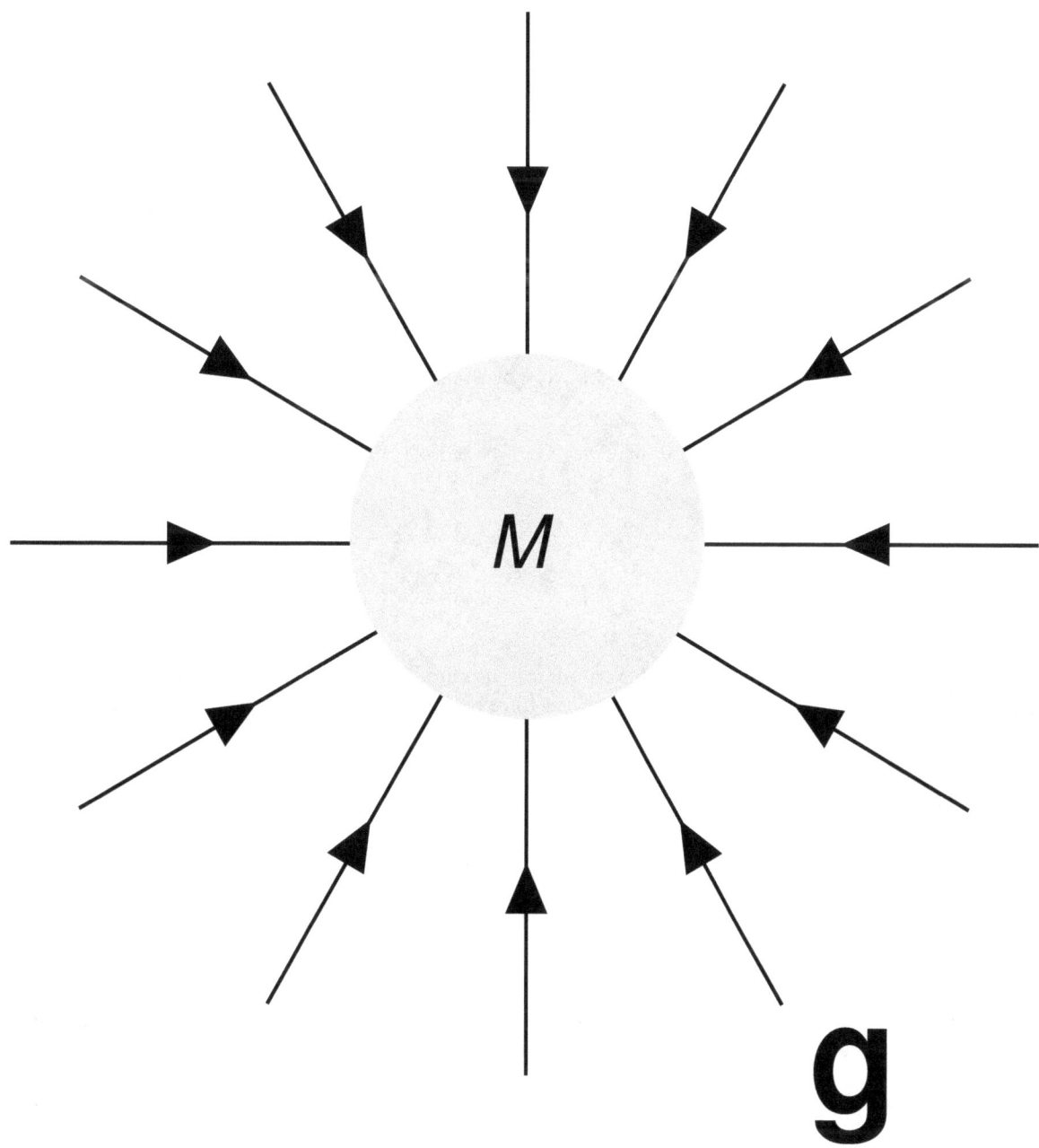

In classical gravitation, mass is the source of an attractive gravitational field **g***.*

5.2.2 Electromagnetism

Main article: Electromagnetism

Michael Faraday first realized the importance of a field as a physical quantity, during his investigations into magnetism. He realized that electric and magnetic fields are not only fields of force which dictate the motion of particles, but also have an independent physical reality because they carry energy.

These ideas eventually led to the creation, by James Clerk Maxwell, of the first unified field theory in physics with the introduction of equations for the electromagnetic field. The modern version of these equations is called Maxwell's equations.

Electrostatics

Main article: Electrostatics

A charged test particle with charge q experiences a force $\mathbf{F}$ based solely on its charge. We can similarly describe the electric field $\mathbf{E}$ so that $\mathbf{F} = q\mathbf{E}$. Using this and Coulomb's law tells us that the electric field due to a single charged particle as

$$\mathbf{E} = \frac{1}{4\pi\epsilon_0} \frac{q}{r^2} \hat{\mathbf{r}}.$$

The electric field is conservative, and hence can be described by a scalar potential, $V(\mathbf{r})$:

$$\mathbf{E}(\mathbf{r}) = -\nabla V(\mathbf{r}).$$

Magnetostatics

Main article: Magnetostatics

A steady current I flowing along a path ℓ will exert a force on nearby moving charged particles that is quantitatively different from the electric field force described above. The force exerted by I on a nearby charge q with velocity $\mathbf{v}$ is

$$\mathbf{F}(\mathbf{r}) = q\mathbf{v} \times \mathbf{B}(\mathbf{r}),$$

where $\mathbf{B}(\mathbf{r})$ is the magnetic field, which is determined from I by the Biot–Savart law:

$$\mathbf{B}(\mathbf{r}) = \frac{\mu_0 I}{4\pi} \int \frac{d\boldsymbol{\ell} \times d\hat{\mathbf{r}}}{r^2}.$$

The magnetic field is not conservative in general, and hence cannot usually be written in terms of a scalar potential. However, it can be written in terms of a vector potential, $\mathbf{A}(\mathbf{r})$:

$$\mathbf{B}(\mathbf{r}) = \nabla \times \mathbf{A}(\mathbf{r})$$

Electrodynamics

Main article: Electrodynamics

In general, in the presence of both a charge density $\rho(\mathbf{r}, t)$ and current density $\mathbf{J}(\mathbf{r}, t)$, there will be both an electric and a magnetic field, and both will vary in time. They are determined by Maxwell's equations, a set of differential equations which directly relate $\mathbf{E}$ and $\mathbf{B}$ to ρ and $\mathbf{J}$.[13]

Alternatively, one can describe the system in terms of its scalar and vector potentials V and $\mathbf{A}$. A set of integral equations known as *retarded potentials* allow one to calculate V and $\mathbf{A}$ from ρ and $\mathbf{J}$,[note 1] and from there the electric and magnetic fields are determined via the relations[14]

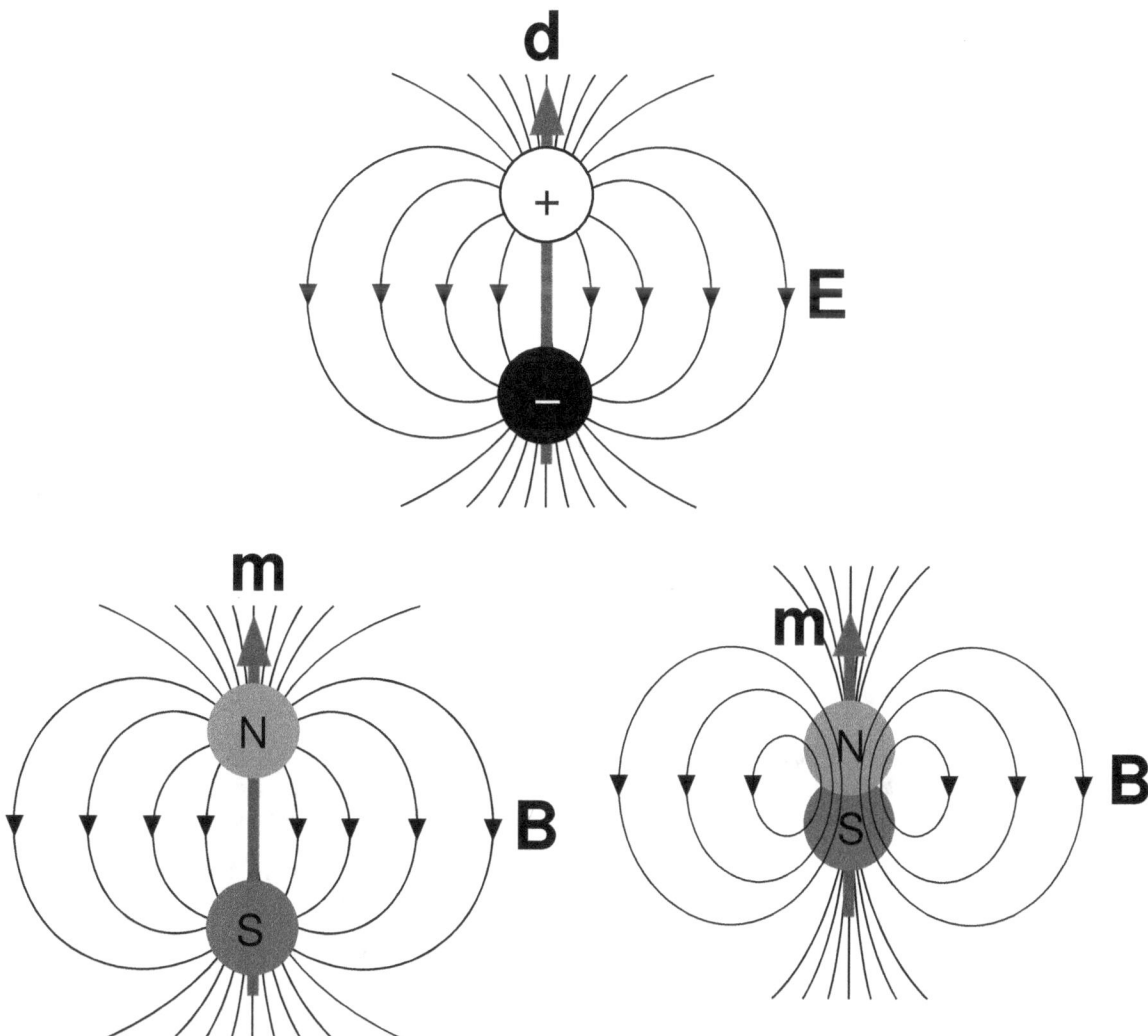

The E fields and B fields due to electric charges (black/white) and magnetic poles (red/blue).[11][12] **Top: E** *field due to an electric dipole moment* **d**. **Bottom left: B** *field due to a* mathematical *magnetic dipole* **m** *formed by two magnetic monopoles.* **Bottom right: B** *field due to a pure magnetic dipole moment* **m** *found in ordinary matter (not from monopoles).*

$$\mathbf{E} = -\boldsymbol{\nabla}V - \frac{\partial \mathbf{A}}{\partial t}$$

$$\mathbf{B} = \boldsymbol{\nabla} \times \mathbf{A}.$$

At the end of the 19th century, the electromagnetic field was understood as a collection of two vector fields in space. Nowadays, one recognizes this as a single antisymmetric 2nd-rank tensor field in spacetime.

5.2.3   Gravitation in general relativity

Einstein's theory of gravity, called general relativity, is another example of a field theory. Here the principal field is the metric tensor, a symmetric 2nd-rank tensor field in spacetime. This replaces Newton's law of universal gravitation.

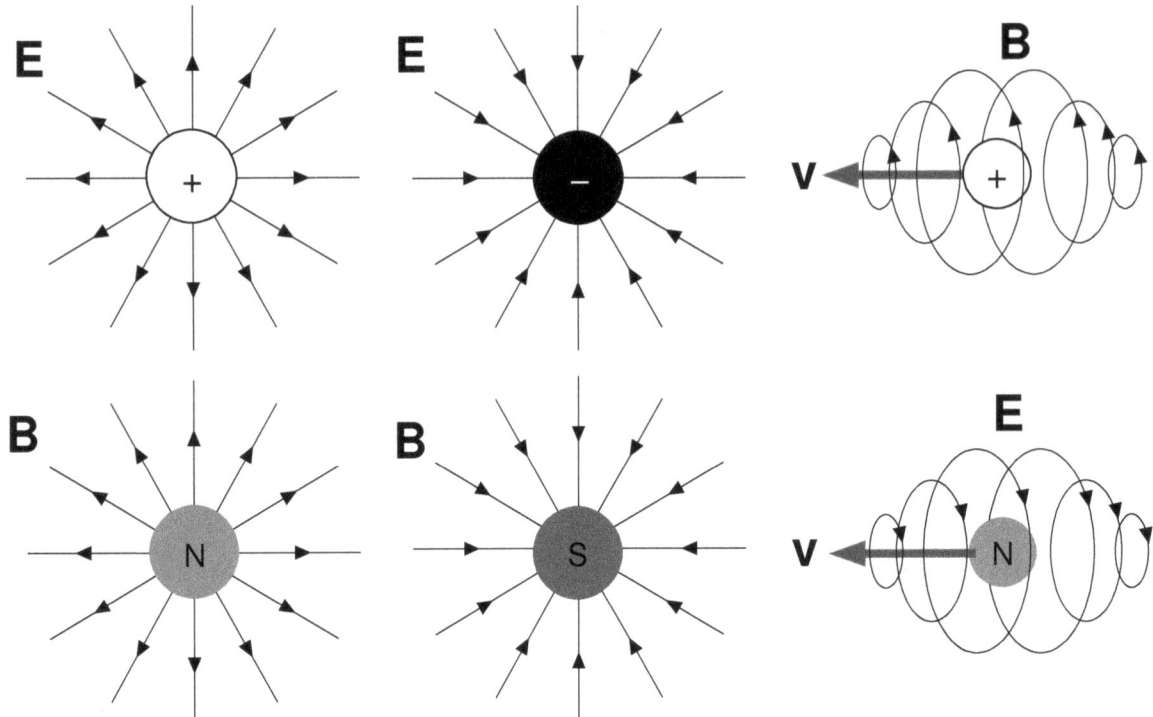

The **E** fields and **B** fields due to electric charges (black/white) and magnetic poles (red/blue).[11][12] **E** fields due to stationary electric charges and **B** fields due to stationary magnetic charges (note in nature N and S monopoles do not exist). In motion (velocity **v**), an electric charge induces a **B** field while a magnetic charge (not found in nature) would induce an **E** field. Conventional current is used.

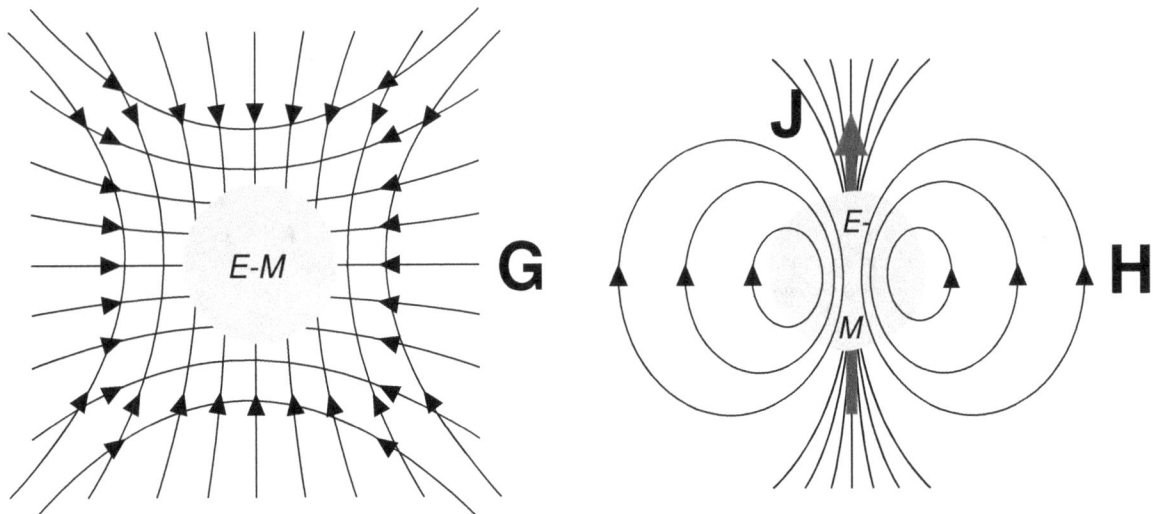

In general relativity, mass-energy warps space time (Einstein tensor **G**),[15] and rotating asymmetric mass-energy distributions with angular momentum **J** generate GEM fields **H**[16]

5.2.4 Waves as fields

Waves can be constructed as physical fields, due to their finite propagation speed and causal nature when a simplified physical model of an isolated closed system is set . They are also subject to the inverse-square law.

For electromagnetic waves, there are optical fields, and terms such as near- and far-field limits for diffraction. In practice, though the field theories of optics are superseded by the electromagnetic field theory of Maxwell.

5.3   Quantum fields

Main article: Quantum field theory

It is now believed that quantum mechanics should underlie all physical phenomena, so that a classical field theory should, at least in principle, permit a recasting in quantum mechanical terms; success yields the corresponding quantum field theory. For example, quantizing classical electrodynamics gives quantum electrodynamics. Quantum electrodynamics is arguably the most successful scientific theory; experimental data confirm its predictions to a higher precision (to more significant digits) than any other theory.[17] The two other fundamental quantum field theories are quantum chromodynamics and the electroweak theory.

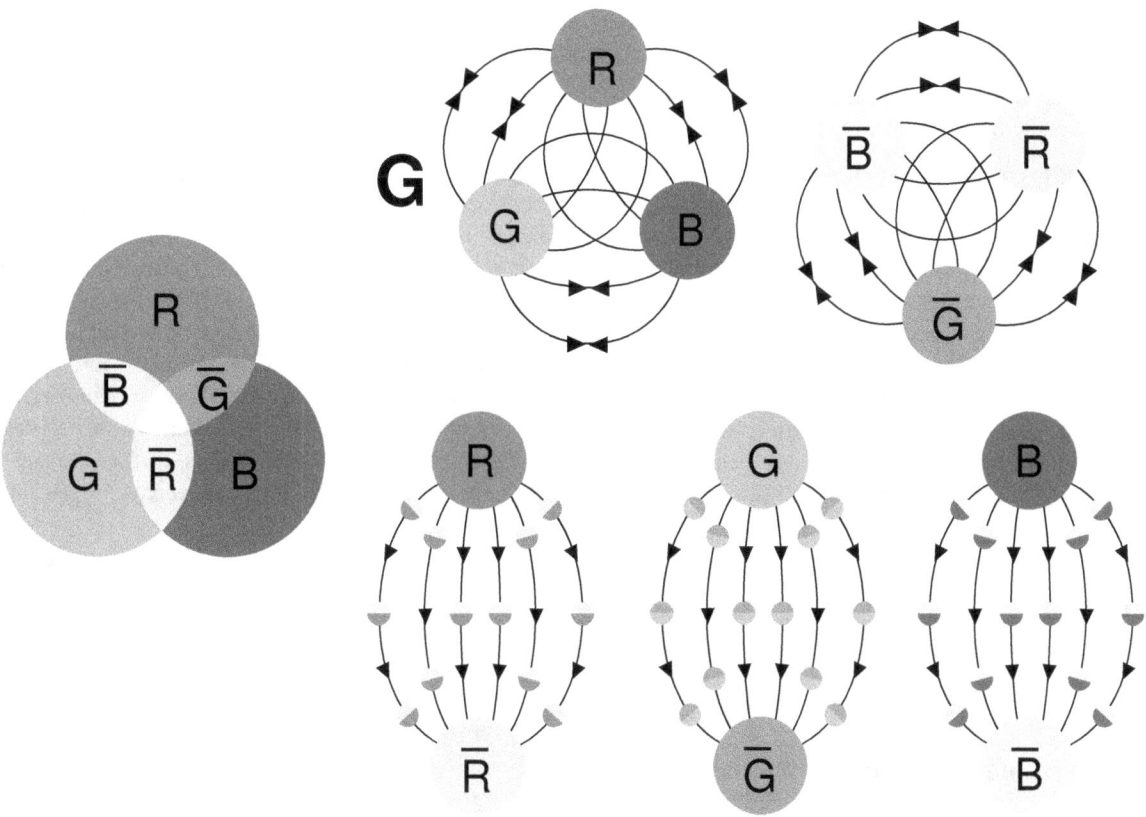

*Fields due to color charges, like in quarks (**G** is the gluon field strength tensor). These are "colorless" combinations. **Top:** Color charge has "ternary neutral states" as well as binary neutrality (analogous to electric charge). **Bottom:** The quark/antiquark combinations.[11][12]*

In quantum chromodynamics, the color field lines are coupled at short distances by gluons, which are polarized by the field and line up with it. This effect increases within a short distance (around 1 fm from the vicinity of the quarks) making the color force increase within a short distance, confining the quarks within hadrons. As the field lines are pulled together tightly by gluons, they do not "bow" outwards as much as an electric field between electric charges.[18]

These three quantum field theories can all be derived as special cases of the so-called standard model of particle physics. General relativity, the Einsteinian field theory of gravity, has yet to be successfully quantized. However an extension, thermal field theory, deals with quantum field theory at *finite temperatures*, something seldom considered in quantum field theory.

In BRST theory one deals with odd fields, e.g. Faddeev–Popov ghosts. There are different descriptions of odd classical fields both on graded manifolds and supermanifolds.

As above with classical fields, it is possible to approach their quantum counterparts from a purely mathematical view using similar techniques as before. The equations governing the quantum fields are in fact PDEs (specifically, relativistic wave equations (RWEs)). Thus one can speak of Yang–Mills, Dirac, Klein–Gordon and Schrödinger fields as being solutions to their respective equations. A possible problem is that these RWEs can deal with complicated mathematical objects with exotic algebraic properties (e.g. spinors are not tensors, so may need calculus over spinor fields), but these in theory can still be subjected to analytical methods given appropriate mathematical generalization.

5.4   Field theory

Field theory usually refers to a construction of the dynamics of a field, i.e. a specification of how a field changes with time or with respect to other independent physical variables on which the field depends. Usually this is done by writing a Lagrangian or a Hamiltonian of the field, and treating it as the classical mechanics (or quantum mechanics) of a system with an infinite number of degrees of freedom. The resulting field theories are referred to as classical or quantum field theories.

The dynamics of a classical field are usually specified by the Lagrangian density in terms of the field components; the dynamics can be obtained by using the action principle.

It is possible to construct simple fields without any a priori knowledge of physics using only mathematics from several variable calculus, potential theory and partial differential equations (PDEs). For example, scalar PDEs might consider quantities such as amplitude, density and pressure fields for the wave equation and fluid dynamics; temperature/concentration fields for the heat/diffusion equations. Outside of physics proper (e.g., radiometry and computer graphics), there are even light fields. All these previous examples are scalar fields. Similarly for vectors, there are vector PDEs for displacement, velocity and vorticity fields in (applied mathematical) fluid dynamics, but vector calculus may now be needed in addition, being calculus over vector fields (as are these three quantities, and those for vector PDEs in general). More generally problems in continuum mechanics may involve for example, directional elasticity (from which comes the term *tensor*, derived from the Latin word for stretch), complex fluid flows or anisotropic diffusion, which are framed as matrix-tensor PDEs, and then require matrices or tensor fields, hence matrix or tensor calculus. It should be noted that the scalars (and hence the vectors, matrices and tensors) can be real or complex as both are fields in the abstract-algebraic/ring-theoretic sense.

In a general setting, classical fields are described by sections of fiber bundles and their dynamics is formulated in the terms of jet manifolds (covariant classical field theory).[19]

In modern physics, the most often studied fields are those that model the four fundamental forces which one day may lead to the Unified Field Theory.

5.4.1   Symmetries of fields

A convenient way of classifying a field (classical or quantum) is by the symmetries it possesses. Physical symmetries are usually of two types:

Spacetime symmetries

Main articles: Global symmetry and Spacetime symmetries

Fields are often classified by their behaviour under transformations of spacetime. The terms used in this classification are:

- scalar fields (such as temperature) whose values are given by a single variable at each point of space. This value does not change under transformations of space.

- vector fields (such as the magnitude and direction of the force at each point in a magnetic field) which are specified by attaching a vector to each point of space. The components of this vector transform between themselves contravariantly under rotations in space. Similarly, a dual (or co-) vector field attaches a dual vector to each point of space, and the components of each dual vector transform covariantly.

- tensor fields, (such as the stress tensor of a crystal) specified by a tensor at each point of space. Under rotations in space, the components of the tensor transform in a more general way which depends on the number of covariant indices and contravariant indices.

- spinor fields (such as the Dirac spinor) arise in quantum field theory to describe particles with spin which transform like vectors except for the one of their component; in other words, when one rotates a vector field 360 degrees around a specific axis, the vector field turns to itself; however, spinors in same case turn to their negatives.

Internal symmetries

Main article: Local symmetry

Fields may have internal symmetries in addition to spacetime symmetries. For example, in many situations one needs fields which are a list of space-time scalars: $(\varphi_1, \varphi_2, ... \varphi N)$. For example, in weather prediction these may be temperature, pressure, humidity, etc. In particle physics, the color symmetry of the interaction of quarks is an example of an internal symmetry of the strong interaction, as is the isospin or flavour symmetry.

If there is a symmetry of the problem, not involving spacetime, under which these components transform into each other, then this set of symmetries is called an *internal symmetry*. One may also make a classification of the charges of the fields under internal symmetries.

5.4.2   Statistical field theory

Main article: Statistical field theory

Statistical field theory attempts to extend the field-theoretic paradigm toward many-body systems and statistical mechanics. As above, it can be approached by the usual infinite number of degrees of freedom argument.

Much like statistical mechanics has some overlap between quantum and classical mechanics, statistical field theory has links to both quantum and classical field theories, especially the former with which it shares many methods. One important example is mean field theory.

5.4.3   Continuous random fields

Classical fields as above, such as the electromagnetic field, are usually infinitely differentiable functions, but they are in any case almost always twice differentiable. In contrast, generalized functions are not continuous. When dealing carefully with classical fields at finite temperature, the mathematical methods of continuous random fields are used, because thermally fluctuating classical fields are nowhere differentiable. Random fields are indexed sets of random variables; a continuous random field is a random field that has a set of functions as its index set. In particular, it is often mathematically convenient to take a continuous random field to have a Schwartz space of functions as its index set, in which case the continuous random field is a tempered distribution.

We can think about a continuous random field, in a (very) rough way, as an ordinary function that is $\pm\infty$ almost everywhere, but such that when we take a weighted average of all the infinities over any finite region, we get a finite result. The infinities are not well-defined; but the finite values can be associated with the functions used as the weight functions to get the finite values, and that can be well-defined. We can define a continuous random field well enough as a linear map from a space of functions into the real numbers.

5.5   See also

- Field strength

- Lagrangian and Eulerian specification of a field

- Covariant Hamiltonian field theory

- Scalar field theory

5.6 Notes

[1] This is contingent on the correct choice of gauge. V and **A** are not completely determined by ρ and **J**; rather, they are only determined up to some scalar function $f(\mathbf{r}, t)$ known as the gauge. The retarded potential formalism requires one to choose the Lorenz gauge.

5.7 References

[1] John Gribbin (1998). *Q is for Quantum: Particle Physics from A to Z*. London: Weidenfeld & Nicolson. p. 138. ISBN 0-297-81752-3.

[2] Richard Feynman (1970). *The Feynman Lectures on Physics Vol II*. Addison Wesley Longman. ISBN 978-0-201-02115-8. A "field" is any physical quantity which takes on different values at different points in space.

[3] Ernan McMullin (2002). "The Origins of the Field Concept in Physics" (PDF). *Phys. Perspect.* **4**: 13–39.

[4] Richard P. Feynman (1970). *The Feynman Lectures on Physics Vol II*. Addison Wesley Longman.

[5] Richard P. Feynman (1970). *The Feynman Lectures on Physics Vol I*. Addison Wesley Longman.

[6] John Archibald Wheeler (1998). *Geons, Black Holes, and Quantum Foam: A Life in Physics*. London: Norton. p. 163.

[7] Richard P. Feynman (1970). *The Feynman Lectures on Physics Vol I*. Addison Wesley Longman.

[8] Steven Weinberg (November 7, 2013). "Physics: What We Do and Don't Know". *New York Review of Books*.

[9] Weinberg, Steven (1977). "The Search for Unity: Notes for a History of Quantum Field Theory". *Daedalus* **106** (4): 17–35. JSTOR 20024506.

[10] Kleppner, David; Kolenkow, Robert. *An Introduction to Mechanics*. p. 85.

[11] Parker, C.B. (1994). *McGraw Hill Encyclopaedia of Physics* (2nd ed.). Mc Graw Hill. ISBN 0-07-051400-3.

[12] M. Mansfield, C. O'Sullivan (2011). *Understanding Physics* (4th ed.). John Wiley & Sons. ISBN 978-0-47-0746370.

[13] Griffiths, David. *Introduction to Electrodynamics* (3rd ed.). p. 326.

[14] Wangsness, Roald. *Electromagnetic Fields* (2nd ed.). p. 469.

[15] J.A. Wheeler, C. Misner, K.S. Thorne (1973). *Gravitation*. W.H. Freeman & Co. ISBN 0-7167-0344-0.

[16] I. Ciufolini and J.A. Wheeler (1995). *Gravitation and Inertia*. Princeton Physics Series. ISBN 0-691-03323-4.

[17] Peskin, Michael E.; Schroeder, Daniel V. (1995). *An Introduction to Quantum Fields*. Westview Press. p. 198. ISBN 0-201-50397-2.. Also see precision tests of QED.

[18] R. Resnick, R. Eisberg (1985). *Quantum Physics of Atoms, Molecules, Solids, Nuclei and Particles* (2nd ed.). John Wiley & Sons. p. 684. ISBN 978-0-471-87373-0.

[19] Giachetta, G., Mangiarotti, L., Sardanashvily, G. (2009) *Advanced Classical Field Theory*. Singapore: World Scientific, ISBN 978-981-283-895-7 (arXiv: 0811.0331v2)

5.8   Further reading

- "Fields". *Principles of Physical Science. Encyclopaedia Britannica (Macropaedia)* **25** (fifteenth ed.). 1994. p. 815.

- Landau, Lev D. and Lifshitz, Evgeny M. (1971). *Classical Theory of Fields* (3rd ed.). London: Pergamon. ISBN 0-08-016019-0. Vol. 2 of the Course of Theoretical Physics.

- Jepsen, Kathryn (July 18, 2013). "Real talk: Everything is made of fields" (PDF). *Symmetry Magazine*.

5.9   External links

- Particle and Polymer Field Theories

Chapter 6

Vector field

In vector calculus, a **vector field** is an assignment of a vector to each point in a subset of space.[1] A vector field in the plane, for instance, can be visualized as a collection of arrows with a given magnitude and direction each attached to a point in the plane. Vector fields are often used to model, for example, the speed and direction of a moving fluid throughout space, or the strength and direction of some force, such as the magnetic or gravitational force, as it changes from point to point.

The elements of differential and integral calculus extend to vector fields in a natural way. When a vector field represents force, the line integral of a vector field represents the work done by a force moving along a path, and under this interpretation conservation of energy is exhibited as a special case of the fundamental theorem of calculus. Vector fields can usefully be thought of as representing the velocity of a moving flow in space, and this physical intuition leads to notions such as the divergence (which represents the rate of change of volume of a flow) and curl (which represents the rotation of a flow).

In coordinates, a vector field on a domain in n-dimensional Euclidean space can be represented as a vector-valued function that associates an n-tuple of real numbers to each point of the domain. This representation of a vector field depends on the coordinate system, and there is a well-defined transformation law in passing from one coordinate system to the other. Vector fields are often discussed on open subsets of Euclidean space, but also make sense on other subsets such as surfaces, where they associate an arrow tangent to the surface at each point (a tangent vector).

More generally, vector fields are defined on differentiable manifolds, which are spaces that look like Euclidean space on small scales, but may have more complicated structure on larger scales. In this setting, a vector field gives a tangent vector at each point of the manifold (that is, a section of the tangent bundle to the manifold). Vector fields are one kind of tensor field.

6.1 Definition

6.1.1 Vector fields on subsets of Euclidean space

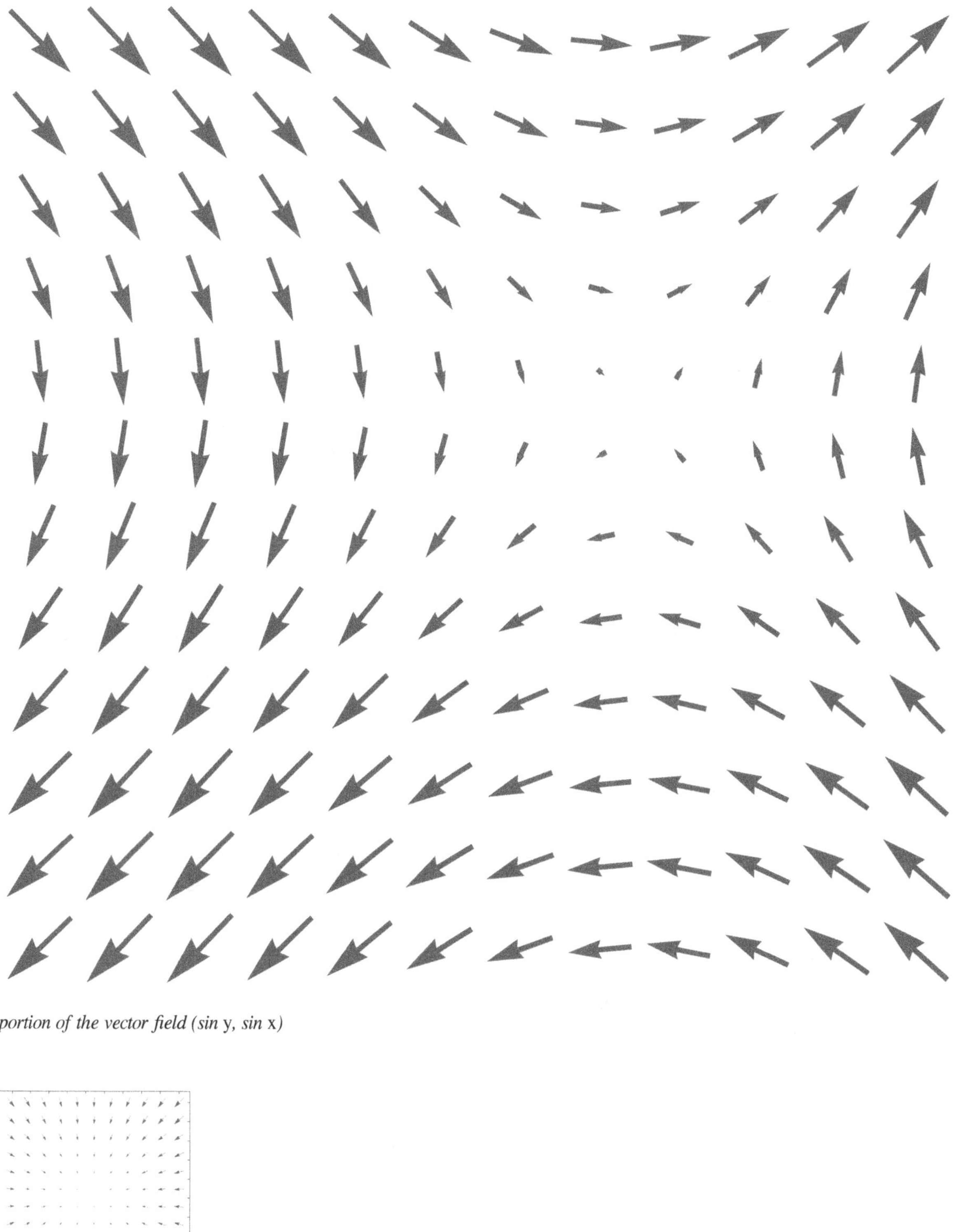

A portion of the vector field (sin y, sin x)

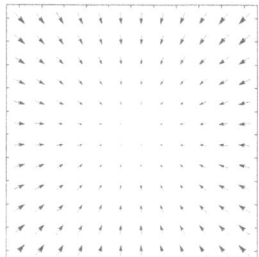

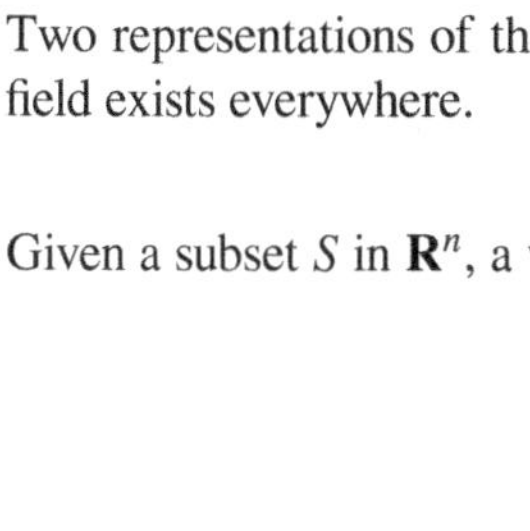

Two representations of the same vector field: $\mathbf{v}(x, y) = -\mathbf{r}$. The arrows depict the field at discrete points, however, the field exists everywhere.

Given a subset S in $\mathbf{R}^n$, a **vector field** is represented by a vector-valued function $V: S \to \mathbf{R}^n$ in standard Cartesian coordinates $(x_1, ..., xn)$. If each component of V is continuous, then V is a continuous vector field, and more generally V is a C^k vector field if each component of V is k times continuously differentiable.

A vector field can be visualized as assigning a vector to individual points within an n-dimensional space.[1]

Given two C^k-vector fields V, W defined on S and a real valued C^k-function f defined on S, the two operations scalar multiplication and vector addition

$$(fV)(p) := f(p)V(p)$$

$$(V + W)(p) := V(p) + W(p)$$

define the module of C^k-vector fields over the ring of C^k-functions.

6.1.2   Coordinate transformation law

In physics, a vector is additionally distinguished by how its coordinates change when one measures the same vector with respect to a different background coordinate system. The transformation properties of vectors distinguish a vector as a geometrically distinct entity from a simple list of scalars, or from a covector.

Thus, suppose that $(x_1,...,xn)$ is a choice of Cartesian coordinates, in terms of which the components of the vector V are

$$V_x = (V_{1,x}, \ldots, V_{n,x})$$

and suppose that $(y_1,...,yn)$ are n functions of the xi defining a different coordinate system. Then the components of the vector V in the new coordinates are required to satisfy the transformation law

Such a transformation law is called contravariant. A similar transformation law characterizes vector fields in physics: specifically, a vector field is a specification of n functions in each coordinate system subject to the transformation law (**1**) relating the different coordinate systems.

Vector fields are thus contrasted with scalar fields, which associate a number or *scalar* to every point in space, and are also contrasted with simple lists of scalar fields, which do not transform under coordinate changes.

6.1.3   Vector fields on manifolds

Given a differentiable manifold M, a **vector field** on M is an assignment of a tangent vector to each point in M.[2] More precisely, a vector field F is a mapping from M into the tangent bundle TM so that $p \circ F$ is the identity mapping where p denotes the projection from TM to M. In other words, a vector field is a section of the tangent bundle.

If the manifold M is smooth or analytic—that is, the change of coordinates is smooth (analytic)—then one can make sense of the notion of smooth (analytic) vector fields. The collection of all smooth vector fields on a smooth manifold M is often denoted by $\Gamma(TM)$ or $C^\infty(M,TM)$ (especially when thinking of vector fields as sections); the collection of all smooth vector fields is also denoted by $\mathfrak{x}(M)$ (a fraktur "X").

6.2   Examples

- A vector field for the movement of air on Earth will associate for every point on the surface of the Earth a vector with the wind speed and direction for that point. This can be drawn using arrows to represent the wind; the length (magnitude) of the arrow will be an indication of the wind speed. A "high" on the usual barometric pressure map would then act as a source (arrows pointing away), and a "low" would be a sink (arrows pointing towards), since air tends to move from high pressure areas to low pressure areas.

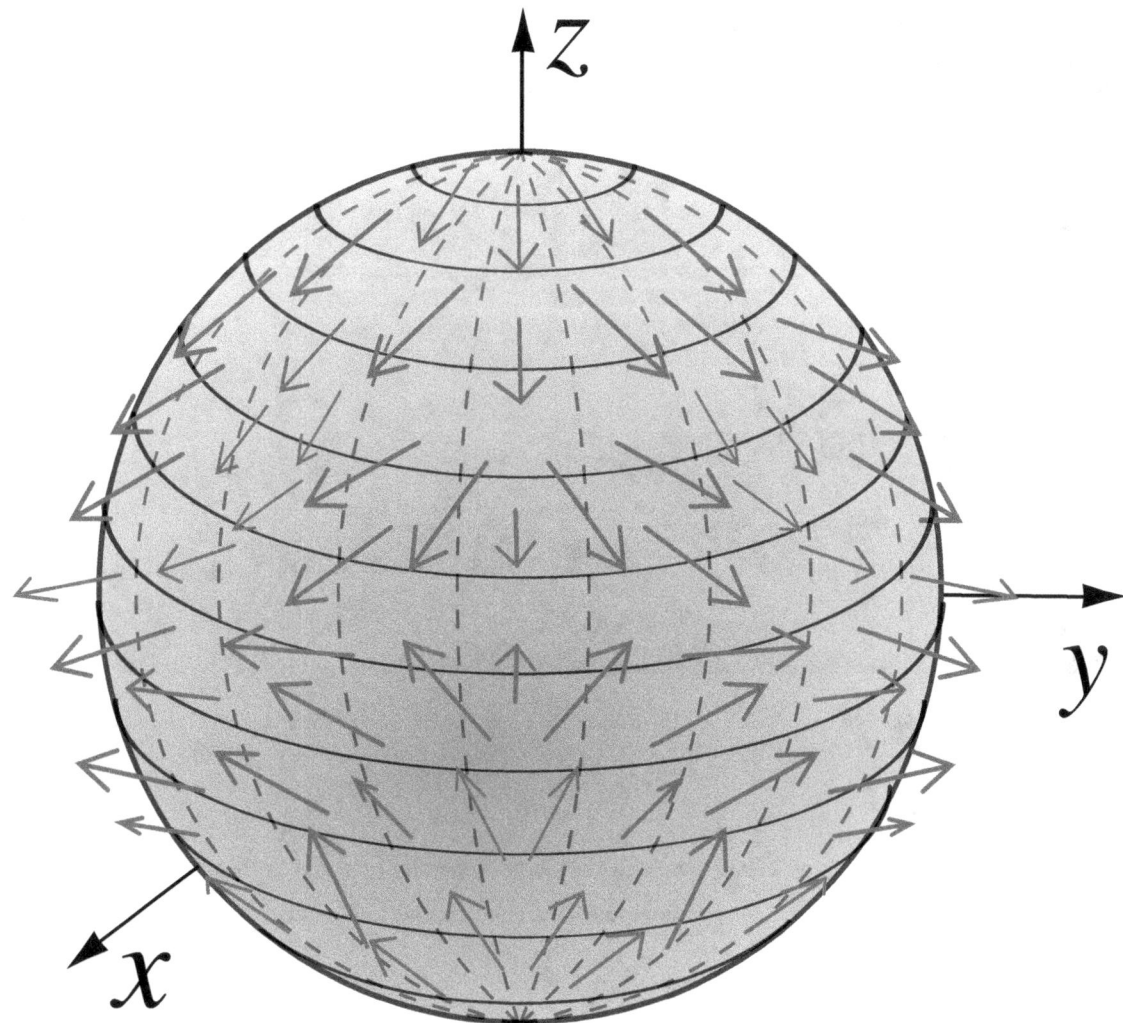

A vector field on a sphere

- Velocity field of a moving fluid. In this case, a velocity vector is associated to each point in the fluid.

- Streamlines, Streaklines and Pathlines are 3 types of lines that can be made from vector fields. They are :

    streaklines — as revealed in wind tunnels using smoke.
    streamlines (or fieldlines)— as a line depicting the instantaneous field at a given time.
    pathlines — showing the path that a given particle (of zero mass) would follow.

- Magnetic fields. The fieldlines can be revealed using small iron filings.

- Maxwell's equations allow us to use a given set of initial conditions to deduce, for every point in Euclidean space, a magnitude and direction for the force experienced by a charged test particle at that point; the resulting vector field is the electromagnetic field.

- A gravitational field generated by any massive object is also a vector field. For example, the gravitational field vectors for a spherically symmetric body would all point towards the sphere's center with the magnitude of the vectors reducing as radial distance from the body increases.

The flow field around an airplane is a vector field in $\mathbf{R}^3$, here visualized by bubbles that follow the streamlines showing a wingtip vortex.

6.2.1 Gradient field

Vector fields can be constructed out of scalar fields using the gradient operator (denoted by the del: ∇).[3]

A vector field V defined on a set S is called a **gradient field** or a **conservative field** if there exists a real-valued function (a scalar field) f on S such that

$$V = \nabla f = \left(\frac{\partial f}{\partial x_1}, \frac{\partial f}{\partial x_2}, \frac{\partial f}{\partial x_3}, \ldots, \frac{\partial f}{\partial x_n} \right).$$

The associated flow is called the **gradient flow**, and is used in the method of gradient descent.

The path integral along any closed curve γ ($\gamma(0) = \gamma(1)$) in a conservative field is zero:

$$\oint_\gamma \langle V(x), \mathrm{d}x \rangle = \oint_\gamma \langle \nabla f(x), \mathrm{d}x \rangle = f(\gamma(1)) - f(\gamma(0)).$$

where the angular brackets and comma: $\langle, \rangle$ denotes the inner product of two vectors (strictly speaking – the integrand $V(x)$ is a 1-form rather than a vector in the elementary sense).[4]

6.2.2 Central field

A C^∞-vector field over $\mathbf{R}^n \setminus \{0\}$ is called a **central field** if

$$V(T(p)) = T(V(p)) \qquad (T \in \mathrm{O}(n, \mathbf{R}))$$

where $\mathrm{O}(n, \mathbf{R})$ is the orthogonal group. We say central fields are invariant under orthogonal transformations around 0.

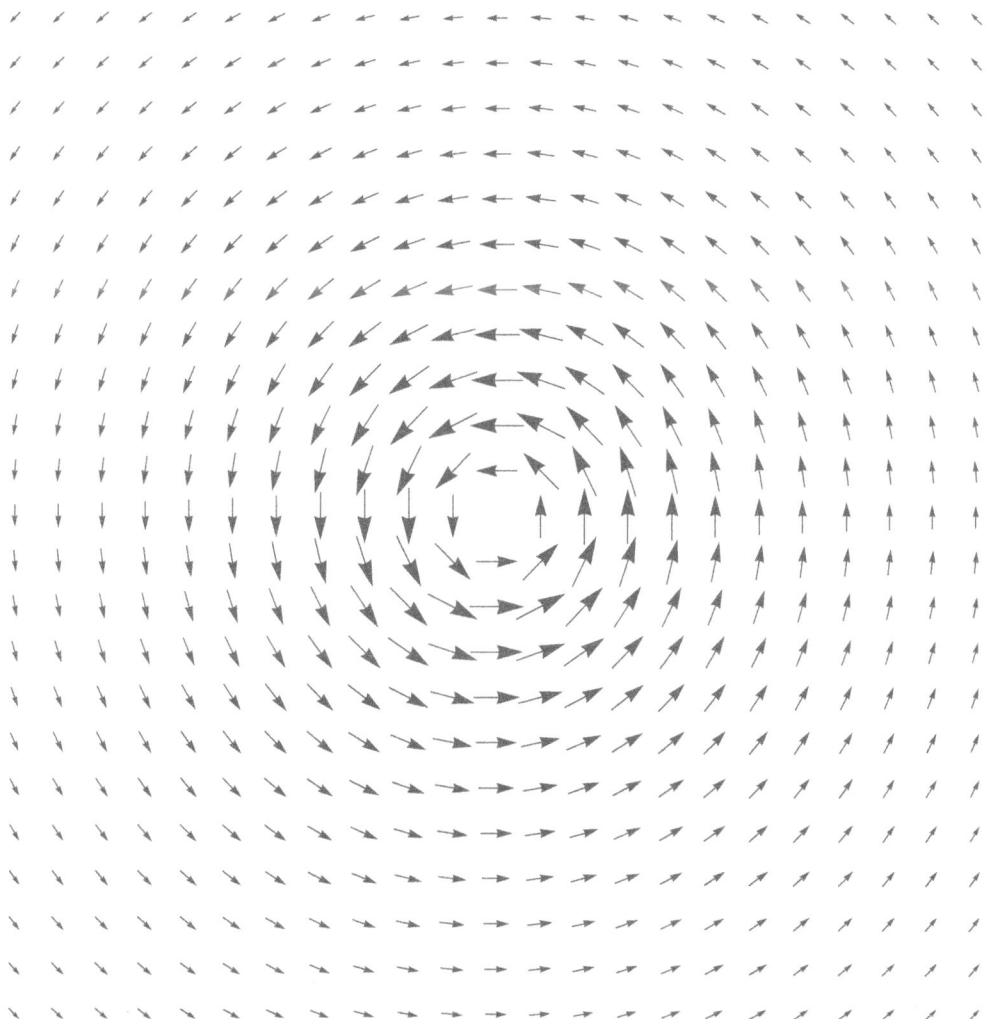

A vector field that has circulation about a point cannot be written as the gradient of a function.

The point 0 is called the **center** of the field.

Since orthogonal transformations are actually rotations and reflections, the invariance conditions mean that vectors of a central field are always directed towards, or away from, 0; this is an alternate (and simpler) definition. A central field is always a gradient field, since defining it on one semiaxis and integrating gives an antigradient.

6.3   Operations on vector fields

6.3.1   Line integral

Main article: Line integral

A common technique in physics is to integrate a vector field along a curve, i.e. to determine its line integral. Given a particle in a gravitational vector field, where each vector represents the force acting on the particle at a given point in space, the line integral is the work done on the particle when it travels along a certain path.

The line integral is constructed analogously to the Riemann integral and it exists if the curve is rectifiable (has finite length) and the vector field is continuous.

Given a vector field V and a curve γ parametrized by $[a, b]$ (where a and b are real) the line integral is defined as

$$\int_\gamma \langle V(x), \mathrm{d}x \rangle = \int_a^b \langle V(\gamma(t)), \gamma'(t) \, \mathrm{d}t \rangle.$$

6.3.2 Divergence

Main article: Divergence

The divergence of a vector field on Euclidean space is a function (or scalar field). In three-dimensions, the divergence is defined by

$$\operatorname{div} \mathbf{F} = \nabla \cdot \mathbf{F} = \frac{\partial F_1}{\partial x} + \frac{\partial F_2}{\partial y} + \frac{\partial F_3}{\partial z},$$

with the obvious generalization to arbitrary dimensions. The divergence at a point represents the degree to which a small volume around the point is a source or a sink for the vector flow, a result which is made precise by the divergence theorem.

The divergence can also be defined on a Riemannian manifold, that is, a manifold with a Riemannian metric that measures the length of vectors.

6.3.3 Curl

Main article: Curl (mathematics)

The curl is an operation which takes a vector field and produces another vector field. The curl is defined only in three-dimensions, but some properties of the curl can be captured in higher dimensions with the exterior derivative. In three-dimensions, it is defined by

$$\operatorname{curl} \mathbf{F} = \nabla \times \mathbf{F} = \left(\frac{\partial F_3}{\partial y} - \frac{\partial F_2}{\partial z} \right) \mathbf{e}_1 - \left(\frac{\partial F_3}{\partial x} - \frac{\partial F_1}{\partial z} \right) \mathbf{e}_2 + \left(\frac{\partial F_2}{\partial x} - \frac{\partial F_1}{\partial y} \right) \mathbf{e}_3.$$

The curl measures the density of the angular momentum of the vector flow at a point, that is, the amount to which the flow circulates around a fixed axis. This intuitive description is made precise by Stokes' theorem.

6.3.4 Index of a vector field

The index of a vector field is a way of describing the behaviour of a vector field around an isolated zero (i.e. non-singular point) which can distinguish saddles from sources and sinks. Take a small sphere around the zero so that no other zeros are included. A map from this sphere to a unit sphere of dimensions $n - 1$ can be constructed by dividing each vector by its length to form a unit length vector which can then be mapped to the unit sphere. The index of the vector field at the point is the degree of this map. The index of the vector field is the sum of the indices of each zero.

The index will be zero around any non singular point, it is +1 around sources and sinks and −1 around saddles. In two dimensions the index is equivalent to the winding number. For an ordinary sphere in three dimension space it can be shown that the index of any vector field on the sphere must be two, this leads to the hairy ball theorem which shows that every such vector field must have a zero. This theorem generalises to the Poincaré–Hopf theorem which relates the index to the Euler characteristic of the space.

6.4 History

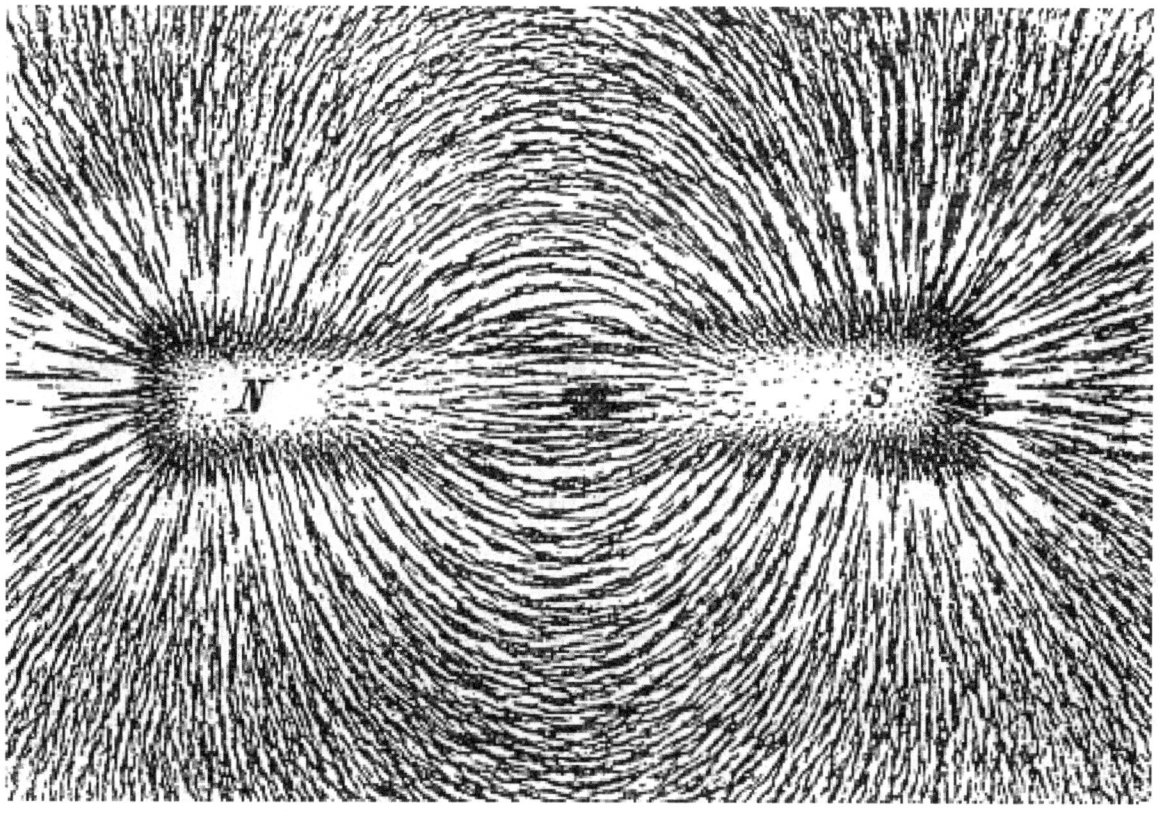

Magnetic field lines of an iron bar (magnetic dipole)

Vector fields arose originally in classical field theory in 19th century physics, specifically in magnetism. They were formalized by Michael Faraday, in his concept of *lines of force,* who emphasized that the field *itself* should be an object of study, which it has become throughout physics in the form of field theory.

In addition to the magnetic field, other phenomena that were modeled as vector fields by Faraday include the electrical field and light field.

6.5 Flow curves

Main article: Integral curve

Consider the flow of a fluid through a region of space. At any given time, any point of the fluid has a particular velocity associated with it; thus there is a vector field associated to any flow. The converse is also true: it is possible to associate a flow to a vector field having that vector field as its velocity.

Given a vector field V defined on S, one defines curves $\gamma(t)$ on S such that for each t in an interval I

$$\gamma'(t) = V(\gamma(t)).$$

By the Picard–Lindelöf theorem, if V is Lipschitz continuous there is a *unique C^1-curve* γx for each point x in S so that

$$\gamma_x(0) = x$$

$$\gamma_x'(t) = V(\gamma_x(t)) \qquad (t \in (-\varepsilon, +\varepsilon) \subset \mathbf{R}).$$

The curves γx are called **flow curves** of the vector field V and partition S into equivalence classes. It is not always possible to extend the interval $(-\varepsilon, +\varepsilon)$ to the whole real number line. The flow may for example reach the edge of S in a finite time. In two or three dimensions one can visualize the vector field as giving rise to a flow on S. If we drop a particle into this flow at a point p it will move along the curve γp in the flow depending on the initial point p. If p is a stationary point of V then the particle will remain at p.

Typical applications are streamline in fluid, geodesic flow, and one-parameter subgroups and the exponential map in Lie groups.

6.5.1 Complete vector fields

A vector field is **complete** if its flow curves exist for all time.[5] In particular, compactly supported vector fields on a manifold are complete. If X is a complete vector field on M, then the one-parameter group of diffeomorphisms generated by the flow along X exists for all time.

6.6 Difference between scalar and vector field

The difference between a scalar and vector field is not that a scalar is just one number while a vector is several numbers. The difference is in how their coordinates respond to coordinate transformations. A scalar *is* a coordinate whereas a vector *can be described* by coordinates, but it *is not* the collection of its coordinates.

6.6.1 Example 1

This example is about 2-dimensional Euclidean space ($\mathbf{R}^2$) where we examine Euclidean (x, y) and polar (r, θ) coordinates (which are undefined at the origin). Thus $x = r \cos \theta$ and $y = r \sin \theta$ and also $r^2 = x^2 + y^2$, $\cos \theta = x/(x^2 + y^2)^{1/2}$ and $\sin \theta = y/(x^2 + y^2)^{1/2}$. Suppose we have a scalar field which is given by the constant function 1, and a vector field which attaches a vector in the r-direction with length 1 to each point. More precisely, they are given by the functions

$$s_{\text{polar}} : (r, \theta) \mapsto 1, \quad v_{\text{polar}} : (r, \theta) \mapsto (1, 0).$$

Let us convert these fields to Euclidean coordinates. The vector of length 1 in the r-direction has the x coordinate $\cos \theta$ and the y coordinate $\sin \theta$. Thus in Euclidean coordinates the same fields are described by the functions

$$s_{\text{Euclidean}} : (x, y) \mapsto 1,$$

$$v_{\text{Euclidean}} : (x, y) \mapsto (\cos \theta, \sin \theta) = \left(\frac{x}{\sqrt{x^2 + y^2}}, \frac{y}{\sqrt{x^2 + y^2}} \right).$$

We see that while the scalar field remains the same, the vector field now looks different. The same holds even in the 1-dimensional case, as illustrated by the next example.

6.6.2 Example 2

Consider the 1-dimensional Euclidean space $\mathbf{R}$ with its standard Euclidean coordinate x. Suppose we have a scalar field and a vector field which are both given in the x coordinate by the constant function 1,

$$s_{\text{Euclidean}} : x \mapsto 1, \quad v_{\text{Euclidean}} : x \mapsto 1.$$

Thus, we have a scalar field which has the value 1 everywhere and a vector field which attaches a vector in the *x*-direction with magnitude 1 unit of *x* to each point.

Now consider the coordinate $\xi := 2x$. If *x* changes one unit then ξ changes 2 units. But since we wish the integral of v along a path to be independent of coordinate, this means v*dx=v'*dξ. So from x increase by 1 unit, ξ increases by 1/2 unit, so v' must be 2. Thus this vector field has a magnitude of 2 in units of ξ. Therefore, in the ξ coordinate the scalar field and the vector field are described by the functions

$$s_{\text{unusual}} : \xi \mapsto 1, \quad v_{\text{unusual}} : \xi \mapsto 2$$

which are different.

6.7   f-relatedness

Given a smooth function between manifolds, $f \colon M \to N$, the derivative is an induced map on tangent bundles, $f^* \colon TM \to TN$. Given vector fields $V \colon M \to TM$ and $W \colon N \to TN$, we say that W is *f*-related to V if the equation $W \circ f^* = f^* \circ V$ holds.

If V_i is *f*-related to W_i, $i = 1, 2$, then the Lie bracket $[V_1, V_2]$ is *f*-related to $[W_1, W_2]$.

6.8   Generalizations

Replacing vectors by *p*-vectors (*p*th exterior power of vectors) yields *p*-vector fields; taking the dual space and exterior powers yields differential *k*-forms, and combining these yields general tensor fields.

Algebraically, vector fields can be characterized as derivations of the algebra of smooth functions on the manifold, which leads to defining a vector field on a commutative algebra as a derivation on the algebra, which is developed in the theory of differential calculus over commutative algebras.

6.9   See also

- Eisenbud–Levine–Khimshiashvili signature formula

- Field line

- Field strength

- Lie derivative

- Scalar field

- Time-dependent vector field

- Vector fields in cylindrical and spherical coordinates

- Tensor fields

6.10   References

[1] Galbis, Antonio & Maestre, Manuel (2012). *Vector Analysis Versus Vector Calculus*. Springer. p. 12. ISBN 978-1-4614-2199-3.

[2] Tu, Loring W. (2010). "Vector fields". *An Introduction to Manifolds*. Springer. p. 149. ISBN 978-1-4419-7399-3.

[3] Dawber, P.G. (1987). *Vectors and Vector Operators*. CRC Press. p. 29. ISBN 978-0-85274-585-4.

[4] T. Frankel (2012), *The Geometry of Physics* (3rd ed.), Cambridge University Press, p. xxxviii, ISBN 978-1107-602601

[5] Sharpe, R. (1997). *Differential geometry*. Springer-Verlag. ISBN 0-387-94732-9.

6.11 Bibliography

- Hubbard, J. H.; Hubbard, B. B. (1999). *Vector calculus, linear algebra, and differential forms. A unified approach.* Upper Saddle River, NJ: Prentice Hall. ISBN 0-13-657446-7.

- Warner, Frank (1983) [1971]. *Foundations of differentiable manifolds and Lie groups.* New York-Berlin: Springer-Verlag. ISBN 0-387-90894-3.

- Boothby, William (1986). *An introduction to differentiable manifolds and Riemannian geometry.* Pure and Applied Mathematics, volume 120 (second ed.). Orlando, FL: Academic Press. ISBN 0-12-116053-X.

6.12 External links

- Hazewinkel, Michiel, ed. (2001), "Vector field", *Encyclopedia of Mathematics*, Springer, ISBN 978-1-55608-010-4

- Vector field — Mathworld

- Vector field — PlanetMath

- 3D Magnetic field viewer

- Vector fields and field lines

- Vector field simulation An interactive application to show the effects of vector fields

- Vector Fields Software 2d & 3d electromagnetic design software that can be used to visualise vector fields and field lines

Chapter 7

Generating set of a group

In abstract algebra, a **generating set of a group** is a subset such that every element of the group can be expressed as the combination (under the group operation) of finitely many elements of the subset and their inverses.

In other words, if S is a subset of a group G, then $\langle S \rangle$, the **subgroup generated by S**, is the smallest subgroup of G containing every element of S, meaning the intersection over all subgroups containing the elements of S; equivalently, $\langle S \rangle$ is the subgroup of all elements of G that can be expressed as the finite product of elements in S and their inverses.

If $G = \langle S \rangle$, then we say S **generates** G; and the elements in S are called **generators** or **group generators**. If S is the empty set, then $\langle S \rangle$ is the trivial group $\{e\}$, since we consider the empty product to be the identity.

When there is only a single element x in S, $\langle S \rangle$ is usually written as $\langle x \rangle$. In this case, $\langle x \rangle$ is the **cyclic subgroup** of the powers of x, a cyclic group, and we say this group is generated by x. Equivalent to saying an element x generates a group is saying that $\langle x \rangle$ equals the entire group G. For finite groups, it is also equivalent to saying that x has order |G|.

7.1   Finitely generated group

If S is finite, then a group $G = \langle S \rangle$ is called **finitely generated**. The structure of finitely generated abelian groups in particular is easily described. Many theorems that are true for finitely generated groups fail for groups in general. It has been proven that if a finite group is generated by a subset S, then each group element may be expressed as a word from the alphabet S of length less than or equal to the order of the group.

Every finite group is finitely generated since $\langle G \rangle = G$. The integers under addition are an example of an infinite group which is finitely generated by both 1 and −1, but the group of rationals under addition cannot be finitely generated. No uncountable group can be finitely generated. For example, the group of real numbers under addition, $(\mathbf{R}, +)$.

Different subsets of the same group can be generating subsets; for example, if p and q are integers with $\gcd(p, q) = 1$, then $\{p, q\}$ also generates the group of integers under addition (by Bézout's identity).

While it is true that every quotient of a finitely generated group is finitely generated (simply take the images of the generators in the quotient), a subgroup of a finitely generated group need not be finitely generated. For example, let G be the free group in two generators, x and y (which is clearly finitely generated, since $G = \langle \{x,y\} \rangle$), and let S be the subset consisting of all elements of G of the form $y^n x y^{-n}$, for n a natural number. Since $\langle S \rangle$ is clearly isomorphic to the free group in countable generators, it cannot be finitely generated. However, every subgroup of a finitely generated abelian group is in itself finitely generated. In fact, more can be said: the class of all finitely generated groups is closed under extensions. To see this, take a generating set for the (finitely generated) normal subgroup and quotient: then the generators for the normal subgroup, together with preimages of the generators for the quotient, generate the group.

7.2 Free group

The most general group generated by a set S is the group **freely generated** by S. Every group generated by S is isomorphic to a quotient of this group, a feature which is utilized in the expression of a group's presentation.

7.3 Frattini subgroup

An interesting companion topic is that of **non-generators**. An element x of the group G is a non-generator if every set S containing x that generates G, still generates G when x is removed from S. In the integers with addition, the only non-generator is 0. The set of all non-generators forms a subgroup of G, the Frattini subgroup.

7.4 Examples

The group of units $U(\mathbf{Z}_9)$ is the group of all integers relatively prime to 9 under multiplication mod 9 ($U_9 = \{1, 2, 4, 5, 7, 8\}$). All arithmetic here is done modulo 9. Seven is not a generator of $U(\mathbf{Z}_9)$, since

$$\{7^i \mod 9 \mid i \in \mathbb{N}\} = \{7, 4, 1\}.$$

while 2 is, since:

$$\{2^i \mod 9 \mid i \in \mathbb{N}\} = \{2, 4, 8, 7, 5, 1\}.$$

On the other hand, for $n > 2$ the symmetric group of degree n is not cyclic, so it is not generated by any one element. However, it is generated by the two permutations (1 2) and (1 2 3 ... n). For example, for S_3 we have:

$$e = (1\ 2)(1\ 2)$$
$$(1\ 2) = (1\ 2)$$
$$(2\ 3) = (1\ 2)(1\ 2\ 3)$$
$$(1\ 3) = (1\ 2\ 3)(1\ 2)$$
$$(1\ 2\ 3) = (1\ 2\ 3)$$
$$(1\ 3\ 2) = (1\ 2)(1\ 2\ 3)(1\ 2)$$

Infinite groups can also have finite generating sets. The additive group of integers has 1 as a generating set. The element 2 is not a generating set, as the odd numbers will be missing. The two-element subset $\{3, 5\}$ is a generating set, since $(-5) + 3 + 3 = 1$ (in fact, any pair of coprime numbers is, as a consequence of Bézout's identity).

7.5 See also

- Cayley graph

- Generating set for related meanings in other structures

- Presentation of a group

7.6   References

- Lang, Serge (2002), *Algebra*, Graduate Texts in Mathematics **211** (Revised third ed.), New York: Springer-Verlag, ISBN 978-0-387-95385-4, Zbl 0984.00001, MR 1878556

- Coxeter, H. S. M. and Moser, W. O. J. (1980). *Generators and Relations for Discrete Groups*. New York: Springer-Verlag. ISBN 0-387-09212-9.

7.7   External links

- Weisstein, Eric W., "Group generators", *MathWorld*.

Chapter 8

Quantum

For other uses, see Quantum (disambiguation).

In physics, a **quantum** (plural: **quanta**) is the minimum amount of any physical entity involved in an interaction. Behind this, one finds the fundamental notion that a physical property may be "quantized," referred to as "the hypothesis of quantization".[1] This means that the magnitude can take on only certain discrete values.

A photon is a single quantum of light, and is referred to as a "light quantum". The energy of an electron bound to an atom is quantized, which results in the stability of atoms, and hence of matter in general.

As incorporated into the theory of quantum mechanics, this is regarded by physicists as part of the fundamental framework for understanding and describing nature at the smallest length-scales.

8.1   Etymology and discovery

The word "quantum" comes from the Latin "quantus", meaning "how much". "Quanta", short for "quanta of electricity" (electrons) was used in a 1902 article on the photoelectric effect by Philipp Lenard, who credited Hermann von Helmholtz for using the word in the area of electricity. However, the word quantum in general was well known before 1900.[2] It was often used by physicians, such as in the term quantum satis. Both Helmholtz and Julius von Mayer were physicians as well as physicists. Helmholtz used "quantum" with reference to heat in his article[3] on Mayer's work, and indeed, the word "quantum" can be found in the formulation of the first law of thermodynamics by Mayer in his letter[4] dated July 24, 1841. Max Planck used "quanta" to mean "quanta of matter and electricity",[5] gas, and heat.[6] In 1905, in response to Planck's work and the experimental work of Lenard (who explained his results by using the term "quanta of electricity"), Albert Einstein suggested that radiation existed in spatially localized packets which he called "quanta of light" ("Lichtquanta").[7]

The concept of quantization of radiation was discovered in 1900 by Max Planck, who had been trying to understand the emission of radiation from heated objects, known as black-body radiation. By assuming that energy can only be absorbed or released in tiny, differential, discrete packets he called "bundles" or "energy elements",[8] Planck accounted for the fact that certain objects change colour when heated.[9] On December 14, 1900, Planck reported his revolutionary findings to the German Physical Society, and introduced the idea of quantization for the first time as a part of his research on black-body radiation.[10] As a result of his experiments, Planck deduced the numerical value of h, known as the Planck constant, and could also report a more precise value for the Avogadro–Loschmidt number, the number of real molecules in a mole and the unit of electrical charge, to the German Physical Society. After his theory was validated, Planck was awarded the Nobel Prize in Physics in 1918 for his discovery.

8.2   Beyond electromagnetic radiation

While quantization was first discovered in electromagnetic radiation, it describes a fundamental aspect of energy not just restricted to photons.[11] In the attempt to bring experiment into agreement with theory, Max Planck postulated that electromagnetic energy is absorbed or emitted in discrete packets, or quanta.[12]

8.3   See also

- Elementary particle
- Graviton
- Introduction to quantum mechanics
- Magnetic flux quantum
- Photon
- Photon polarization
- Quantal analysis
- Quantization (physics)
- Quantum cellular automata
- Quantum channel
- Quantum coherence
- Quantum chromodynamics
- Quantum computer
- Quantum cryptography
- Quantum dot
- Quantum electronics
- Quantum entanglement
- Quantum immortality
- Quantum lithography
- Quantum mechanics
- Quantum number
- Quantum sensor
- Quantum state
- Subatomic particle

8.4 References

[1] Wiener, N. (1966). *Differential Space, Quantum Systems, and Prediction*. Cambridge: The Massachusetts Institute of Technology Press

[2] E. Cobham Brewer 1810–1897. Dictionary of Phrase and Fable. 1898.

[3] E. Helmholtz, Robert Mayer's Priorität (German)

[4] Herrmann,A. Weltreich der Physik, GNT-Verlag (1991) (German)

[5] Planck, M. (1901). "Ueber die Elementarquanta der Materie und der Elektricität". *Annalen der Physik* (in German) **309** (3): 564–566. Bibcode:1901AnP...309..564P. doi:10.1002/andp.19013090311.

[6] Planck, Max (1883). "Ueber das thermodynamische Gleichgewicht von Gasgemengen". *Annalen der Physik* (in German) **255** (6): 358. Bibcode:1883AnP...255..358P. doi:10.1002/andp.18832550612.

[7] Einstein, A. (1905). "Über einen die Erzeugung und Verwandlung des Lichtes betreffenden heuristischen Gesichtspunkt" (PDF). *Annalen der Physik* (in German) **17** (6): 132–148. Bibcode:1905AnP...322..132E. doi:10.1002/andp.19053220607.. A partial English translation is available from Wikisource.

[8] Max Planck (1901). "Ueber das Gesetz der Energieverteilung im Normalspectrum (On the Law of Distribution of Energy in the Normal Spectrum)". *Annalen der Physik* **309** (3): 553. Bibcode:1901AnP...309..553P. doi:10.1002/andp.19013090310. Archived from the original on 2008-04-18.

[9] Brown, T., LeMay, H., Bursten, B. (2008). *Chemistry: The Central Science* Upper Saddle River, NJ: Pearson Education ISBN 0-13-600617-5

[10] Klein, Martin J. (1961). "Max Planck and the beginnings of the quantum theory". *Archive for History of Exact Sciences* **1** (5): 459. doi:10.1007/BF00327765.

[11] Melville, K. (2005, February 11). Real-World Quantum Effects Demonstrated

[12] Modern Applied Physics-Tippens third edition; McGraw-Hill.

8.5 Further reading

- B. Hoffmann, *The Strange Story of the Quantum*, Pelican 1963.

- Lucretius, *On the Nature of the Universe*, transl. from the Latin by R.E. Latham, Penguin Books Ltd., Harmondsworth 1951. There are, of course, many translations, and the translation's title varies. Some put emphasis on how things work, others on what things are found in nature.

- J. Mehra and H. Rechenberg, *The Historical Development of Quantum Theory*, Vol.1, Part 1, Springer-Verlag New York Inc., New York 1982.

- M. Planck, *A Survey of Physical Theory*, transl. by R. Jones and D.H. Williams, Methuen & Co., Ltd., London 1925 (Dover editions 1960 and 1993) including the Nobel lecture.

- Rodney, Brooks (2011) *Fields of Color: The theory that escaped Einstein*. Allegra Print & Imaging.

Chapter 9

Lagrangian

This article is about the Lagrangian function in Lagrangian mechanics. For other uses, see Lagrangian (disambiguation).

The **Lagrangian**, L, of a dynamical system is a mathematical function that summarizes the dynamics of the system. For a simple mechanical system, it is the value given by the kinetic energy of the particle minus the potential energy of the particle but it may be generalized to more complex systems. It is used primarily as a key component in the Euler-Lagrange equations to find the path of a particle according to the principle of least action.

The Lagrangian is named after Italian mathematician and astronomer Joseph Louis Lagrange. The concept of a Lagrangian was introduced in a reformulation of classical mechanics introduced by Lagrange known as Lagrangian mechanics in 1788. This reformulation was needed in order to explore mechanics in alternative systems to cartesian coordinates such as polar, cylindrical and spherical coordinates, and where the coordinates do not necessarily refer to position, for which Newton's formulation of classical mechanics is not convenient.

The Lagrangian has since been used in a method to find the acceleration of a particle in a Newtonian gravitational field and to derive the Einstein field equations. This led to its use in applying electromagnetism to curved spacetime and in describing charged black holes. It also has additional uses in Mathematical formalism to find the functional derivative of an action, and in engineering for the analysis and optimisation of dynamic systems.

9.1 Definition

In classical mechanics, the natural form of the Lagrangian is defined as the kinetic energy, T, of the system minus its potential energy, V.[1] In symbols,

$$L = T - V.$$

If the Lagrangian of a system is known, then the equations of motion of the system may be obtained by a direct substitution of the expression for the Lagrangian into the Euler–Lagrange equation. The Lagrangian of a given system is not unique, and two Lagrangians describing the same system can differ by the total derivative with respect to time of some function $f(q,t)$, but solving any equivalent Lagrangians will give the same equations of motion.[2][3]

9.2 The Lagrangian formulation

9.2.1 Simple example

The trajectory of a thrown ball is characterized by the sum of the Lagrangian values at each time being a (local) minimum.

The Lagrangian L can be calculated at several instants of time t, and a graph of L against t can be drawn. The area under the curve is the action. Any different path between the initial and final positions leads to a larger action than that chosen by nature. Nature chooses the smallest action – this is the Principle of Least Action.

> If Nature has defined the mechanics problem of the thrown ball in so elegant a fashion, might She have defined other problems similarly. So it seems now. Indeed, at the present time it appears that we can describe all the fundamental forces in terms of a Lagrangian. The search for Nature's One Equation, which rules all of the universe, has been largely a search for an adequate Lagrangian.
> —Robert Adair, The Great Design: Particles, Fields, and Creation[4]

Using only the principle of least action and the Lagrangian we can deduce the correct trajectory, by trial and error or the calculus of variations.

9.2.2 Importance

The Lagrangian formulation of mechanics is important not just for its broad applications, but also for its role in advancing deep understanding of physics. Although Lagrange only sought to describe classical mechanics, the *action principle* that is used to derive the Lagrange equation was later recognized to be applicable to quantum mechanics as well.

Physical action and quantum-mechanical phase are related via Planck's constant, and the principle of stationary action can be understood in terms of constructive interference of wave functions.

The same principle, and the Lagrangian formalism, are tied closely to Noether's theorem, which connects physical conserved quantities to continuous symmetries of a physical system.

Lagrangian mechanics and Noether's theorem together yield a natural formalism for first quantization by including commutators between certain terms of the Lagrangian equations of motion for a physical system.

9.2.3 Advantages over other methods

- The formulation is not tied to any one coordinate system – rather, any convenient variables may be used to describe the system; these variables are called "generalized coordinates" qi and may be any quantitative attributes of the system (for example, strength of the magnetic field at a particular location; angle of a pulley; position of a particle in space; or degree of excitation of a particular eigenmode in a complex system) which are functions of the independent variable(s). This trait makes it easy to incorporate constraints into a theory by defining coordinates that only describe states of the system that satisfy the constraints.

- If the Lagrangian is invariant under a symmetry, then the resulting equations of motion are also invariant under that symmetry. This characteristic is very helpful in showing that theories are consistent with either special relativity or general relativity.

9.2.4 Cyclic coordinates and conservation laws

An important property of the Lagrangian is that conservation laws can easily be read off from it. For example, if the Lagrangian L does *not* depend on q_i itself (q_i representing a set of generalized coordinates), then the *generalized momentum* (p_i), given by:

$$p_i = \frac{\partial L}{\partial \dot{q}_i},$$

is a *conserved* quantity, because of Lagrange's equations:

$$\dot{p}_i = \frac{d}{dt}\frac{\partial L}{\partial \dot{q}_i} = \frac{\partial L}{\partial q_i} = 0.$$

It doesn't matter if L depends on the *time derivative* $\dot{q}_i$ of that generalized coordinate, since the Lagrangian independence of the coordinate always makes the above partial derivative zero. This is a special case of Noether's theorem. Such coordinates are called "cyclic" or "ignorable".

For example, the conservation of the generalized momentum,

$$p_2 = \frac{\partial L}{\partial \dot{q}_2},$$

say, can be directly seen if the Lagrangian of the system is of the form

$$L(q_1, q_3, q_4, \ldots; \dot{q}_1, \dot{q}_2, \dot{q}_3, \dot{q}_4, \ldots; t).$$

Also, if the time t, does not appear in L, then the Hamiltonian, which is related to the Lagrangian by a Legendre transformation, is conserved. This is the energy conservation unless the potential energy depends on velocity, as in electrodynamics.[5][6]

9.3 Explanation

The Lagrangian in many classical systems is a function of generalized coordinates qi and their velocities dqi/dt. These coordinates (and velocities) are, in their turn, parametric functions of time. In the classical view, time is an independent variable and qi (and dqi/dt) are dependent variables as is often seen in phase space explanations of systems. This formalism was generalized further to handle field theory. In field theory, the independent variable is replaced by an event in spacetime (x, y, z, t), or more generally still by a point s on a manifold. The dependent variables (q) are replaced by the value of a field at that point in spacetime $\varphi(x,y,z,t)$ so that the equations of motion are obtained by means of an action principle, written as:

$$\frac{\delta S}{\delta \varphi_i} = 0,$$

where the *action*, S, is a functional of the dependent variables $\varphi i(s)$ with their derivatives and s itself

$$S[\varphi_i] = \int \mathcal{L}\left(\varphi_i(s), \frac{\partial \varphi_i(s)}{\partial s^\alpha}, s^\alpha\right) d^n s$$

and where $s = \{s^\alpha\}$ denotes the set of n independent variables of the system, indexed by $\alpha = 1, 2, 3,..., n$. Notice L is used in the case of one independent variable (t) and $\mathcal{L}$ is used in the case of multiple independent variables (usually four: x, y, z, t).

The equations of motion obtained from this functional derivative are the Euler–Lagrange equations of this action. For example, in the classical mechanics of particles, the only independent variable is time, t. So the Euler–Lagrange equations are

$$\frac{d}{dt}\frac{\partial L}{\partial \dot{\varphi}_i} = \frac{\partial L}{\partial \varphi_i}.$$

Dynamical systems whose equations of motion are obtainable by means of an action principle on a suitably chosen Lagrangian are known as *Lagrangian dynamical systems*. Examples of Lagrangian dynamical systems range from the classical version of the Standard Model, to Newton's equations, to purely mathematical problems such as geodesic equations and Plateau's problem.

9.4 An example from classical mechanics

9.4.1 In Cartesian coordinates

Suppose we have a three-dimensional space in which a particle of mass m moves under the influence of a conservative force $\mathbf{F}$. Since the force is conservative, it corresponds to a potential energy function $V(\mathbf{x})$ given by $\mathbf{F} = -\nabla V(\mathbf{x})$. The Lagrangian of the particle can be written

$$L(\mathbf{x}, \dot{\mathbf{x}}) = \frac{1}{2} m \dot{\mathbf{x}}^2 - V(\mathbf{x}).$$

The equations of motion for the particle are found by applying the Euler–Lagrange equation

$$\frac{\mathrm{d}}{\mathrm{d}t} \left(\frac{\partial L}{\partial \dot{x}_i} \right) - \frac{\partial L}{\partial x_i} = 0,$$

where $i = 1, 2, 3$.

Then

$$\frac{\partial L}{\partial x_i} = -\frac{\partial V}{\partial x_i},$$

$$\frac{\partial L}{\partial \dot{x}_i} = \frac{\partial}{\partial \dot{x}_i} \left(\frac{1}{2} m \dot{\mathbf{x}}^2 \right) = \frac{1}{2} m \frac{\partial}{\partial \dot{x}_i} \left(\dot{x}_i^2 \right) = m \dot{x}_i,$$

and

$$\frac{\mathrm{d}}{\mathrm{d}t} \left(\frac{\partial L}{\partial \dot{x}_i} \right) = m \ddot{x}_i.$$

Thus

$$m \ddot{\mathbf{x}} + \nabla V = 0,$$

which is Newton's second law of motion for a particle subject to a conservative force. Here the time derivative is written conventionally as a dot above the quantity being differentiated, and ∇ is the del operator.

9.4.2 In spherical coordinates

Suppose we have a three-dimensional space using spherical coordinates (r, θ, φ) with the Lagrangian

$$L = \frac{m}{2} (\dot{r}^2 + r^2 \dot{\theta}^2 + r^2 \sin^2 \theta \, \dot{\varphi}^2) - V(r).$$

Then the Euler–Lagrange equations are:

$$m\ddot{r} - mr(\dot{\theta}^2 + \sin^2 \theta \, \dot{\varphi}^2) + V' = 0,$$

$$\frac{\mathrm{d}}{\mathrm{d}t}(mr^2\dot{\theta}) - mr^2\sin\theta\cos\theta\,\dot{\varphi}^2 = 0,$$

$$\frac{\mathrm{d}}{\mathrm{d}t}(mr^2\sin^2\theta\,\dot{\varphi}) = 0.$$

Here the set of parameters s_i is just the time t, and the dynamical variables $\phi_i(s)$ are the trajectories $\vec{x}(t)$ of the particle.

Despite the use of standard variables such as x, the Lagrangian allows the use of any coordinates, which do not need to be orthogonal. These are "generalized coordinates".

9.5   Lagrangian of a test particle

A test particle is a particle whose mass and charge are assumed to be so small that its effect on external system is insignificant. It is often a hypothetical simplified point particle with no properties other than mass and charge. Real particles like electrons and up quarks are more complex and have additional terms in their Lagrangians.

9.5.1   Classical test particle with Newtonian gravity

Suppose we are given a particle with mass m kilograms, and position $\mathbf{x}$ meters in a Newtonian gravitation field with potential Φ in J·kg^{-1}. The particle's world line is parameterized by time t seconds. The particle's kinetic energy is:

$$T(t) = \tfrac{1}{2}m\,|\dot{\mathbf{x}}(t)|^2$$

and the particle's gravitational potential energy is:

$$V(t) = m\Phi(\mathbf{x}(t), t).$$

Then its Lagrangian is L joules, where

$$L(t) = T(t) - V(t) = \tfrac{1}{2}m\,|\dot{\mathbf{x}}(t)|^2 - m\Phi(\mathbf{x}(t), t).$$

Varying $\mathbf{x}$ in the integral (equivalent to the Euler–Lagrange differential equation), we get

$$0 = \delta\int L(t)\,\mathrm{d}t = \int \delta L(t)\,\mathrm{d}t$$

$$= \int \left(m\dot{\mathbf{x}}(t)\cdot\dot{\delta\mathbf{x}}(t) - m\nabla\Phi(\mathbf{x}(t),t)\cdot\delta\mathbf{x}(t)\right)\mathrm{d}t.$$

Integrate the first term by parts and discard the total integral. Then divide out the variation to get

$$0 = -m\ddot{\mathbf{x}}(t) - m\nabla\Phi(\mathbf{x}(t), t)$$

and thus

is the equation of motion – two different expressions for the force.

9.5.2   Special relativistic test particle with electromagnetism

In special relativity, the energy (rest energy plus kinetic energy) of a free test particle is

$$mc^2 \frac{dt}{d\tau} = \frac{mc^2}{\sqrt{1 - \frac{v^2(t)}{c^2}}} = +mc^2 + \frac{1}{2}mv^2(t) + \frac{3}{8}m\frac{v^4(t)}{c^2} + \dots .$$

However, the term in the Lagrangian that gives rise to the derivative of the momentum is no longer the kinetic energy.

One possible Lagrangian

$$-mc^2 \frac{d\tau(t)}{dt} = -mc^2 \sqrt{1 - \frac{v^2(t)}{c^2}} = -mc^2 + \frac{1}{2}mv^2(t) + \frac{1}{8}m\frac{v^4(t)}{c^2} + \dots$$

where c is the vacuum speed of light in m·s^{-1}, τ is the proper time in seconds (i.e. time measured by a clock moving with the particle) and $v^2(t)=\dot{\mathbf{x}}(t)\cdot\dot{\mathbf{x}}(t)$. The second term in the series is just the classical kinetic energy. Suppose the particle has electrical charge q coulombs and is in an electromagnetic field with scalar potential ϕ volts (a volt is a joule per coulomb) and vector potential A V·s·m^{-1}. The Lagrangian of a special relativistic test particle in an electromagnetic field is:

$$L(t) = -mc^2 \sqrt{1 - \frac{v^2(t)}{c^2}} - q\phi(\mathbf{x}(t), t) + q\dot{\mathbf{x}}(t) \cdot \mathbf{A}(\mathbf{x}(t), t).$$

Varying this with respect to **x** , we get

$$0 = -\frac{d}{dt}\left(\frac{m\dot{\mathbf{x}}(t)}{\sqrt{1 - \frac{v^2(t)}{c^2}}}\right) - q\nabla\phi(\mathbf{x}(t), t) - q\frac{\partial \mathbf{A}(\mathbf{x}(t), t)}{\partial t} - q\dot{\mathbf{x}}(t) \cdot \nabla \mathbf{A}(\mathbf{x}(t), t) + q\nabla \mathbf{A}(\mathbf{x}(t), t) \cdot \dot{\mathbf{x}}(t)$$

which is

$$\frac{d}{dt}\left(\frac{m\dot{\mathbf{x}}(t)}{\sqrt{1 - \frac{v^2(t)}{c^2}}}\right) = q\mathbf{E}(\mathbf{x}(t), t) + q\dot{\mathbf{x}}(t) \times \mathbf{B}(\mathbf{x}(t), t)$$

which is the equation for the Lorentz force, where:

$$\mathbf{E}(\mathbf{x}, t) = -\nabla\phi(\mathbf{x}, t) - \frac{\partial \mathbf{A}(\mathbf{x}, t)}{\partial t}$$

$$\mathbf{B}(\mathbf{x}, t) = \nabla \times \mathbf{A}(\mathbf{x}, t)$$

are the fields and potentials.

An alternative Lagrangian for a special relativistic test particle is

$$L(\tau) = \frac{1}{2}mu^\mu(\tau)u_\mu(\tau) + qu^\mu(\tau)A_\mu(x)$$

where $u^\mu = dx^\mu/d\tau$ is the four-velocity of the test particle.

The Euler-Lagrange equations

$$\frac{\partial L}{\partial x^\nu} - \frac{d}{d\tau}\frac{\partial L}{\partial u^\nu} = 0$$

become

$$qu^\mu \frac{\partial A_\mu}{\partial x^\nu} = \frac{d}{d\tau}(mu_\nu + qA_\nu)$$

9.5.3   General relativistic test particle

In general relativity, the first term generalizes (includes) both the classical kinetic energy and the interaction with the gravitational field. It becomes:[7][8]

$$-mc^2 \frac{d\tau(t)}{dt} = -mc^2 \sqrt{-c^{-2}g_{\mu\nu}(x(t))\frac{dx^\mu(t)}{dt}\frac{dx^\nu(t)}{dt}}.$$

The Lagrangian of a general relativistic test particle in an electromagnetic field is:

$$L(t) = -mc^2 \sqrt{-c^{-2}g_{\mu\nu}(x(t))\frac{dx^\mu(t)}{dt}\frac{dx^\nu(t)}{dt}} + q\frac{dx^\mu(t)}{dt}A_\mu(x(t)).$$

If the four spacetime coordinates x^μ are given in arbitrary units (i.e. unitless), then gμv in m^2 is the rank 2 symmetric metric tensor which is also the gravitational potential. Also, Aμ in V·s is the electromagnetic 4-vector potential.

More generally, suppose the Lagrangian is that of a single particle plus an interaction term LI

$$L = -mc^2 \frac{d\tau}{dt} + L_I.$$

Varying this with respect to the position of the particle x^α as a function of time t gives

$$\delta L = m\frac{dt}{2d\tau}\delta\left(g_{\mu\nu}\frac{dx^\mu}{dt}\frac{dx^\nu}{dt}\right) + \delta L_I.$$

$$\delta L = m\frac{dt}{2d\tau}\left(g_{\mu\nu,\alpha}\delta x^\alpha \frac{dx^\mu}{dt}\frac{dx^\nu}{dt} + 2g_{\alpha\nu}\frac{d\delta x^\alpha}{dt}\frac{dx^\nu}{dt}\right) + \frac{\partial L_I}{\partial x^\alpha}\delta x^\alpha + \frac{\partial L_I}{\partial \frac{dx^\alpha}{dt}}\frac{d\delta x^\alpha}{dt}.$$

$$\delta L = \frac{1}{2}mg_{\mu\nu,\alpha}\delta x^\alpha \frac{dx^\mu}{d\tau}\frac{dx^\nu}{dt} - \frac{d}{dt}\left(mg_{\alpha\nu}\frac{dx^\nu}{d\tau}\right)\delta x^\alpha + \frac{\partial L_I}{\partial x^\alpha}\delta x^\alpha - \frac{d}{dt}\left(\frac{\partial L_I}{\partial \frac{dx^\alpha}{dt}}\right)\delta x^\alpha + \frac{d(...)}{dt}.$$

This gives the equation of motion

$$0 = \frac{1}{2}mg_{\mu\nu,\alpha}\frac{dx^\mu}{d\tau}\frac{dx^\nu}{dt} - \frac{d}{dt}\left(mg_{\alpha\nu}\frac{dx^\nu}{d\tau}\right) + f_\alpha$$

where

$$f_\alpha = \frac{\partial L_I}{\partial x^\alpha} - \frac{d}{dt}\left(\frac{\partial L_I}{\partial \frac{dx^\alpha}{dt}}\right)$$

is the non-gravitational force on the particle. (For m to be independent of time, we must have $f_\alpha \frac{dx^\alpha}{dt} = 0$.)

Rearranging gets the force equation

$$\frac{d}{dt}\left(m\frac{dx^\nu}{d\tau}\right) = -m\Gamma^\nu_{\mu\sigma}\frac{dx^\mu}{d\tau}\frac{dx^\sigma}{dt} + g^{\nu\alpha}f_\alpha$$

where Γ is the Christoffel symbol which is the gravitational force field.

If we let

$$p^\nu = m\frac{dx^\nu}{d\tau}$$

be the (kinetic) linear momentum for a particle with mass, then

$$\frac{dp^\nu}{dt} = -\Gamma^\nu_{\mu\sigma}p^\mu\frac{dx^\sigma}{dt} + g^{\nu\alpha}f_\alpha$$

and

$$\frac{dx^\nu}{dt} = \frac{p^\nu}{p^0}$$

hold even for a massless particle.

9.6 Lagrangians and Lagrangian densities in field theory

The time integral of the Lagrangian is called the action denoted by S. In field theory, a distinction is occasionally made between the Lagrangian L, of which the action is the time integral:

$$\mathcal{S} = \int L \, dt$$

and the *Lagrangian density* $\mathcal{L}$, which one integrates over all spacetime to get the action:

$$\mathcal{S}[\varphi_i] = \int \mathcal{L}(\varphi_i(x)) \, d^4x.$$

- General form of Lagrangian density: $\mathcal{L} = \mathcal{L}(\varphi_i, \varphi_{i,\mu})$ [9] where $\varphi_{i,\mu} \equiv \frac{\partial \varphi_i}{\partial x^\mu} \equiv \partial_\mu \varphi_i$ (see 4-gradient)
- The relationship between $\mathcal{L}$ and L: $L = \int \mathcal{L} \, d^{n-1}x$, where n is the space-time dimension [9] similar to $q = \int \rho \, dV$.

- In field theory, the independent variable t was replaced by an event in spacetime (x, y, z, t) or still more generally by a point s on a manifold.

The Lagrangian is then the spatial integral of the Lagrangian density. However, $\mathcal{L}$ is also frequently simply called the Lagrangian, especially in modern use; it is far more useful in relativistic theories since it is a locally defined, Lorentz scalar field. Both definitions of the Lagrangian can be seen as special cases of the general form, depending on whether the spatial variable x is incorporated into the index i or the parameters s in $\varphi i(s)$. Quantum field theories in particle physics, such as quantum electrodynamics, are usually described in terms of $\mathcal{L}$, and the terms in this form of the Lagrangian translate quickly to the rules used in evaluating Feynman diagrams.

Notice that, in the presence of gravity or when using general curvilinear coordinates, the Lagrangian density $\mathcal{L}$ will include a factor of $\sqrt{|g|}$ or its equivalent to ensure that it is a scalar density so that the integral will be invariant.

9.7　Selected fields

To go with the section on test particles above, here are the equations for the fields in which they move. The equations below pertain to the fields in which the test particles described above move and allow the calculation of those fields. The equations below will not give you the equations of motion of a test particle in the field but will instead give you the potential (field) induced by quantities such as mass or charge density at any point (x,t) . For example, in the case of Newtonian gravity, the Lagrangian density integrated over spacetime gives you an equation which, if solved, would yield $\Phi(x,t)$. This $\Phi(x,t)$, when substituted back in equation (**1**), the Lagrangian equation for the test particle in a Newtonian gravitational field, provides the information needed to calculate the acceleration of the particle.

9.7.1　Newtonian gravity

The Lagrangian (density) is $\mathcal{L}$ in J·m^{-3}. The interaction term $m\Phi$ is replaced by a term involving a continuous mass density ϱ in kg·m^{-3}. This is necessary because using a point source for a field would result in mathematical difficulties. The resulting Lagrangian for the classical gravitational field is:

$$\mathcal{L}(\mathbf{x}, t) = -\rho(\mathbf{x}, t)\Phi(\mathbf{x}, t) - \frac{1}{8\pi G}(\nabla\Phi(\mathbf{x}, t))^2$$

where G in m^3·kg^{-1}·s^{-2} is the gravitational constant. Variation of the integral with respect to Φ gives:

$$\delta\mathcal{L}(\mathbf{x}, t) = -\rho(\mathbf{x}, t)\delta\Phi(\mathbf{x}, t) - \frac{2}{8\pi G}(\nabla\Phi(\mathbf{x}, t)) \cdot (\nabla\delta\Phi(\mathbf{x}, t)).$$

Integrate by parts and discard the total integral. Then divide out by $\delta\Phi$ to get:

$$0 = -\rho(\mathbf{x}, t) + \frac{1}{4\pi G}\nabla \cdot \nabla\Phi(\mathbf{x}, t)$$

and thus

$$4\pi G\rho(\mathbf{x}, t) = \nabla^2\Phi(\mathbf{x}, t)$$

which yields Gauss's law for gravity.

9.7.2　Einstein Gravity

Main article: Einstein–Hilbert action

The Lagrange density for general relativity in the presence of matter fields is

$$\mathcal{L}_{\text{GR}} = \mathcal{L}_{\text{EH}} + \mathcal{L}_{\text{matter}} = \frac{c^4}{16\pi G}(R - 2\Lambda) + \mathcal{L}_{\text{matter}}$$

R is the curvature scalar, which is the Ricci tensor contracted with the metric tensor, and the Ricci tensor is the Riemann tensor contracted with a Kronecker delta. The integral of $\mathcal{L}_{\text{EH}}$ is known as the Einstein-Hilbert action. The Riemann tensor is the tidal force tensor, and is constructed out of Christoffel symbols and derivatives of Christoffel symbols, which are the gravitational force field. Plugging this Lagrangian into the Euler-Lagrange equation and taking the metric tensor $g_{\mu\nu}$ as the field, we obtain the Einstein field equations

$$R_{\mu\nu} - \frac{1}{2}Rg_{\mu\nu} + g_{\mu\nu}\Lambda = \frac{8\pi G}{c^4}T_{\mu\nu}$$

The last tensor is the energy momentum tensor and is defined by

$$T_{\mu\nu} \equiv \frac{-2}{\sqrt{-g}}\frac{\delta(\mathcal{L}_{\text{matter}}\sqrt{-g})}{\delta g^{\mu\nu}} = -2\frac{\delta\mathcal{L}_{\text{matter}}}{\delta g^{\mu\nu}} + g_{\mu\nu}\mathcal{L}_{\text{matter}}.$$

g is the determinant of the metric tensor when regarded as a matrix. Λ is the Cosmological constant. Generally, in general relativity, the integration measure of the action of Lagrange density is $\sqrt{-g}d^4x$. This makes the integral coordinate independent, as the root of the metric determinant is equivalent to the Jacobian determinant. The minus sign is a consequence of the metric signature (the determinant by itself is negative).[10]

9.7.3   Electromagnetism in special relativity

The interaction terms

$$-q\phi(\mathbf{x}(t),t) + q\dot{\mathbf{x}}(t)\cdot\mathbf{A}(\mathbf{x}(t),t)$$

are replaced by terms involving a continuous charge density ρ in A·s·m^{-3} and current density j in A·m^{-2}. The resulting Lagrangian for the electromagnetic field is:

$$\mathcal{L}(\mathbf{x},t) = -\rho(\mathbf{x},t)\phi(\mathbf{x},t) + \mathbf{j}(\mathbf{x},t)\cdot\mathbf{A}(\mathbf{x},t) + \frac{\epsilon_0}{2}E^2(\mathbf{x},t) - \frac{1}{2\mu_0}B^2(\mathbf{x},t).$$

Varying this with respect to φ, we get

$$0 = -\rho(\mathbf{x},t) + \epsilon_0\nabla\cdot\mathbf{E}(\mathbf{x},t)$$

which yields Gauss' law.

Varying instead with respect to A , we get

$$0 = \mathbf{j}(\mathbf{x},t) + \epsilon_0\dot{\mathbf{E}}(\mathbf{x},t) - \frac{1}{\mu_0}\nabla\times\mathbf{B}(\mathbf{x},t)$$

which yields Ampère's law.

Using tensor notation, we can write all this more compactly. The term $-\rho\phi(\mathbf{x},t) + \mathbf{j}\cdot\mathbf{A}$ is actually the inner product of two four-vectors. We package the charge density into the current 4-vector and the potential into the potential 4-vector. These two new vectors are

$$j^\mu = (\rho,\mathbf{j}) \quad \text{and} \quad A_\mu = (-\phi,\mathbf{A})$$

We can then write the interaction term as

$$-\rho\phi + \mathbf{j}\cdot\mathbf{A} = j^\mu A_\mu$$

Additionally, we can package the E and B fields into what is known as the electromagnetic tensor $F_{\mu\nu}$. We define this tensor as

$$F_{\mu\nu} = \partial_\mu A_\nu - \partial_\nu A_\mu$$

The term we are looking out for turns out to be

$$\frac{\epsilon_0}{2}E^2 - \frac{1}{2\mu_0}B^2 = -\frac{1}{4\mu_0}F_{\mu\nu}F^{\mu\nu} = -\frac{1}{4\mu_0}F_{\mu\nu}F_{\rho\sigma}\eta^{\mu\rho}\eta^{\nu\sigma}$$

We have made use of the Minkowski metric to raise the indices on the EMF tensor. In this notation, Maxwell's equations are

$$\partial_\mu F^{\mu\nu} = -\mu_0 j^\nu \quad \text{and} \quad \epsilon^{\mu\nu\lambda\sigma}\partial_\nu F_{\lambda\sigma} = 0$$

where ϵ is the Levi-Civita tensor. So the Lagrange density for electromagnetism in special relativity written in terms of Lorentz vectors and tensors is

$$\mathcal{L}(x) = j^\mu(x)A_\mu(x) - \frac{1}{4\mu_0}F_{\mu\nu}(x)F^{\mu\nu}(x)$$

In this notation it is apparent that classical electromagnetism is a Lorentz-invariant theory. By the equivalence principle, it becomes simple to extend the notion of electromagnetism to curved spacetime.[11][12]

9.7.4 Electromagnetism in general relativity

The Lagrange density of electromagnetism in general relativity also contains the Einstein-Hilbert action from above. The pure electromagnetic Lagrangian is precisely a matter Lagrangian $\mathcal{L}_{\text{matter}}$. The Lagrangian is

$$\mathcal{L}(x) = j^\mu(x)A_\mu(x) - \frac{1}{4\mu_0}F_{\mu\nu}(x)F_{\rho\sigma}(x)g^{\mu\rho}(x)g^{\nu\sigma}(x) + \frac{c^4}{16\pi G}R(x)$$
$$= \mathcal{L}_{\text{Maxwell}} + \mathcal{L}_{\text{Einstein-Hilbert}}.$$

This Lagrangian is obtained by simply replacing the Minkowski metric in the above flat Lagrangian with a more general (possibly curved) metric $g_{\mu\nu}(x)$. We can generate the Einstein Field Equations in the presence of an EM field using this lagrangian. The energy-momentum tensor is

$$T^{\mu\nu}(x) = \frac{2}{\sqrt{-g(x)}}\frac{\delta}{\delta g_{\mu\nu}(x)}\mathcal{S}_{\text{Maxwell}} = \frac{1}{\mu_0}\left(F^\mu_\lambda(x)F^{\nu\lambda}(x) - \frac{1}{4}g^{\mu\nu}(x)F_{\rho\sigma}(x)F^{\rho\sigma}(x)\right)$$

It can be shown that this energy momentum tensor is traceless, i.e. that

$$T = g_{\mu\nu}T^{\mu\nu} = 0$$

If we take the trace of both sides of the Einstein Field Equations, we obtain

$$R = -\frac{8\pi G}{c^4}T$$

So the tracelessness of the energy momentum tensor implies that the curvature scalar in an electromagnetic field vanishes. The Einstein equations are then

$$R^{\mu\nu} = \frac{8\pi G}{c^4} \frac{1}{\mu_0} \left(F^{\mu}_{\lambda}(x) F^{\nu\lambda}(x) - \frac{1}{4} g^{\mu\nu}(x) F_{\rho\sigma}(x) F^{\rho\sigma}(x) \right)$$

Additionally, Maxwell's equations are

$$D_\mu F^{\mu\nu} = -\mu_0 j^\nu$$

where D_μ is the covariant derivative. For free space, we can set the current tensor equal to zero, $j^\mu = 0$. Solving both Einstein and Maxwell's equations around a spherically symmetric mass distribution in free space leads to the Reissner-Nordstrom charged black hole, with the defining line element (written in natural units and with charge Q):[13]

$$ds^2 = \left(1 - \frac{2M}{r} + \frac{Q^2}{r^2}\right) dt^2 - \left(1 - \frac{2M}{r} + \frac{Q^2}{r^2}\right)^{-1} dr^2 - r^2 d\Omega^2$$

9.7.5 Electromagnetism using differential forms

Using differential forms, the electromagnetic action S in vacuum on a (pseudo-) Riemannian manifold $\mathcal{M}$ can be written (using natural units, $c = \varepsilon_0 = 1$) as

$$\mathcal{S}[\mathbf{A}] = \int_{\mathcal{M}} \left(-\frac{1}{2} \mathbf{F} \wedge \star \mathbf{F} + \mathbf{A} \wedge \star \mathbf{J} \right).$$

Here, $\mathbf{A}$ stands for the electromagnetic potential 1-form, $\mathbf{J}$ is the current 1-form, $\mathbf{F}$ is the field strength 2-form and the star denotes the Hodge star operator. This is exactly the same Lagrangian as in the section above, except that the treatment here is coordinate-free; expanding the integrand into a basis yields the identical, lengthy expression. Note that with forms, an additional integration measure is not necessary because forms have coordinate differentials built in. Variation of the action leads to

$$\mathrm{d}\star \mathbf{F} = \mathbf{J}.$$

These are Maxwell's equations for the electromagnetic potential. Substituting $\mathbf{F} = \mathrm{d}\mathbf{A}$ immediately yields the equation for the fields,

$$\mathrm{d}\mathbf{F} = 0$$

because $\mathbf{F}$ is an exact form.

9.7.6 Dirac Lagrangian

The Lagrangian density for a Dirac field is:[14]

$$\mathcal{L} = i\hbar c \bar{\psi} \slashed{\partial} \psi - mc^2 \bar{\psi}\psi$$

where ψ is a Dirac spinor (annihilation operator), $\bar{\psi} = \psi^\dagger \gamma^0$ is its Dirac adjoint (creation operator), and $\slashed{\partial}$ is Feynman slash notation for $\gamma^\sigma \partial_\sigma$.

9.7.7   Quantum electrodynamic Lagrangian

The Lagrangian density for QED is:

$$\mathcal{L}_{\text{QED}} = i\hbar c \bar{\psi} \slashed{D} \psi - mc^2 \bar{\psi}\psi - \frac{1}{4\mu_0} F_{\mu\nu} F^{\mu\nu}$$

where $F^{\mu\nu}$ is the electromagnetic tensor, D is the gauge covariant derivative, and $\slashed{D}$ is Feynman notation for $\gamma^\sigma D_\sigma$.

9.7.8   Quantum chromodynamic Lagrangian

The Lagrangian density for quantum chromodynamics is:[15][16][17]

$$\mathcal{L}_{\text{QCD}} = \sum_n \left(i\hbar c \bar{\psi}_n \slashed{D} \psi_n - m_n c^2 \bar{\psi}_n \psi_n \right) - \frac{1}{4} G^\alpha{}_{\mu\nu} G_\alpha{}^{\mu\nu}$$

where D is the QCD gauge covariant derivative, $n = 1, 2, ...6$ counts the quark types, and $G^\alpha{}_{\mu\nu}$ is the gluon field strength tensor.

9.8   Mathematical formalism

Suppose we have an n-dimensional manifold, M, and a target manifold, T. Let $\mathcal{C}$ be the configuration space of smooth functions from M to T.

9.8.1   Examples

- In classical mechanics, in the Hamiltonian formalism, M is the one-dimensional manifold $\mathbb{R}$, representing time and the target space is the cotangent bundle of space of generalized positions.

- In field theory, M is the spacetime manifold and the target space is the set of values the fields can take at any given point. For example, if there are m real-valued scalar fields, $\phi_1, ..., \phi m$, then the target manifold is $\mathbb{R}^m$. If the field is a real vector field, then the target manifold is isomorphic to $\mathbb{R}^n$. There is actually a much more elegant way using tangent bundles over M, but we will just stick to this version.

9.8.2   Mathematical development

Consider a functional,

$$\mathcal{S} : \mathcal{C} \to \mathbb{R}$$

called the action. Physical considerations require it be a mapping to $\mathbb{R}$ (the set of all real numbers), not $\mathbb{C}$ (the set of all complex numbers).

In order for the action to be local, we need additional restrictions on the action. If $\varphi \in \mathcal{C}$, we assume $\mathcal{S}[\varphi]$ is the integral over M of a function of φ , its derivatives and the position called the **Lagrangian**, $\mathcal{L}(\varphi, \partial\varphi, \partial\partial\varphi, ..., x)$. In other words,

$$\forall \varphi \in \mathcal{C}, \ \ \mathcal{S}[\varphi] \equiv \int_M \mathrm{d}^n x \mathcal{L}\big(\varphi(x), \partial\varphi(x), \partial\partial\varphi(x), ..., x\big).$$

It is assumed below, in addition, that the Lagrangian depends on only the field value and its first derivative but not the higher derivatives.

Given boundary conditions, basically a specification of the value of φ at the boundary if M is compact or some limit on φ as $x \to \infty$ (this will help in doing integration by parts), the subspace of c consisting of functions, φ, such that all functional derivatives of S at φ are zero and φ satisfies the given boundary conditions is the subspace of on shell solutions.

The solution is given by the Euler–Lagrange equations (thanks to the boundary conditions),

$$\frac{\delta S}{\delta \varphi} = -\partial_\mu \left(\frac{\partial \mathcal{L}}{\partial(\partial_\mu \varphi)} \right) + \frac{\partial \mathcal{L}}{\partial \varphi} = 0.$$

The left hand side is the functional derivative of the action with respect to φ.

9.9   Uses in Engineering

Circa 1963 Lagrangians were a general part of the engineering curriculum, but a quarter of a century later, even with the ascendency of dynamical systems, they were dropped as requirements for some engineering programs, and are generally considered to be the domain of theoretical dynamics. Circa 2003 this changed dramatically, and Lagrangians are not only a required part of many ME and EE graduate-level curricula, but also find applications in finance, economics, and biology, mainly as the basis of the formulation of various path integral schemes to facilitate the solution of parabolic partial differential equations via random walks.

Circa 2013, Lagrangians find their way into hundreds of direct engineering solutions, including robotics, turbulent flow analysis (Lagrangian and Eulerian specification of the flow field), signal processing, microscopic component contact and nanotechnology (superlinear convergent augmented Lagrangians), gyroscopic forcing and dissipation, semi-infinite supercomputing (which also involve Lagrange multipliers in the subfield of semi-infinite programming), chemical engineering (specific heat linear Lagrangian interpolation in reaction planning), civil engineering (dynamic analysis of traffic flows), optics engineering and design (Lagrangian and Hamiltonian optics) aerospace (Lagrangian interpolation), force stepping integrators, and even airbag deployment (coupled Eulerian-Lagrangians as well as SELM—the stochastic Eulerian Lagrangian method).[18]

9.10   See also

9.11   Notes

[1] Torby, Bruce (1984). "Energy Methods". *Advanced Dynamics for Engineers.* HRW Series in Mechanical Engineering. United States of America: CBS College Publishing. ISBN 0-03-063366-4.

[2] Goldstein, Herbert; Poole, Charles P.; Safko, John L. (2002). *Classical Mechanics* (3rd ed.). Addison-Wesley. p. 21. ISBN 978-0-201-65702-9.

[3] Bell, L.D. Landau and E.M. Lifshitz ; translated from the Russian by J.B. Sykes and J.S. (1999). *Mechanics* (3rd ed. ed.). Oxford: Butterworth-Heinemann. p. 4. ISBN 978-0-7506-2896-9.

[4] The Great Design: Particles, Fields, and Creation (New York: Oxford University Press, 1989), ROBERT K. ADAIR, p.22–24

[5] Classical Mechanics, T.W.B. Kibble, European Physics Series, McGraw-Hill (UK), 1973, ISBN 0-07-084018-0

[6] Analytical Mechanics, L.N. Hand, J.D. Finch, Cambridge University Press, 2008, ISBN 978-0-521-57572-0

[7] Lev Davidovich Landau & Evgeny Mikhailovich Lifshitz, *The Classical Theory of Fields*, (1975), Elsevier Ltd., ISBN 978-0-7506-2768-9, page 26

[8] Noel A. Doughty, *Lagrangian Interaction*, (1990), Addison-Wesley Publishers Ltd., ISBN 0-201-41625-5, pages 310

[9] Mandl F., Shaw G., *Quantum Field Theory*, chapter 2

[10] Zee, A. (2013). *Einstein gravity in a nutshell*. Princeton: Princeton University Press. pp. 344–390. ISBN 9780691145587.

[11] Zee, A. (2013). *Einstein gravity in a nutshell*. Princeton: Princeton University Press. pp. 244–253. ISBN 9780691145587.

[12] Mexico, Kevin Cahill, University of New (2013). *Physical mathematics* (Repr. ed.). Cambridge: Cambridge University Press. ISBN 9781107005211.

[13] Zee, A. (2013). *Einstein gravity in a nutshell*. Princeton: Princeton University Press. pp. 381–383, 477–478. ISBN 9780691145587.

[14] Itzykson-Zuber, eq. 3-152

[15] http://www.fuw.edu.pl/~{}dobaczew/maub-42w/node9.html

[16] http://smallsystems.isn-oldenburg.de/Docs/THEO3/publications/semiclassical.qcd.prep.pdf

[17] http://www-zeus.physik.uni-bonn.de/~{}brock/teaching/jets_ws0405/seminar09/sluka_quark_gluon_jets.pdf

[18] Roger F Gans (2013). *Engineering Dynamics: From the Lagrangian to Simulation*. New York: Springer. ISBN 978-1-4614-3929-5.

9.12   References

- David Tong Classical Dynamics (Cambridge lecture notes)

Chapter 10

Invariant (physics)

In mathematics and theoretical physics, an **invariant** is a property of a system which remains unchanged under some transformation.

Invariance does not imply not varying, it pertains to a condition where there is no variation of the system under observation, and the only applicable condition is the instantaneous condition. Invariance pertains to now(). Now(+1), to a condition where all variations are solely due the internal variables, with no external aspects imparting nor removing energy (Newton's law of motion: a system in motion continues in motion, unless an external force imparts or removes energy). That condition is met by using the partial derivative function, ∂f(internal)xf(external) and presuming/setting f(external)=constant, leading to ∂f(external)=1 using the chain rule. Obviously, this is a model used solely for calculations, and not a reality. Reality is, that at all and every instance, energy is both removed and added to any system in observation.

10.1   Examples

See also: Lorentz scalar

In the current era, the immobility of Polaris (the North Star) under the diurnal motion of the celestial sphere is a classical illustration of physical invariance.

> For example the rule describing Newton's force of gravity between two chunks of matter is the same whether they are in this galaxy or another (translational invariance in space). It is also the same today as it was a million years ago (translational invariance in time). The law does not work differently depending on whether one chunk is east or north of the other one (rotational invariance). Nor does the law have to be changed depending on whether you measure the force between the two chunks in a railroad station, or do the same experiment with the two chunks on a uniformly moving train (principle of relativity).
> — David Mermin: *It's About Time - Understanding Einstein's Relativity*, Chapter 1

Another example of a physical invariant is the speed of light under a Lorentz transformation[1] and time under a Galilean transformation. Such spacetime transformations represent shifts between the reference frames of different observers, and so by Noether's theorem invariance under a transformation represents a fundamental conservation law. For example, invariance under translation leads to conservation of momentum, and invariance in time leads to conservation of energy.

Quantities can be invariant under some common transformations but not under others. For example, the velocity of a particle is invariant when switching from rectangular coordinates to curvilinear coordinates, but is not invariant when transforming between frames of reference that are moving with respect to each other. Other quantities, like the speed of light, are always invariant.

10.2   Importance

Invariants are important in modern theoretical physics, and many theories are expressed in terms of their symmetries and invariants.

Covariance and contravariance generalize the mathematical properties of invariance in tensor mathematics, and are frequently used in electromagnetism, special relativity, and general relativity.

10.3   See also

- General covariance

- Invariant (mathematics)

- Physical constant

- Eigenvalues and eigenvectors

10.4   References

[1] French, A.P. (1968). *Special Relativity*. W. W. Norton & Company. ISBN 0-393-09793-5.

Chapter 11

Lie algebra

"Lie bracket" redirects here. For the operation on vector fields, see Lie bracket of vector fields.

In mathematics, a **Lie algebra** (/ˈliː/, not /ˈlaɪ/) is a vector space together with a non-associative multiplication called "Lie bracket" $[x, y]$. It was introduced to study the concept of infinitesimal transformations. Hermann Weyl introduced the term "Lie algebra" (after Sophus Lie) in the 1930s. In older texts, the name "**infinitesimal group**" is used.

Lie algebras are closely related to Lie groups which are groups that are also smooth manifolds, with the property that the group operations of multiplication and inversion are smooth maps. Any Lie group gives rise to a Lie algebra. Conversely, to any finite-dimensional Lie algebra over real or complex numbers, there is a corresponding connected Lie group unique up to covering (Lie's third theorem). This correspondence between Lie groups and Lie algebras allows one to study Lie groups in terms of Lie algebras.

11.1 Definitions

A **Lie algebra** is a vector space $\mathfrak{g}$ over some field F together with a binary operation $[\cdot, \cdot] : \mathfrak{g} \times \mathfrak{g} \to \mathfrak{g}$ called the **Lie bracket**, which satisfies the following axioms:

- Bilinearity,

$$[ax + by, z] = a[x, z] + b[y, z], \quad [z, ax + by] = a[z, x] + b[z, y]$$

for all scalars a, b in F and all elements x, y, z in $\mathfrak{g}$.

- Alternativity on $\mathfrak{g}$,

$$[x, x] = 0$$

for all x in $\mathfrak{g}$.

- The Jacobi identity,

$$[x, [y, z]] + [z, [x, y]] + [y, [z, x]] = 0$$

for all x, y, z in $\mathfrak{g}$.

Note that the bilinearity and alternating properties imply

- Anticommutativity,

$$[x,y] = -[y,x],$$

  for all elements x, y in $\mathfrak{g}$, while anticommutativity only implies the alternating property if the field's characteristic is not 2.[1]

It is customary to express a Lie algebra in lower-case fraktur, like $\mathfrak{g}$. If a Lie algebra is associated with a Lie group, then the spelling of the Lie algebra is the same as that Lie group. For example, the Lie algebra of $\mathrm{SU}(n)$ is written as $\mathfrak{su}(n)$.

11.1.1 Generators and dimension

Elements of a Lie algebra $\mathfrak{g}$ are said to be **generators** of the Lie algebra if the smallest subalgebra of $\mathfrak{g}$ containing them is $\mathfrak{g}$ itself. The **dimension** of a Lie algebra is its dimension as a vector space over F. The cardinality of a minimal generating set of a Lie algebra is always less than or equal to its dimension.

11.1.2 Subalgebras, ideals, and homomorphisms

The Lie bracket is not associative in general, meaning that $[[x, y], z]$ need not equal $[x, [y, z]]$. Nonetheless, much of the terminology that was developed in the theory of associative rings or associative algebras is commonly applied to Lie algebras. A subspace $\mathfrak{h} \subseteq \mathfrak{g}$ that is closed under the Lie bracket is called a **Lie subalgebra**. If a subspace $I \subseteq \mathfrak{g}$ satisfies a stronger condition that

$$[\mathfrak{g}, I] \subseteq I,$$

then I is called an **ideal** in the Lie algebra $\mathfrak{g}$.[2] A **homomorphism** between two Lie algebras (over the same base field) is a linear map that is compatible with the respective Lie brackets:

$$f : \mathfrak{g} \to \mathfrak{g}', \quad f([x,y]) = [f(x), f(y)],$$

for all elements x and y in $\mathfrak{g}$. As in the theory of associative rings, ideals are precisely the kernels of homomorphisms, given a Lie algebra $\mathfrak{g}$ and an ideal I in it, one constructs the **factor algebra** $\mathfrak{g}/I$, and the first isomorphism theorem holds for Lie algebras.

Let S be a subset of $\mathfrak{g}$. The set of elements x such that $[x, s] = 0$ for all s in S forms a subalgebra called the centralizer of S. The centralizer of $\mathfrak{g}$ itself is called the center of $\mathfrak{g}$. Similar to centralizers, if S is a subspace,[3] then the set of x such that $[x, s]$ is in S for all s in S forms a subalgebra called the normalizer of S.

11.1.3 Direct sum and semidirect product

Given two Lie algebras $\mathfrak{g}$ and $\mathfrak{g}'$, their direct sum is the Lie algebra consisting of the vector space $\mathfrak{g} \oplus \mathfrak{g}'$, of the pairs (x, x'), $x \in \mathfrak{g}, x' \in \mathfrak{g}'$, with the operation

$$[(x, x'), (y, y')] = ([x, y], [x', y']), \quad x, y \in \mathfrak{g}, \ x', y' \in \mathfrak{g}'.$$

Let $\mathfrak{g}$ be a Lie algebra and $\mathfrak{i}$ its ideal. If the canonical map $\mathfrak{g} \to \mathfrak{g}/\mathfrak{i}$ splits (i.e., admits a section), then $\mathfrak{g}$ is said to be a semidirect product of $\mathfrak{i}$ and $\mathfrak{g}/\mathfrak{i}$.

Levi's theorem says that a finite-dimensional Lie algebra is a semidirect product of its radical and the complementary subalgebra (Levi subalgebra).

11.2 Properties

11.2.1 Admits an enveloping algebra

See also: Universal enveloping algebra

For any associative algebra A with multiplication $*$, one can construct a Lie algebra $L(A)$. As a vector space, $L(A)$ is the same as A. The Lie bracket of two elements of $L(A)$ is defined to be their commutator in A:

$$[a, b] = a * b - b * a.$$

The associativity of the multiplication $*$ in A implies the Jacobi identity of the commutator in $L(A)$. For example, the associative algebra of $n \times n$ matrices over a field F gives rise to the general linear Lie algebra $\mathfrak{gl}_n(F)$. The associative algebra A is called an **enveloping algebra** of the Lie algebra $L(A)$. Every Lie algebra can be embedded into one that arises from an associative algebra in this fashion; see universal enveloping algebra.

11.2.2 Representation

Given a vector space V, let $\mathfrak{gl}(V)$ denote the Lie algebra enveloped by the associative algebra of all linear endomorphisms of V. A representation of a Lie algebra $\mathfrak{g}$ on V is a Lie algebra homomorphism

$$\pi : \mathfrak{g} \to \mathfrak{gl}(V).$$

A representation is said to be faithful if its kernel is trivial. Every finite-dimensional Lie algebra has a faithful representation on a finite-dimensional vector space (Ado's theorem).[4]

For example,

$$\mathrm{ad} : \mathfrak{g} \to \mathfrak{gl}(\mathfrak{g})$$

given by $\mathrm{ad}(x)(y) = [x, y]$ is a representation of $\mathfrak{g}$ on the vector space $\mathfrak{g}$ called the adjoint representation. A derivation on the Lie algebra $\mathfrak{g}$ (in fact on any non-associative algebra) is a linear map $\delta : \mathfrak{g} \to \mathfrak{g}$ that obeys the Leibniz' law, that is,

$$\delta([x, y]) = [\delta(x), y] + [x, \delta(y)]$$

for all x and y in the algebra. For any x, $\mathrm{ad}(x)$ is a derivation; a consequence of the Jacobi identity. Thus, the image of ad lies in the subalgebra of $\mathfrak{gl}(\mathfrak{g})$ consisting of derivations on $\mathfrak{g}$. A derivation that happens to be in the image of ad is called an inner derivation. If $\mathfrak{g}$ is semisimple, every derivation on $\mathfrak{g}$ is inner.

11.3 Examples

11.3.1 Vector spaces

- Any vector space V endowed with the identically zero Lie bracket becomes a Lie algebra. Such Lie algebras are called abelian, cf. below. Any one-dimensional Lie algebra over a field is abelian, by the antisymmetry of the Lie bracket.

- The real vector space of all $n \times n$ skew-hermitian matrices is closed under the commutator and forms a real Lie algebra denoted $\mathfrak{u}(n)$. This is the Lie algebra of the unitary group $U(n)$.

11.3.2   Subspaces

- The subspace of the general linear Lie algebra $\mathfrak{gl}_n(F)$ consisting of matrices of trace zero is a subalgebra,[5] the special linear Lie algebra, denoted $\mathfrak{sl}_n(F)$.

11.3.3   Real matrix groups

- Any Lie group G defines an associated real Lie algebra $\mathfrak{g}$ =Lie(G). The definition in general is somewhat technical, but in the case of real matrix groups, it can be formulated via the exponential map, or the matrix exponent. The Lie algebra $\mathfrak{g}$ consists of those matrices X for which $\exp(tX) \in G$, $\forall$ real numbers t.

  The Lie bracket of $\mathfrak{g}$ is given by the commutator of matrices. As a concrete example, consider the special linear group SL(*n*,**R**), consisting of all $n \times n$ matrices with real entries and determinant 1. This is a matrix Lie group, and its Lie algebra consists of all $n \times n$ matrices with real entries and trace 0.

11.3.4   Three dimensions

- The three-dimensional Euclidean space $\mathbf{R}^3$ with the Lie bracket given by the cross product of vectors becomes a three-dimensional Lie algebra.

- The Heisenberg algebra $H_3(\mathrm{R})$ is a three-dimensional Lie algebra generated by elements x, y and z with Lie brackets

$$[x, y] = z, \quad [x, z] = 0, \quad [y, z] = 0$$

It is explicitly realized as the space of 3×3 strictly upper-triangular matrices, with the Lie bracket given by the matrix commutator,

$$x = \begin{pmatrix} 0 & 1 & 0 \\ 0 & 0 & 0 \\ 0 & 0 & 0 \end{pmatrix}, \quad y = \begin{pmatrix} 0 & 0 & 0 \\ 0 & 0 & 1 \\ 0 & 0 & 0 \end{pmatrix}, \quad z = \begin{pmatrix} 0 & 0 & 1 \\ 0 & 0 & 0 \\ 0 & 0 & 0 \end{pmatrix}.$$

Any element of the Heisenberg group is thus representable as a product of group generators, i.e., matrix exponentials of these Lie algebra generators,

$$\begin{pmatrix} 1 & a & c \\ 0 & 1 & b \\ 0 & 0 & 1 \end{pmatrix} = e^{by} e^{cz} e^{ax} .$$

- The commutation relations between the *x*, *y*, and *z* components of the angular momentum operator in quantum mechanics are the same as those of $\mathfrak{su}(2)$ and $\mathfrak{so}(3)$,

$$[L_x, L_y] = i\hbar L_z$$

$$[L_y, L_z] = i\hbar L_x$$

$$[L_z, L_x] = i\hbar L_y$$

(The physicist convention for Lie algebras is used in the above equations, hence the factor of *i*.) The Lie algebra formed by these operators have, in fact, representations of all finite dimensions.

11.3.5 Infinite dimensions

- An important class of infinite-dimensional real Lie algebras arises in differential topology. The space of smooth vector fields on a differentiable manifold *M* forms a Lie algebra, where the Lie bracket is defined to be the commutator of vector fields. One way of expressing the Lie bracket is through the formalism of Lie derivatives, which identifies a vector field *X* with a first order partial differential operator *LX* acting on smooth functions by letting *LX*(*f*) be the directional derivative of the function *f* in the direction of *X*. The Lie bracket [*X*,*Y*] of two vector fields is the vector field defined through its action on functions by the formula:

$$L_{[X,Y]}f = L_X(L_Y f) - L_Y(L_X f).$$

- A Kac–Moody algebra is an example of an infinite-dimensional Lie algebra.

- The Moyal algebra is an infinite-dimensional Lie algebra which contains all classical Lie algebras as subalgebras.

11.4 Structure theory and classification

Lie algebras can be classified to some extent. In particular, this has an application to the classification of Lie groups.

11.4.1 Abelian, nilpotent, and solvable

Analogously to abelian, nilpotent, and solvable groups, defined in terms of the derived subgroups, one can define abelian, nilpotent, and solvable Lie algebras.

A Lie algebra $\mathfrak{g}$ is **abelian** if the Lie bracket vanishes, i.e. $[x,y] = 0$, for all x and y in $\mathfrak{g}$. Abelian Lie algebras correspond to commutative (or abelian) connected Lie groups such as vector spaces K^n or tori T^n, and are all of the form $\mathfrak{k}^n$, meaning an n-dimensional vector space with the trivial Lie bracket.

A more general class of Lie algebras is defined by the vanishing of all commutators of given length. A Lie algebra $\mathfrak{g}$ is **nilpotent** if the lower central series

$$\mathfrak{g} > [\mathfrak{g},\mathfrak{g}] > [[\mathfrak{g},\mathfrak{g}],\mathfrak{g}] > [[[\mathfrak{g},\mathfrak{g}],\mathfrak{g}],\mathfrak{g}] > \cdots$$

becomes zero eventually. By Engel's theorem, a Lie algebra is nilpotent if and only if for every u in $\mathfrak{g}$ the adjoint endomorphism

$$\mathrm{ad}(u) : \mathfrak{g} \to \mathfrak{g}, \quad \mathrm{ad}(u)v = [u,v]$$

is nilpotent.

More generally still, a Lie algebra $\mathfrak{g}$ is said to be **solvable** if the derived series:

$$\mathfrak{g} > [\mathfrak{g},\mathfrak{g}] > [[\mathfrak{g},\mathfrak{g}],[\mathfrak{g},\mathfrak{g}]] > [[[\mathfrak{g},\mathfrak{g}],[\mathfrak{g},\mathfrak{g}]],[[\mathfrak{g},\mathfrak{g}],[\mathfrak{g},\mathfrak{g}]]] > \cdots$$

becomes zero eventually.

Every finite-dimensional Lie algebra has a unique maximal solvable ideal, called its radical. Under the Lie correspondence, nilpotent (respectively, solvable) connected Lie groups correspond to nilpotent (respectively, solvable) Lie algebras.

11.4.2   Simple and semisimple

A Lie algebra is "simple" if it has no non-trivial ideals and is not abelian. A Lie algebra $\mathfrak{g}$ is called **semisimple** if its radical is zero. Equivalently, $\mathfrak{g}$ is semisimple if it does not contain any non-zero abelian ideals. In particular, a simple Lie algebra is semisimple. Conversely, it can be proven that any semisimple Lie algebra is the direct sum of its minimal ideals, which are canonically determined simple Lie algebras.

The concept of semisimplicity for Lie algebras is closely related with the complete reducibility (semisimplicity) of their representations. When the ground field F has characteristic zero, any finite-dimensional representation of a semisimple Lie algebra is semisimple (i.e., direct sum of irreducible representations.) In general, a Lie algebra is called reductive if the adjoint representation is semisimple. Thus, a semisimple Lie algebra is reductive.

11.4.3   Cartan's criterion

Cartan's criterion gives conditions for a Lie algebra to be nilpotent, solvable, or semisimple. It is based on the notion of the Killing form, a symmetric bilinear form on $\mathfrak{g}$ defined by the formula

$$K(u, v) = \mathrm{tr}(\mathrm{ad}(u)\,\mathrm{ad}(v)),$$

where tr denotes the trace of a linear operator. A Lie algebra $\mathfrak{g}$ is semisimple if and only if the Killing form is nondegenerate. A Lie algebra $\mathfrak{g}$ is solvable if and only if $K(\mathfrak{g}, [\mathfrak{g}, \mathfrak{g}]) = 0$.

11.4.4   Classification

The Levi decomposition expresses an arbitrary Lie algebra as a semidirect sum of its solvable radical and a semisimple Lie algebra, almost in a canonical way. Furthermore, semisimple Lie algebras over an algebraically closed field have been completely classified through their root systems. However, the classification of solvable Lie algebras is a 'wild' problem, and cannot be accomplished in general.

11.5   Relation to Lie groups

See also: Lie group–Lie algebra correspondence

Although Lie algebras are often studied in their own right, historically they arose as a means to study Lie groups.

Lie's fundamental theorems describe a relation between Lie groups and Lie algebras. In particular, any Lie group gives rise to a canonically determined Lie algebra (concretely, *the tangent space at the identity*); and, conversely, for any Lie algebra there is a corresponding connected Lie group (Lie's third theorem; see the Baker–Campbell–Hausdorff formula). This Lie group is not determined uniquely; however, any two connected Lie groups with the same Lie algebra are *locally isomorphic*, and in particular, have the same universal cover. For instance, the special orthogonal group SO(3) and the special unitary group SU(2) give rise to the same Lie algebra, which is isomorphic to $\mathbf{R}^3$ with the cross-product, while SU(2) is a simply-connected twofold cover of SO(3).

Given a Lie group, a Lie algebra can be associated to it either by endowing the tangent space to the identity with the differential of the adjoint map, or by considering the left-invariant vector fields as mentioned in the examples. In the case of real matrix groups, the Lie algebra $\mathfrak{g}$ consists of those matrices X for which exp(tX) ∈ G for all real numbers t, where exp is the exponential map.

Some examples of Lie algebras corresponding to Lie groups are the following:

- The Lie algebra $\mathfrak{gl}_n(\mathbb{C})$ for the group $\mathrm{GL}_n(\mathbb{C})$ is the algebra of complex $n{\times}n$ matrices

- The Lie algebra $\mathfrak{sl}_n(\mathbb{C})$ for the group $SL_n(\mathbb{C})$ is the algebra of complex $n \times n$ matrices with trace 0

- The Lie algebras $\mathfrak{o}(n)$ for the group $O(n)$ and $\mathfrak{so}(n)$ for $SO(n)$ are both the algebra of real anti-symmetric $n \times n$ matrices (See Antisymmetric matrix: Infinitesimal rotations for a discussion)

- The Lie algebra $\mathfrak{u}(n)$ for the group $U(n)$ is the algebra of skew-Hermitian complex $n \times n$ matrices while the Lie algebra $\mathfrak{su}(n)$ for $SU(n)$ is the algebra of skew-Hermitian, traceless complex $n \times n$ matrices.

In the above examples, the Lie bracket $[X, Y]$ (for X and Y matrices in the Lie algebra) is defined as $[X, Y] = XY - YX$.

Given a set of generators T^a, the **structure constants** f^{abc} express the Lie brackets of pairs of generators as linear combinations of generators from the set, i.e., $[T^a, T^b] = f^{abc} T^c$. The structure constants determine the Lie brackets of elements of the Lie algebra, and consequently nearly completely determine the group structure of the Lie group. The structure of the Lie group near the identity element is displayed explicitly by the Baker–Campbell–Hausdorff formula, an expansion in Lie algebra elements X, Y and their Lie brackets, all nested together within a single exponent, $\exp(tX)$ $\exp(tY) = \exp(tX + tY + \frac{1}{2} t^2 [X, Y] + O(t^3))$).

The mapping from Lie groups to Lie algebras is functorial, which implies that homomorphisms of Lie groups lift to homomorphisms of Lie algebras, and various properties are satisfied by this lifting: it commutes with composition, it maps Lie subgroups, kernels, quotients and cokernels of Lie groups to subalgebras, kernels, quotients and cokernels of Lie algebras, respectively.

The functor **L** which takes each Lie group to its Lie algebra and each homomorphism to its differential is faithful and exact. It is however not an equivalence of categories: different Lie groups may have isomorphic Lie algebras (for example $SO(3)$ and $SU(2)$), and there are (infinite dimensional) Lie algebras that are not associated to any Lie group.[6]

However, when the Lie algebra $\mathfrak{g}$ is finite-dimensional, one can associate to it a simply connected Lie group having $\mathfrak{g}$ as its Lie algebra. More precisely, the Lie algebra functor **L** has a left adjoint functor Γ from finite-dimensional (real) Lie algebras to Lie groups, factoring through the full subcategory of simply connected Lie groups.[7] In other words, there is a natural isomorphism of bifunctors

$$\text{Hom}(\Gamma(\mathfrak{g}), H) \cong \text{Hom}(\mathfrak{g}, \text{L}(H)).$$

The adjunction $\mathfrak{g} \to \text{L}(\Gamma(\mathfrak{g}))$ (corresponding to the identity on $\Gamma(\mathfrak{g})$) is an isomorphism, and the other adjunction $\Gamma(\text{L}(H)) \to H$ is the projection homomorphism from the universal cover group of the identity component of H to H. It follows immediately that if G is simply connected, then the Lie algebra functor establishes a bijective correspondence between Lie group homomorphisms $G \to H$ and Lie algebra homomorphisms $\text{L}(G) \to \text{L}(H)$.

The universal cover group above can be constructed as the image of the Lie algebra under the exponential map. More generally, we have that the Lie algebra is homeomorphic to a neighborhood of the identity. But globally, if the Lie group is compact, the exponential will not be injective, and if the Lie group is not connected, simply connected or compact, the exponential map need not be surjective.

If the Lie algebra is infinite-dimensional, the issue is more subtle. In many instances, the exponential map is not even locally a homeomorphism (for example, in $\text{Diff}(\mathbf{S}^1)$, one may find diffeomorphisms arbitrarily close to the identity that are not in the image of exp). Furthermore, some infinite-dimensional Lie algebras are not the Lie algebra of any group.

The correspondence between Lie algebras and Lie groups is used in several ways, including in the classification of Lie groups and the related matter of the representation theory of Lie groups. Every representation of a Lie algebra lifts uniquely to a representation of the corresponding connected, simply connected Lie group, and conversely every representation of any Lie group induces a representation of the group's Lie algebra; the representations are in one to one correspondence. Therefore, knowing the representations of a Lie algebra settles the question of representations of the group.

As for classification, it can be shown that any connected Lie group with a given Lie algebra is isomorphic to the universal cover mod a discrete central subgroup. So classifying Lie groups becomes simply a matter of counting the discrete subgroups of the center, once the classification of Lie algebras is known (solved by Cartan et al. in the semisimple case).

11.6   Category theoretic definition

Using the language of category theory, a **Lie algebra** can be defined as an object A in **Vec**k, the category of vector spaces over a field k of characteristic not 2, together with a morphism $[.,.]: A \otimes A \rightarrow A$, where $\otimes$ refers to the monoidal product of **Vec**k, such that

- $[\cdot,\cdot] \circ (\mathrm{id} + \tau_{A,A}) = 0$

- $[\cdot,\cdot] \circ ([\cdot,\cdot] \otimes \mathrm{id}) \circ (\mathrm{id} + \sigma + \sigma^2) = 0$

where $\tau\,(a \otimes b) := b \otimes a$ and σ is the cyclic permutation braiding $(\mathrm{id} \otimes \tau A,A) \,^\circ\, (\tau A,A \otimes \mathrm{id})$. In diagrammatic form:

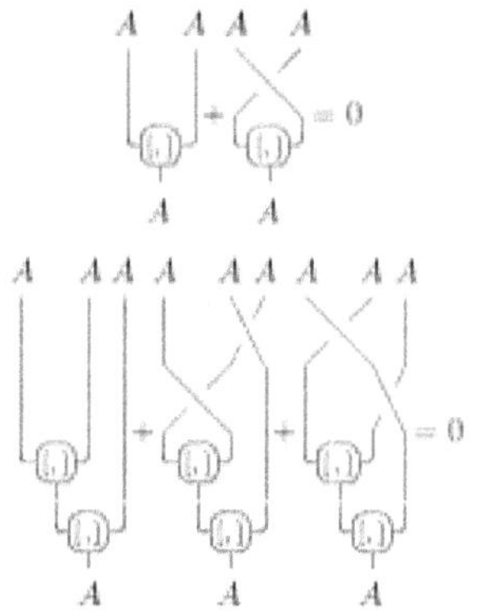

11.7   See also

11.8   Notes

[1] Humphreys p. 1

[2] Due to the anticommutativity of the commutator, the notions of a left and right ideal in a Lie algebra coincide.

[3] Jacobson 1962, pg. 28

[4] Jacobson 1962, Ch. VI

[5] Humphreys p.2

[6] Beltita 2005, pg. 75

[7] Adjoint property is discussed in more general context in Hofman & Morris (2007) (e.g., page 130) but is a straightforward consequence of, e.g., Bourbaki (1989) Theorem 1 of page 305 and Theorem 3 of page 310.

11.9   References

- Beltita, Daniel. *Smooth Homogeneous Structures in Operator Theory*, CRC Press, 2005. ISBN 978-1-4200-3480-6

- Boza, Luis; Fedriani, Eugenio M. & Núñez, Juan. *A new method for classifying complex filiform Lie algebras*, Applied Mathematics and Computation, 121 (2-3): 169–175, 2001

- Bourbaki, Nicolas. "Lie Groups and Lie Algebras - Chapters 1-3", Springer, 1989, ISBN 3-540-64242-0

- Erdmann, Karin & Wildon, Mark. *Introduction to Lie Algebras*, 1st edition, Springer, 2006. ISBN 1-84628-040-0

- Hall, Brian C. *Lie Groups, Lie Algebras, and Representations: An Elementary Introduction*, Springer, 2003. ISBN 0-387-40122-9

- Hofman, Karl & Morris, Sidney. "The Lie Theory of Connected Pro-Lie Groups", European Mathematical Society, 2007, ISBN 978-3-03719-032-6

- Humphreys, James E. *Introduction to Lie Algebras and Representation Theory*, Second printing, revised. Graduate Texts in Mathematics, 9. Springer-Verlag, New York, 1978. ISBN 0-387-90053-5

- Jacobson, Nathan, *Lie algebras*, Republication of the 1962 original. Dover Publications, Inc., New York, 1979. ISBN 0-486-63832-4

- Kac, Victor G. et al. *Course notes for MIT 18.745: Introduction to Lie Algebras*, math.mit.edu

- O'Connor, J.J. & Robertson, E.F. Biography of Sophus Lie, MacTutor History of Mathematics Archive, www-history.mcs.st-andrews.ac.uk

- O'Connor, J.J. & Robertson, E.F. Biography of Wilhelm Killing, MacTutor History of Mathematics Archive, www-history.mcs.st-andrews.ac.uk

- Serre, Jean-Pierre. "Lie Algebras and Lie Groups", 2nd edition, Springer, 2006. ISBN 3-540-55008-9

- Steeb, W.-H. *Continuous Symmetries, Lie Algebras, Differential Equations and Computer Algebra*, second edition, World Scientific, 2007, ISBN 978-981-270-809-0

- Varadarajan, V.S. *Lie Groups, Lie Algebras, and Their Representations*, 1st edition, Springer, 2004. ISBN 0-387-90969-9.

11.10 External links

- Hazewinkel, Michiel, ed. (2001), "Lie algebra", *Encyclopedia of Mathematics*, Springer, ISBN 978-1-55608-010-4

Chapter 12

Electromagnetic four-potential

An **electromagnetic four-potential** is a relativistic vector function from which the electromagnetic field can be derived. It combines both an electric scalar potential and a magnetic vector potential into a single four-vector. [1]

As measured in a given frame of reference, and for a given gauge, the first component of the electromagnetic four-potential is the electric scalar potential, and the other three components make up the magnetic vector potential. While both the scalar and vector potential depend upon the frame, the electromagnetic four-potential is Lorentz covariant.

Like other potentials, many different electromagnetic four-potentials correspond to the same electromagnetic field, depending upon the choice of gauge.

In this article, index notation and the Minkowski metric (+−−−) will be used, see also Ricci calculus, covariance and contravariance of vectors and raising and lowering indices for more details on notation. Formulae are given in SI units and Gaussian-cgs units.

12.1 Definition

The **electromagnetic four-potential** can be defined as:[2]

in which ϕ is the electric potential, and $\mathbf{A}$ is the magnetic potential (a vector potential). The units of A^α are V·s·m^{-1} in SI, and Mx·cm^{-1} in Gaussian-cgs.

The electric and magnetic fields associated with these four-potentials are:[3]

In special relativity, the electric and magnetic fields must be written in the form of a tensor so they transform correctly under Lorentz transformations - achieved by the electromagnetic tensor. This is written in terms of the electromagnetic four-potential as:

$$F^{\mu\nu} = \partial^\mu A^\nu - \partial^\nu A^\mu.$$

This essentially defines the four-potential in terms of physically observable quantities, as well as reducing to the above definition.

12.2   In the Lorenz gauge

Main articles: mathematical descriptions of the electromagnetic field and Retarded potential

Often, the Lorenz gauge condition $\partial_\alpha A^\alpha = 0$ in an inertial frame of reference is employed to simplify Maxwell's equations as:[4]

where J^α are the components of the four-current, and

$$\Box = \frac{1}{c^2}\frac{\partial^2}{\partial t^2} - \nabla^2$$

is the d'Alembertian operator. In terms of the scalar and vector potentials, this last equation becomes:

For a given charge and current distribution, $\rho(\mathbf{r}, t)$ and $\mathbf{j}(\mathbf{r}, t)$, the solutions to these equations in SI units are:[5]

$$\phi(\mathbf{r}, t) = \frac{1}{4\pi\epsilon_0}\int d^3x' \frac{\rho(\mathbf{r}', t_r)}{|\mathbf{r} - \mathbf{r}'|}$$

$$\mathbf{A}(\mathbf{r}, t) = \frac{\mu_0}{4\pi}\int d^3x' \frac{\mathbf{j}(\mathbf{r}', t_r)}{|\mathbf{r} - \mathbf{r}'|},$$

where

$$t_r = t - \frac{|\mathbf{r} - \mathbf{r}'|}{c}$$

is the retarded time. This is sometimes also expressed with

$$\rho(\mathbf{r}', t_r) = [\rho(\mathbf{r}', t)],$$

where the square brackets are meant to indicate that the time should be evaluated at the retarded time. Of course, since the above equations are simply the solution to an inhomogeneous differential equation, any solution to the homogeneous equation can be added to these to satisfy the boundary conditions. These homogeneous solutions in general represent waves propagating from sources outside the boundary.

When the integrals above are evaluated for typical cases, e.g. of an oscillating current (or charge), they are found to give both a magnetic field component varying according to r^{-2} (the induction field) and a component decreasing as r^{-1} (the radiation field).

12.3   See also

- Covariant formulation of classical electromagnetism
- Jefimenko's equations
- Gluon field

12.4   References

[1] Gravitation, J.A. Wheeler, C. Misner, K.S. Thorne, W.H. Freeman & Co, 1973, ISBN 0-7167-0344-0

[2] Introduction to Electrodynamics (3rd Edition), D.J. Griffiths, Pearson Education, Dorling Kindersley, 2007, ISBN 81-7758-293-3

[3] Electromagnetism (2nd Edition), I.S. Grant, W.R. Phillips, Manchester Physics, John Wiley & Sons, 2008, ISBN 978-0-471-92712-9

[4] Introduction to Electrodynamics (3rd Edition), D.J. Griffiths, Pearson Education, Dorling Kindersley, 2007, ISBN 81-7758-293-3

[5] Electromagnetism (2nd Edition), I.S. Grant, W.R. Phillips, Manchester Physics, John Wiley & Sons, 2008, ISBN 978-0-471-92712-9

- Rindler, Wolfgang (1991). *Introduction to Special Relativity (2nd)*. Oxford: Oxford University Press. ISBN 0-19-853952-5.

- Jackson, J D (1999). *Classical Electrodynamics (3rd)*. New York: Wiley. ISBN ISBN 0-471-30932-X.

Chapter 13

Gauge boson

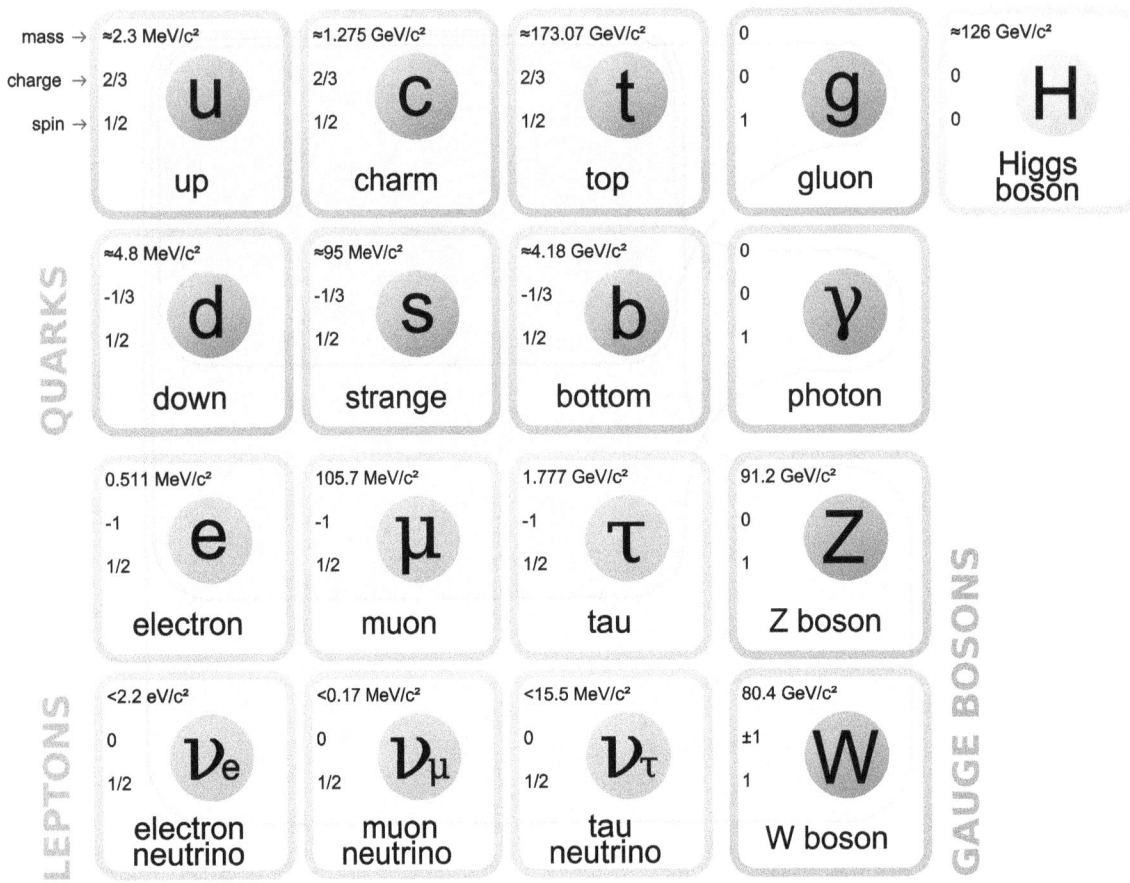

The Standard Model of elementary particles, with the gauge bosons in the fourth column in red

In particle physics, a **gauge boson** is a force carrier, a bosonic particle that carries any of the fundamental interactions of nature.[1][2] Elementary particles, whose interactions are described by a gauge theory, interact with each other by the exchange of gauge bosons—usually as virtual particles.

13.1   Gauge bosons in the Standard Model

The Standard Model of particle physics recognizes four kinds of gauge bosons: photons, which carry the electromagnetic interaction; W and Z bosons, which carry the weak interaction; and gluons, which carry the strong interaction.[3]

Isolated gluons do not occur at low energies because they are color-charged, and subject to color confinement.

13.1.1   Multiplicity of gauge bosons

In a quantized gauge theory, gauge bosons are quanta of the gauge fields. Consequently, there are as many gauge bosons as there are generators of the gauge field. In quantum electrodynamics, the gauge group is $U(1)$; in this simple case, there is only one gauge boson. In quantum chromodynamics, the more complicated group $SU(3)$ has eight generators, corresponding to the eight gluons. The three W and Z bosons correspond (roughly) to the three generators of $SU(2)$ in GWS theory.

13.1.2   Massive gauge bosons

For technical reasons involving gauge invariance, gauge bosons are described mathematically by field equations for massless particles. Therefore, at a naïve theoretical level all gauge bosons are required to be massless, and the forces that they describe are required to be long-ranged. The conflict between this idea and experimental evidence that the weak interaction has a very short range requires further theoretical insight.

According to the Standard Model, the W and Z bosons gain mass via the Higgs mechanism. In the Higgs mechanism, the four gauge bosons (of $SU(2) \times U(1)$ symmetry) of the unified electroweak interaction couple to a Higgs field. This field undergoes spontaneous symmetry breaking due to the shape of its interaction potential. As a result, the universe is permeated by a nonzero Higgs vacuum expectation value (VEV). This VEV couples to three of the electroweak gauge bosons (the Ws and Z), giving them mass; the remaining gauge boson remains massless (the photon). This theory also predicts the existence of a scalar Higgs boson, which has been observed in experiments that were reported on 4 July 2012.[4]

13.2   Beyond the Standard Model

13.2.1   Grand unification theories

A grand unified theory predicts additional gauge bosons named X and Y bosons. The hypothetical X and Y bosons direct interactions between quarks and leptons, hence violating conservation of baryon number and causing proton decay. Such bosons would be even more massive than W and Z bosons due to symmetry breaking. Analysis of data collected from such sources as the Super-Kamiokande neutrino detector has yielded no evidence of X and Y bosons.

13.2.2   Gravitons

The fourth fundamental interaction, gravity, may also be carried by a boson, called the graviton. In the absence of experimental evidence and a mathematically coherent theory of quantum gravity, it is unknown whether this would be a gauge boson or not. The role of gauge invariance in general relativity is played by a similar symmetry: diffeomorphism invariance.

13.2.3   W' and Z' bosons

Main article: W' and Z' bosons

W' and Z' bosons refer to hypothetical new gauge bosons (named in analogy with the Standard Model W and Z bosons).

13.3   See also

- 1964 PRL symmetry breaking papers

- Boson

- Glueball

- Quantum chromodynamics

- Quantum electrodynamics

13.4   References

[1] Gribbin, John (2000). *Q is for Quantum – An Encyclopedia of Particle Physics.* Simon & Schuster. ISBN 0-684-85578-X.

[2] Clark, John, E.O. (2004). *The Essential Dictionary of Science.* Barnes & Noble. ISBN 0-7607-4616-8.

[3] Veltman, Martinus (2003). *Facts and Mysteries in Elementary Particle Physics.* World Scientific. ISBN 981-238-149-X.

[4] "CERN experiments observe particle consistent with long-sought Higgs boson". CERN. Retrieved 4 July 2012.

13.5   External links

- Explanation of gauge boson and gauge fields by Christopher T. Hill

Chapter 14

Elementary particle

This article is about the physics concept. For the novel, see The Elementary Particles.

In particle physics, an **elementary particle** or **fundamental particle** is a particle whose substructure (domain of the

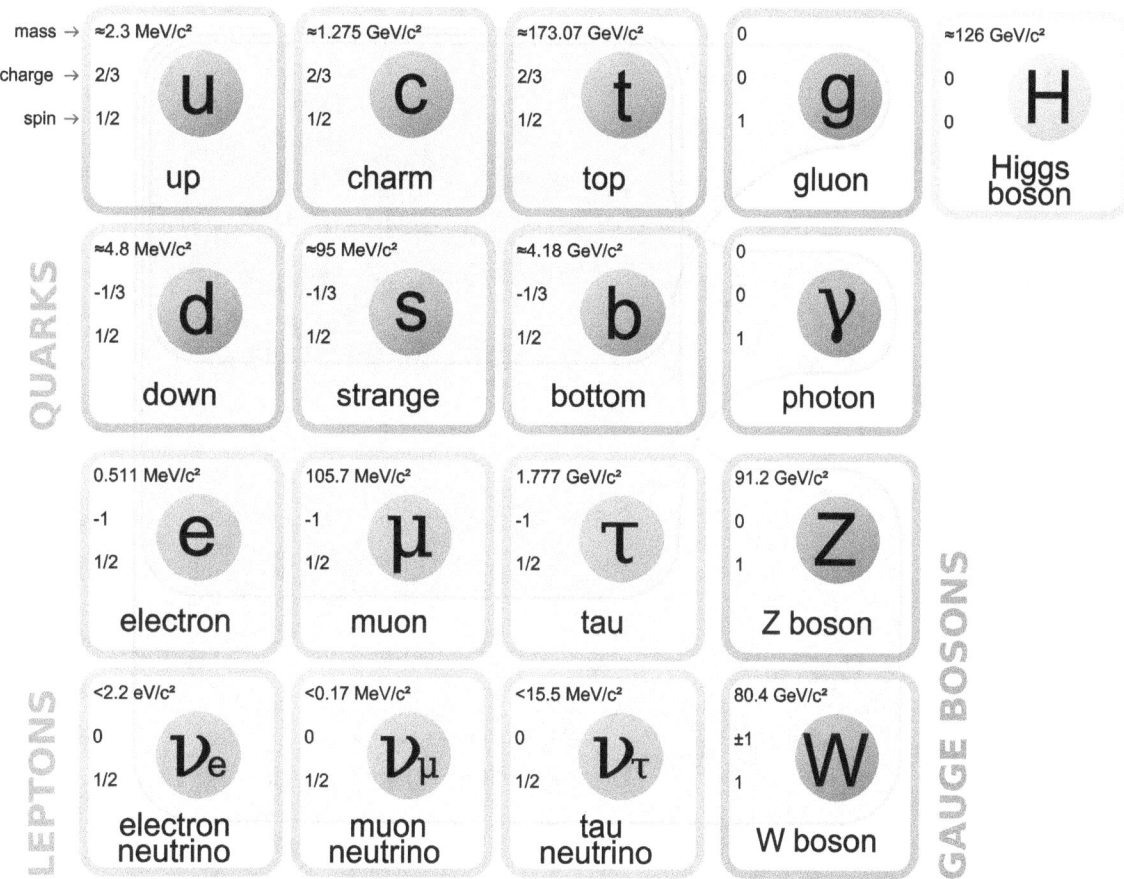

Elementary particles included in the Standard Model

bigger structure which shares the similar characteristics of the domain) is unknown, thus it is unknown whether it is composed of other particles.[1] Known elementary particles include the fundamental fermions (quarks, leptons, antiquarks, and antileptons), which generally are "matter particles" and "antimatter particles", as well as the fundamental bosons (gauge bosons and Higgs boson), which generally are "force particles" that mediate interactions among fermions.[1] A particle containing two or more elementary particles is a *composite particle*.

Everyday matter is composed of atoms, once presumed to be matter's elementary particles—*atom* meaning "indivisible" in Greek—although the atom's existence remained controversial until about 1910, as some leading physicists regarded molecules as mathematical illusions, and matter as ultimately composed of energy.[1][2] Soon, subatomic constituents of the atom were identified. As the 1930s opened, the electron and the proton had been observed, along with the photon, the particle of electromagnetic radiation.[1] At that time, the recent advent of quantum mechanics was radically altering the conception of particles, as a single particle could seemingly span a field as would a wave, a paradox still eluding satisfactory explanation.[3][4][5]

Via quantum theory, protons and neutrons were found to contain quarks—up quarks and down quarks—now considered elementary particles.[1] And within a molecule, the electron's three degrees of freedom (charge, spin, orbital) can separate via wavefunction into three quasiparticles (holon, spinon, orbiton).[6] Yet a free electron—which, not orbiting an atomic nucleus, lacks orbital motion—appears unsplittable and remains regarded as an elementary particle.[6]

Around 1980, an elementary particle's status as indeed elementary—an *ultimate constituent* of substance—was mostly discarded for a more practical outlook,[1] embodied in particle physics' Standard Model, science's most experimentally successful theory.[5][7] Many elaborations upon and theories beyond the Standard Model, including the extremely popular supersymmetry, double the number of elementary particles by hypothesizing that each known particle associates with a "shadow" partner far more massive,[8][9] although all such superpartners remain undiscovered.[7][10] Meanwhile, an elementary boson mediating gravitation—the graviton—is generally presumed, but remains hypothetical.[1]

14.1   Overview

Main article: Standard Model
See also: Physics beyond the Standard Model

All elementary particles are—depending on their *spin*—either bosons or fermions. These are differentiated via the spin–

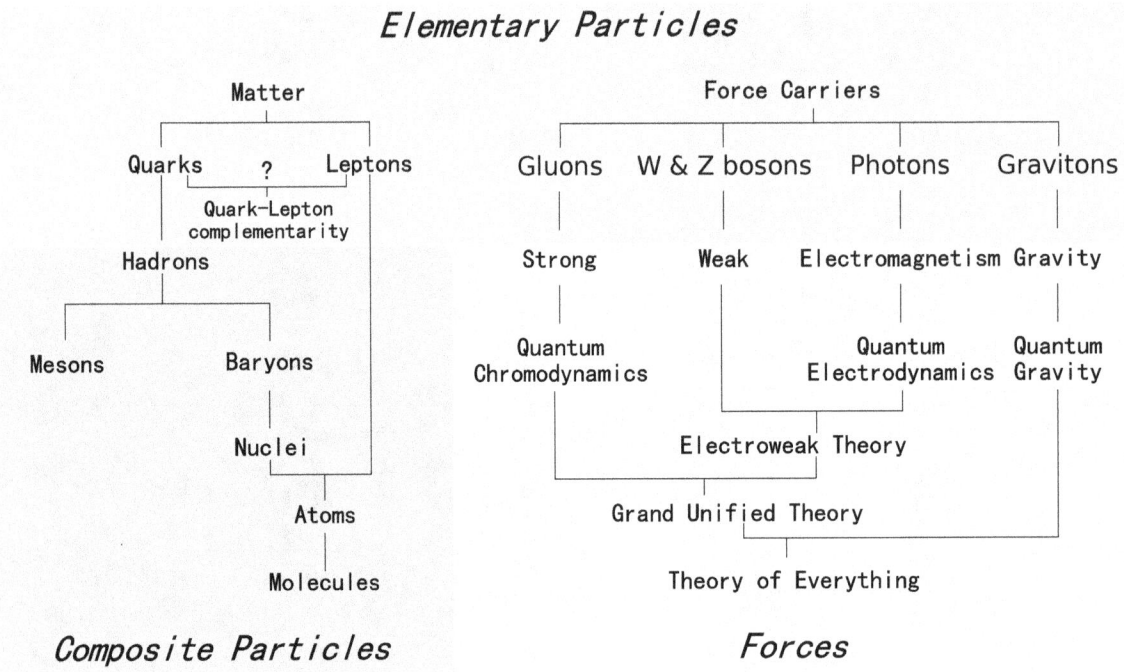

An overview of the various families of elementary and composite particles, and the theories describing their interactions

statistics theorem of quantum statistics. Particles of *half-integer* spin exhibit Fermi–Dirac statistics and are fermions.[1] Particles of *integer* spin, in other words full-integer, exhibit Bose–Einstein statistics and are bosons.[1]

Elementary fermions:

- Matter particles

  - Quarks:

    - up, down
    - charm, strange
    - top, bottom

  - Leptons:

    - electron, electron neutrino (a.k.a., "neutrino")
    - muon, muon neutrino
    - tau, tau neutrino

- Antimatter particles

  - Antiquarks
  - Antileptons

Elementary bosons:

- Force particles (gauge bosons):

  - photon
  - gluon (numbering eight)[1]
  - W^+, W^-, and Z^0 bosons
  - graviton (hypothetical)[1]

- Scalar boson

  - Higgs boson

A particle's mass is quantified in units of energy versus the electron's (electronvolts). Through conversion of energy into mass, any particle can be produced through collision of other particles at high energy,[1][11] although the output particle might not contain the input particles, for instance matter creation from colliding photons. Likewise, the composite fermions protons were collided at nearly light speed to produce a Higgs boson, which elementary boson is far more massive.[11] The most massive elementary particle, the top quark, rapidly decays into, but apparently does not contain, lighter particles.

When probed at energies available in experiments, particles exhibit spherical sizes. In operating particle physics' Standard Model, elementary particles are usually represented for predictive utility as point particles, which, as zero-dimensional, lack spatial extension. Though extremely successful, the Standard Model is limited to the microcosm by its omission of gravitation, and has some parameters arbitrarily added but unexplained.[12] Seeking to resolve those shortcomings, string theory posits that elementary particles are ultimately composed of one-dimensional energy strings whose absolute minimum size is the Planck length.

14.2   Common elementary particles

Main article: cosmic abundance of elements

According to the current models of big bang nucleosynthesis, the primordial composition of visible matter of the universe should be about 75% hydrogen and 25% helium-4 (in mass). Neutrons are made up of one up and two down quark, while protons are made of two up and one down quark. Since the other common elementary particles (such as electrons, neutrinos, or weak bosons) are so light or so rare when compared to atomic nuclei, we can neglect their mass contribution to the observable universe's total mass. Therefore, one can conclude that most of the visible mass of the universe consists of protons and neutrons, which, like all baryons, in turn consist of up quarks and down quarks.

Some estimates imply that there are roughly 10^{80} baryons (almost entirely protons and neutrons) in the observable universe.[13][14][15]

The number of protons in the observable universe is called the Eddington number.

In terms of number of particles, some estimates imply that nearly all the matter, excluding dark matter, occurs in neutrinos, and that roughly 10^{86} elementary particles of matter exist in the visible universe, mostly neutrinos.[15] Other estimates imply that roughly 10^{97} elementary particles exist in the visible universe (not including dark matter), mostly photons, gravitons, and other massless force carriers.[15]

14.3 Standard Model

Main article: Standard Model

The Standard Model of particle physics contains 12 flavors of elementary fermions, plus their corresponding antiparticles, as well as elementary bosons that mediate the forces and the Higgs boson, which was reported on July 4, 2012, as having been likely detected by the two main experiments at the LHC (ATLAS and CMS). However, the Standard Model is widely considered to be a provisional theory rather than a truly fundamental one, since it is not known if it is compatible with Einstein's general relativity. There may be hypothetical elementary particles not described by the Standard Model, such as the graviton, the particle that would carry the gravitational force, and sparticles, supersymmetric partners of the ordinary particles.

14.3.1 Fundamental fermions

Main article: Fermion

The 12 fundamental fermionic flavours are divided into three generations of four particles each. Six of the particles are quarks. The remaining six are leptons, three of which are neutrinos, and the remaining three of which have an electric charge of −1: the electron and its two cousins, the muon and the tau.

Antiparticles

Main article: Antimatter

There are also 12 fundamental fermionic antiparticles that correspond to these 12 particles. For example, the antielectron (positron) $e+$ is the electron's antiparticle and has an electric charge of +1.

Quarks

Main article: Quark

Isolated quarks and antiquarks have never been detected, a fact explained by confinement. Every quark carries one of three color charges of the strong interaction; antiquarks similarly carry anticolor. Color-charged particles interact via gluon exchange in the same way that charged particles interact via photon exchange. However, gluons are themselves color-charged, resulting in an amplification of the strong force as color-charged particles are separated. Unlike the electromagnetic force, which diminishes as charged particles separate, color-charged particles feel increasing force.

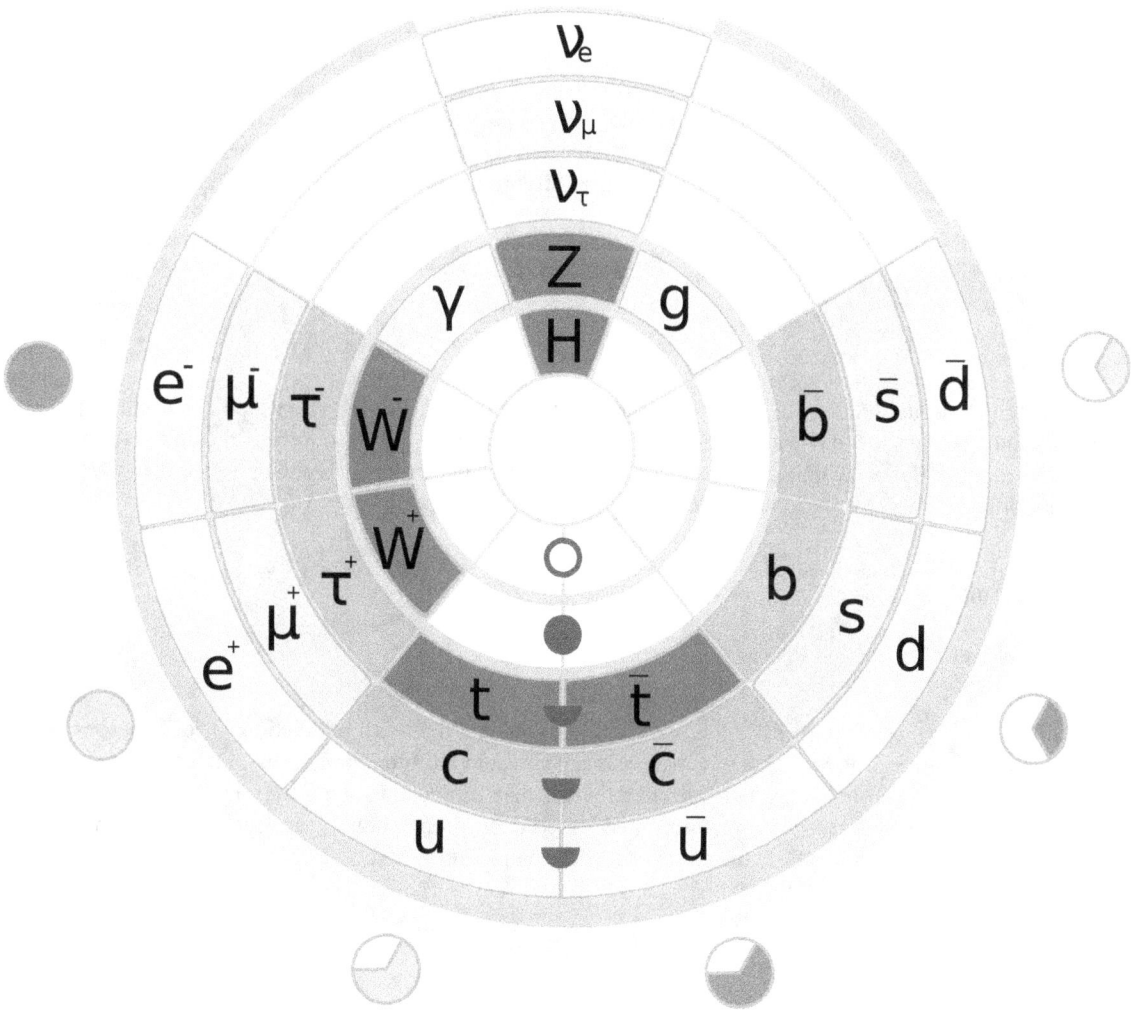

Graphic representation of the standard model. Spin, charge, mass and participation in different force interactions are shown. Click on the image to see the full description

However, color-charged particles may combine to form color neutral composite particles called hadrons. A quark may pair up with an antiquark: the quark has a color and the antiquark has the corresponding anticolor. The color and anticolor cancel out, forming a color neutral meson. Alternatively, three quarks can exist together, one quark being "red", another "blue", another "green". These three colored quarks together form a color-neutral baryon. Symmetrically, three antiquarks with the colors "antired", "antiblue" and "antigreen" can form a color-neutral antibaryon.

Quarks also carry fractional electric charges, but, since they are confined within hadrons whose charges are all integral, fractional charges have never been isolated. Note that quarks have electric charges of either $+2/3$ or $-1/3$, whereas antiquarks have corresponding electric charges of either $-2/3$ or $+1/3$.

Evidence for the existence of quarks comes from deep inelastic scattering: firing electrons at nuclei to determine the distribution of charge within nucleons (which are baryons). If the charge is uniform, the electric field around the proton should be uniform and the electron should scatter elastically. Low-energy electrons do scatter in this way, but, above a particular energy, the protons deflect some electrons through large angles. The recoiling electron has much less energy and a jet of particles is emitted. This inelastic scattering suggests that the charge in the proton is not uniform but split among smaller charged particles: quarks.

14.3.2 Fundamental bosons

Main article: Boson

In the Standard Model, vector (spin−1) bosons (gluons, photons, and the W and Z bosons) mediate forces, whereas the Higgs boson (spin-0) is responsible for the intrinsic mass of particles. Bosons differ from fermions in the fact that multiple bosons can occupy the same quantum state (Pauli exclusion principle). Also, bosons can be either elementary, like photons, or a combination, like mesons. The spin of bosons are integers instead of half integers.

Gluons

Main article: Gluon

Gluons mediate the strong interaction, which join quarks and thereby form hadrons, which are either baryons (three quarks) or mesons (one quark and one antiquark). Protons and neutrons are baryons, joined by gluons to form the atomic nucleus. Like quarks, gluons exhibit colour and anticolour—unrelated to the concept of visual color—sometimes in combinations, altogether eight variations of gluons.

Electroweak bosons

Main articles: W and Z bosons and Photon

There are three weak gauge bosons: W^+, W^-, and Z^0; these mediate the weak interaction. The W bosons are known for their mediation in nuclear decay. The W^- converts a neutron into a proton then decay into an electron and electron antineutrino pair. The Z^0 does not convert charge but rather changes momentum and is the only mechanism for elastically scattering neutrinos. The weak gauge bosons were discovered due to momentum change in electrons from neutrino-Z exchange. The massless photon mediates the electromagnetic interaction. These four gauge bosons form the electroweak interaction among elementary particles.

Higgs boson

Main article: Higgs boson

Although the weak and electromagnetic forces appear quite different to us at everyday energies, the two forces are theorized to unify as a single electroweak force at high energies. This prediction was clearly confirmed by measurements of cross-sections for high-energy electron-proton scattering at the HERA collider at DESY. The differences at low energies is a consequence of the high masses of the W and Z bosons, which in turn are a consequence of the Higgs mechanism. Through the process of spontaneous symmetry breaking, the Higgs selects a special direction in electroweak space that causes three electroweak particles to become very heavy (the weak bosons) and one to remain massless (the photon). On 4 July 2012, after many years of experimentally searching for evidence of its existence, the Higgs boson was announced to have been observed at CERN's Large Hadron Collider. Peter Higgs who first posited the existence of the Higgs boson was present at the announcement.[16] The Higgs boson is believed to have a mass of approximately 125 GeV.[17] The statistical significance of this discovery was reported as 5-sigma, which implies a certainty of roughly 99.99994%. In particle physics, this is the level of significance required to officially label experimental observations as a discovery. Research into the properties of the newly discovered particle continues.

Graviton

Main article: Graviton

The graviton is hypothesized to mediate gravitation, but remains undiscovered and yet is sometimes included in tables of elementary particles.[1] Its spin would be two—thus a boson—and it would lack charge or mass. Besides mediating an extremely feeble force, the graviton would have its own antiparticle and rapidly annihilate, rendering its detection extremely difficult even if it exists.

14.4   Beyond the Standard Model

Although experimental evidence overwhelmingly confirms the predictions derived from the Standard Model, some of its parameters were added arbitrarily, not determined by a particular explanation, which remain mysteries, for instance the hierarchy problem. Theories beyond the Standard Model attempt to resolve these shortcomings.

14.4.1   Grand unification

Main article: Grand Unified Theory

One extension of the Standard Model attempts to combine the electroweak interaction with the strong interaction into a single 'grand unified theory' (GUT). Such a force would be spontaneously broken into the three forces by a Higgs-like mechanism. The most dramatic prediction of grand unification is the existence of X and Y bosons, which cause proton decay. However, the non-observation of proton decay at the Super-Kamiokande neutrino observatory rules out the simplest GUTs, including SU(5) and SO(10).

14.4.2   Supersymmetry

Main article: Supersymmetry

Supersymmetry extends the Standard Model by adding another class of symmetries to the Lagrangian. These symmetries exchange fermionic particles with bosonic ones. Such a symmetry predicts the existence of supersymmetric particles, abbreviated as *sparticles*, which include the sleptons, squarks, neutralinos, and charginos. Each particle in the Standard Model would have a superpartner whose spin differs by 1/2 from the ordinary particle. Due to the breaking of supersymmetry, the sparticles are much heavier than their ordinary counterparts; they are so heavy that existing particle colliders would not be powerful enough to produce them. However, some physicists believe that sparticles will be detected by the Large Hadron Collider at CERN.

14.4.3   String theory

Main article: String theory

String Theory is a model of physics where all "particles" that make up matter are composed of strings (measuring at the Planck length) that exist in an 11-dimensional (according to M-theory, the leading version) universe. These strings vibrate at different frequencies that determine mass, electric charge, color charge, and spin. A string can be open (a line) or closed in a loop (a one-dimensional sphere, like a circle). As a string moves through space it sweeps out something called a *world sheet*. String theory predicts 1- to 10-branes (a 1-brane being a string and a 10-brane being a 10-dimensional object) that prevent tears in the "fabric" of space using the uncertainty principle (E.g., the electron orbiting a hydrogen atom has the probability, albeit small, that it could be anywhere else in the universe at any given moment).

String theory proposes that our universe is merely a 4-brane, inside which exist the 3 space dimensions and the 1 time dimension that we observe. The remaining 6 theoretical dimensions either are very tiny and curled up (and too small to be macroscopically accessible) or simply do not/cannot exist in our universe (because they exist in a grander scheme called the "multiverse" outside our known universe).

Some predictions of the string theory include existence of extremely massive counterparts of ordinary particles due to vibrational excitations of the fundamental string and existence of a massless spin-2 particle behaving like the graviton.

14.4.4 Technicolor

Main article: Technicolor (physics)

Technicolor theories try to modify the Standard Model in a minimal way by introducing a new QCD-like interaction. This means one adds a new theory of so-called Techniquarks, interacting via so called Technigluons. The main idea is that the Higgs-Boson is not an elementary particle but a bound state of these objects.

14.4.5 Preon theory

Main article: Preon

According to preon theory there are one or more orders of particles more fundamental than those (or most of those) found in the Standard Model. The most fundamental of these are normally called preons, which is derived from "pre-quarks". In essence, preon theory tries to do for the Standard Model what the Standard Model did for the particle zoo that came before it. Most models assume that almost everything in the Standard Model can be explained in terms of three to half a dozen more fundamental particles and the rules that govern their interactions. Interest in preons has waned since the simplest models were experimentally ruled out in the 1980s.

14.4.6 Acceleron theory

Accelerons are the hypothetical subatomic particles that integrally link the newfound mass of the neutrino and to the dark energy conjectured to be accelerating the expansion of the universe.[18]

In theory, neutrinos are influenced by a new force resulting from their interactions with accelerons. Dark energy results as the universe tries to pull neutrinos apart.[18]

14.5 See also

- Asymptotic freedom
- List of particles
- Physical ontology
- Quantum field theory
- Quantum gravity
- Quantum triviality
- UV fixed point

14.6 Notes

[1] Sylvie Braibant; Giorgio Giacomelli; Maurizio Spurio (2012). *Particles and Fundamental Interactions: An Introduction to Particle Physics* (2nd ed.). Springer. pp. 1–3. ISBN 978-94-007-2463-1.

[2] Ronald Newburgh; Joseph Peidle; Wolfgang Rueckner (2006). "Einstein, Perrin, and the reality of atoms: 1905 revisited" (PDF). *American Journal of Physics.* **74** (6): 478–481. Bibcode:2006AmJPh..74..478N. doi:10.1119/1.2188962.

[3] Friedel Weinert (2004). *The Scientist as Philosopher: Philosophical Consequences of Great Scientific Discoveries.* Springer. p. 43. ISBN 978-3-540-20580-7.

[4] Friedel Weinert (2004). *The Scientist as Philosopher: Philosophical Consequences of Great Scientific Discoveries.* Springer. pp. 57–59. ISBN 978-3-540-20580-7.

[5] Meinard Kuhlmann (24 Jul 2013). "Physicists debate whether the world is made of particles or fields—or something else entirely". *Scientific American.*

[6] Zeeya Merali (18 Apr 2012). "Not-quite-so elementary, my dear electron: Fundamental particle 'splits' into quasiparticles, including the new 'orbiton'". *Nature.* doi:10.1038/nature.2012.10471.

[7] Ian O'Neill (24 Jul 2013). "LHC discovery maims supersymmetry, again". *Discovery News.* Retrieved 2013-08-28.

[8] Particle Data Group. "Unsolved mysteries—supersymmetry". *The Particle Adventure.* Berkeley Lab. Retrieved 2013-08-28.

[9] National Research Council (2006). *Revealing the Hidden Nature of Space and Time: Charting the Course for Elementary Particle Physics.* National Academies Press. p. 68. ISBN 978-0-309-66039-6.

[10] "CERN latest data shows no sign of supersymmetry—yet". *Phys.Org.* 25 Jul 2013. Retrieved 2013-08-28.

[11] Ryan Avent (19 Jul 2012). "The Q&A: Brian Greene—Life after the Higgs". *The Economist.* Retrieved 2013-08-28.

[12] Sylvie Braibant; Giorgio Giacomelli; Maurizio Spurio (2012). *Particles and Fundamental Interactions: An Introduction to Particle Physics* (2nd ed.). Springer. p. 384. ISBN 978-94-007-2463-1.

[13] Frank Heile. "Is the Total Number of Particles in the Universe Stable Over Long Periods of Time?". 2014.

[14] Jared Brooks. "Galaxies and Cosmology". 2014. p. 4, equation 16.

[15] Robert Munafo (24 Jul 2013). "Notable Properties of Specific Numbers". Retrieved 2013-08-28.

[16] Lizzy Davies (4 July 2014). "Higgs boson announcement live: CERN scientists discover subatomic particle". *The Guardian.* Retrieved 2012-07-06.

[17] Lucas Taylor (4 Jul 2014). "Observation of a new particle with a mass of 125 GeV". CMS. Retrieved 2012-07-06.

[18] "New theory links neutrino's slight mass to accelerating Universe expansion". *ScienceDaily.* 28 Jul 2004. Retrieved 2008-06-05.

14.7   Further reading

14.7.1   General readers

- Feynman, R.P. & Weinberg, S. (1987) *Elementary Particles and the Laws of Physics: The 1986 Dirac Memorial Lectures.* Cambridge Univ. Press.

- Ford, Kenneth W. (2005) *The Quantum World.* Harvard Univ. Press.

- Brian Greene (1999). *The Elegant Universe.* W.W.Norton & Company. ISBN 0-393-05858-1.

- John Gribbin (2000) *Q is for Quantum – An Encyclopedia of Particle Physics.* Simon & Schuster. ISBN 0-684-85578-X.

- Oerter, Robert (2006) *The Theory of Almost Everything: The Standard Model, the Unsung Triumph of Modern Physics.* Plume.

- Schumm, Bruce A. (2004) *Deep Down Things: The Breathtaking Beauty of Particle Physics.* Johns Hopkins University Press. ISBN 0-8018-7971-X.

- Martinus Veltman (2003). *Facts and Mysteries in Elementary Particle Physics*. World Scientific. ISBN 981-238-149-X.

- Frank Close (2004). *Particle Physics: A Very Short Introduction*. Oxford: Oxford University Press. ISBN 0-19-280434-0.

- Seiden, Abraham (2005). *Particle Physics – A Comprehensive Introduction*. Addison Wesley. ISBN 0-8053-8736-6.

14.7.2 Textbooks

- Bettini, Alessandro (2008) *Introduction to Elementary Particle Physics*. Cambridge Univ. Press. ISBN 978-0-521-88021-3

- Coughlan, G. D., J. E. Dodd, and B. M. Gripaios (2006) *The Ideas of Particle Physics: An Introduction for Scientists*, 3rd ed. Cambridge Univ. Press. An undergraduate text for those not majoring in physics.

- Griffiths, David J. (1987) *Introduction to Elementary Particles*. John Wiley & Sons. ISBN 0-471-60386-4.

- Kane, Gordon L. (1987). *Modern Elementary Particle Physics*. Perseus Books. ISBN 0-201-11749-5.

- Perkins, Donald H. (2000) *Introduction to High Energy Physics*, 4th ed. Cambridge Univ. Press.

14.8 External links

The most important address about the current experimental and theoretical knowledge about elementary particle physics is the Particle Data Group, where different international institutions collect all experimental data and give short reviews over the contemporary theoretical understanding.

- Particle Data Group

other pages are:

- Greene, Brian, "*Elementary particles*", The Elegant Universe, NOVA (PBS)

- particleadventure.org, a well-made introduction also for non physicists

- CERNCourier: Season of Higgs and melodrama

- Pentaquark information page

- Interactions.org, particle physics news

- Symmetry Magazine, a joint Fermilab/SLAC publication

- "Sized Matter: perception of the extreme unseen", Michigan University project for artistic visualisation of sub-atomic particles

- Elementary Particles made thinkable, an interactive visualisation allowing physical properties to be compared

Chapter 15

Global symmetry

In physics, a **global symmetry** is a symmetry that holds at all points in the spacetime under consideration, as opposed to a local symmetry which varies from point to point.

Global symmetries require conservation laws, but not forces, in physics.

An example of a global symmetry is the action of the $U(1) = e^{iq\theta}$ (for θ a constant - making it a global transformation) group on the Dirac Lagrangian:

$$\mathcal{L}_D = \bar{\psi}\left(i\gamma^\mu \partial_\mu - m\right)\psi$$

Under this transformation the wavefunction changes as $\psi \to e^{iq\theta}\psi$ and $\bar{\psi} \to e^{-iq\theta}\bar{\psi}$ and so:

$$\mathcal{L} \to \bar{\mathcal{L}} = e^{-iq\theta}\bar{\psi}\left(i\gamma^\mu \partial_\mu - m\right)e^{iq\theta}\psi = e^{-iq\theta}e^{iq\theta}\bar{\psi}\left(i\gamma^\mu \partial_\mu - m\right)\psi = \mathcal{L}$$

15.1 See also

- Field (physics)
- Global spacetime structure
- Local spacetime structure

Chapter 16

Local symmetry

In physics, a **local symmetry** is symmetry of some physical quantity, which smoothly depends on the point of the base manifold. Such quantities can be for example an observable, a tensor or the Lagrangian of a theory. If a symmetry is local in this sense, then one can apply a local transformation (resp. local gauge transformation), which means that the representation of the symmetry group is a function of the manifold and can thus be taken to act differently on different points of spacetime.

The diffeomorphism group is a local symmetry and thus every geometrical or generally covariant theory (i.e. a theory whose equations are tensor equations, for example general relativity) has local symmetries.

Often the term local symmetry is specifically associated with local gauge symmetries in Yang–Mills theory (see also standard model) where the Lagrangian is locally symmetric under some compact Lie group. Local gauge symmetries always come together with some bosonic gauge fields, like the photon or gluon field, which induce a force in addition to requiring conservation laws.[1]

16.1 Examples

- General relativity has a local symmetry (general covariance, diffeomorphisms) which can be seen as generating the gravitational force.[2] Special relativity only has a global symmetry (Lorentz symmetry or more generally Poincaré symmetry)

- There are many global symmetries (such as SU(2) of isospin symmetry) and local symmetries (like SU(2) of weak interactions) in particle physics. The standard model of particle physics consists of Yang-Mills Theories

- The symmetry group of Supergravity is a local symmetry, whereas supersymmetry is a global symmetry.

16.2 See also

- Field (physics)

- Global spacetime structure

- Local spacetime structure

- Gauge theory

- Gravitation (book)

16.3   References

[1] Kaku, Michio (1993). *Quantum Field Theory: A Modern Introduction*. New York: Oxford University Press. ISBN 0-19-507652-4.

[2] Misner, Charles W.; Thorne, Kip S.; Wheeler, John Archibald (1973-09-15). "Gravitation". San Francisco: W. H. Freeman. ISBN 978-0-7167-0344-0.

Chapter 17

Photon

This article is about the elementary particle of light. For other uses, see Photon (disambiguation).

A **photon** is an elementary particle, the quantum of light and all other forms of electromagnetic radiation. It is the force carrier for the electromagnetic force, even when static via virtual photons. The effects of this force are easily observable at the microscopic and at the macroscopic level, because the photon has zero rest mass; this allows long distance interactions. Like all elementary particles, photons are currently best explained by quantum mechanics and exhibit wave–particle duality, exhibiting properties of waves and of particles. For example, a single photon may be refracted by a lens or exhibit wave interference with itself, but also act as a particle giving a definite result when its position is measured.

The modern photon concept was developed gradually by Albert Einstein in the first years of the 20th century to explain experimental observations that did not fit the classical wave model of light. In particular, the photon model accounted for the frequency dependence of light's energy, and explained the ability of matter and radiation to be in thermal equilibrium. It also accounted for anomalous observations, including the properties of black-body radiation, that other physicists, most notably Max Planck, had sought to explain using *semiclassical models*, in which light is still described by Maxwell's equations, but the material objects that emit and absorb light do so in amounts of energy that are *quantized* (i.e., they change energy only by certain particular discrete amounts and cannot change energy in any arbitrary way). Although these semiclassical models contributed to the development of quantum mechanics, many further experiments[2][3] starting with Compton scattering of single photons by electrons, first observed in 1923, validated Einstein's hypothesis that *light itself* is quantized. In 1926 the optical physicist Frithiof Wolfers and the chemist Gilbert N. Lewis coined the name *photon* for these particles, and after 1927, when Arthur H. Compton won the Nobel Prize for his scattering studies, most scientists accepted the validity that quanta of light have an independent existence, and the term *photon* for light quanta was accepted.

In the Standard Model of particle physics, photons and other elementary particles are described as a necessary consequence of physical laws having a certain symmetry at every point in spacetime. The intrinsic properties of particles, such as charge, mass and spin, are determined by the properties of this gauge symmetry. The photon concept has led to momentous advances in experimental and theoretical physics, such as lasers, Bose–Einstein condensation, quantum field theory, and the probabilistic interpretation of quantum mechanics. It has been applied to photochemistry, high-resolution microscopy, and measurements of molecular distances. Recently, photons have been studied as elements of quantum computers and for applications in optical imaging and optical communication such as quantum cryptography.

17.1 Nomenclature

In 1900, the German physicist Max Planck was working on black-body radiation and suggested that the energy in electromagnetic waves could only be released in "packets" of energy. In his 1901 article [4] in Annalen der Physik he called these packets "energy elements". The word *quanta* (singular *quantum*) was used even before 1900 to mean particles or amounts of different quantities, including electricity. Later, in 1905, Albert Einstein went further by suggesting that

111

electromagnetic waves could only exist in these discrete wave-packets.[5] He called such a wave-packet *the light quantum* (German: *das Lichtquant*).[Note 1] The name *photon* derives from the Greek word for light, φῶς (transliterated *phôs*). Arthur Compton used *photon* in 1928, referring to Gilbert N. Lewis.[6] The same name was used earlier, by the American physicist and psychologist Leonard T. Troland, who coined the word in 1916, in 1921 by the Irish physicist John Joly, in 1924 by the French physiologist René Wurmser (1890-1993) and in 1926 by the French physicist Frithiof Wolfers (1891-1971).[7] The name was suggested initially as a unit related to the illumination of the eye and the resulting sensation of light and was used later on in a physiological context. Although Wolfers's and Lewis's theories were never accepted, as they were contradicted by many experiments, the new name was adopted very soon by most physicists after Compton used it.[7][Note 2]

In physics, a photon is usually denoted by the symbol γ (the Greek letter gamma). This symbol for the photon probably derives from gamma rays, which were discovered in 1900 by Paul Villard,[8][9] named by Ernest Rutherford in 1903, and shown to be a form of electromagnetic radiation in 1914 by Rutherford and Edward Andrade.[10] In chemistry and optical engineering, photons are usually symbolized by $h\nu$, the energy of a photon, where h is Planck's constant and the Greek letter ν (nu) is the photon's frequency. Much less commonly, the photon can be symbolized by hf, where its frequency is denoted by f.

17.2   Physical properties

See also: Special relativity and Photonic molecule

A photon is massless,[Note 3] has no electric charge,[11] and is stable. A photon has two possible polarization states. In the momentum representation, which is preferred in quantum field theory, a photon is described by its wave vector, which determines its wavelength λ and its direction of propagation. A photon's wave vector may not be zero and can be represented either as a spatial 3-vector or as a (relativistic) four-vector; in the latter case it belongs to the light cone (pictured). Different signs of the four-vector denote different circular polarizations, but in the 3-vector representation one should account for the polarization state separately; it actually is a spin quantum number. In both cases the space of possible wave vectors is three-dimensional.

The photon is the gauge boson for electromagnetism,[12]:29-30 and therefore all other quantum numbers of the photon (such as lepton number, baryon number, and flavour quantum numbers) are zero.[13] Also, the photon does not obey the Pauli exclusion principle.[14]:1221

Photons are emitted in many natural processes. For example, when a charge is accelerated it emits synchrotron radiation. During a molecular, atomic or nuclear transition to a lower energy level, photons of various energy will be emitted, from radio waves to gamma rays. A photon can also be emitted when a particle and its corresponding antiparticle are annihilated (for example, electron–positron annihilation).[14]:572, 1114, 1172

In empty space, the photon moves at c (the speed of light) and its energy and momentum are related by $E = pc$, where p is the magnitude of the momentum vector **p**. This derives from the following relativistic relation, with $m = 0$:[15]

$$E^2 = p^2 c^2 + m^2 c^4.$$

The energy and momentum of a photon depend only on its frequency (ν) or inversely, its wavelength (λ):

$$E = \hbar\omega = h\nu = \frac{hc}{\lambda}$$

$$\boldsymbol{p} = \hbar\boldsymbol{k},$$

where $\boldsymbol{k}$ is the wave vector (where the wave number $k = |\boldsymbol{k}| = 2\pi/\lambda$), $\omega = 2\pi\nu$ is the angular frequency, and $\hbar = h/2\pi$ is the reduced Planck constant.[16]

Since $\boldsymbol{p}$ points in the direction of the photon's propagation, the magnitude of the momentum is

$$p = \hbar k = \frac{h\nu}{c} = \frac{h}{\lambda}.$$

The photon also carries spin angular momentum that does not depend on its frequency.[17] The magnitude of its spin is $\sqrt{2}\hbar$ and the component measured along its direction of motion, its helicity, must be $\pm\hbar$. These two possible helicities, called right-handed and left-handed, correspond to the two possible circular polarization states of the photon.[18]

To illustrate the significance of these formulae, the annihilation of a particle with its antiparticle in free space must result in the creation of at least *two* photons for the following reason. In the center of momentum frame, the colliding antiparticles have no net momentum, whereas a single photon always has momentum (since it is determined, as we have seen, only by the photon's frequency or wavelength—which cannot be zero). Hence, conservation of momentum (or equivalently, translational invariance) requires that at least two photons are created, with zero net momentum. (However, it is possible if the system interacts with another particle or field for annihilation to produce one photon, as when a positron annihilates with a bound atomic electron, it is possible for only one photon to be emitted, as the nuclear Coulomb field breaks translational symmetry.)[19]:64-65 The energy of the two photons, or, equivalently, their frequency, may be determined from conservation of four-momentum. Seen another way, the photon can be considered as its own antiparticle. The reverse process, pair production, is the dominant mechanism by which high-energy photons such as gamma rays lose energy while passing through matter.[20] That process is the reverse of "annihilation to one photon" allowed in the electric field of an atomic nucleus.

The classical formulae for the energy and momentum of electromagnetic radiation can be re-expressed in terms of photon events. For example, the pressure of electromagnetic radiation on an object derives from the transfer of photon momentum per unit time and unit area to that object, since pressure is force per unit area and force is the change in momentum per unit time.[21]

17.2.1 Experimental checks on photon mass

Current commonly accepted physical theories imply or assume the photon to be strictly massless, but this should be also checked experimentally. If the photon is not a strictly massless particle, it would not move at the exact speed of light in vacuum, c. Its speed would be lower and depend on its frequency. Relativity would be unaffected by this; the so-called speed of light, c, would then not be the actual speed at which light moves, but a constant of nature which is the maximum speed that any object could theoretically attain in space-time.[22] Thus, it would still be the speed of space-time ripples (gravitational waves and gravitons), but it would not be the speed of photons.

If a photon did have non-zero mass, there would be other effects as well. Coulomb's law would be modified and the electromagnetic field would have an extra physical degree of freedom. These effects yield more sensitive experimental probes of the photon mass than the frequency dependence of the speed of light. If Coulomb's law is not exactly valid, then that would cause the presence of an electric field inside a hollow conductor when it is subjected to an external electric field. This thus allows one to test Coulomb's law to very high precision.[23] A null result of such an experiment has set a limit of $m \lesssim 10^{-14}$ eV/c^2.[24]

Sharper upper limits have been obtained in experiments designed to detect effects caused by the galactic vector potential. Although the galactic vector potential is very large because the galactic magnetic field exists on very long length scales, only the magnetic field is observable if the photon is massless. In case of a massive photon, the mass term $\frac{1}{2}m^2 A_\mu A^\mu$ would affect the galactic plasma. The fact that no such effects are seen implies an upper bound on the photon mass of $m < 3\times10^{-27}$ eV/c^2.[25] The galactic vector potential can also be probed directly by measuring the torque exerted on a magnetized ring.[26] Such methods were used to obtain the sharper upper limit of 10^{-18}eV/c^2 (the equivalent of 1.07×10^{-27} atomic mass units) given by the Particle Data Group.[27]

These sharp limits from the non-observation of the effects caused by the galactic vector potential have been shown to be model dependent.[28] If the photon mass is generated via the Higgs mechanism then the upper limit of $m\lesssim10^{-14}$ eV/c^2 from the test of Coulomb's law is valid.

Photons inside superconductors do develop a nonzero effective rest mass; as a result, electromagnetic forces become short-range inside superconductors.[29]

See also: Supernova/Acceleration Probe

17.3   Historical development

Main article: Light

In most theories up to the eighteenth century, light was pictured as being made up of particles. Since particle models cannot easily account for the refraction, diffraction and birefringence of light, wave theories of light were proposed by René Descartes (1637),[30] Robert Hooke (1665),[31] and Christiaan Huygens (1678);[32] however, particle models remained dominant, chiefly due to the influence of Isaac Newton.[33] In the early nineteenth century, Thomas Young and August Fresnel clearly demonstrated the interference and diffraction of light and by 1850 wave models were generally accepted.[34] In 1865, James Clerk Maxwell's prediction[35] that light was an electromagnetic wave—which was confirmed experimentally in 1888 by Heinrich Hertz's detection of radio waves[36]—seemed to be the final blow to particle models of light.

The Maxwell wave theory, however, does not account for *all* properties of light. The Maxwell theory predicts that the energy of a light wave depends only on its intensity, not on its frequency; nevertheless, several independent types of experiments show that the energy imparted by light to atoms depends only on the light's frequency, not on its intensity. For example, some chemical reactions are provoked only by light of frequency higher than a certain threshold; light of frequency lower than the threshold, no matter how intense, does not initiate the reaction. Similarly, electrons can be ejected from a metal plate by shining light of sufficiently high frequency on it (the photoelectric effect); the energy of the ejected electron is related only to the light's frequency, not to its intensity.[37][Note 4]

At the same time, investigations of blackbody radiation carried out over four decades (1860–1900) by various researchers[38] culminated in Max Planck's hypothesis[4][39] that the energy of *any* system that absorbs or emits electromagnetic radiation of frequency ν is an integer multiple of an energy quantum $E = h\nu$. As shown by Albert Einstein,[5][40] some form of energy quantization *must* be assumed to account for the thermal equilibrium observed between matter and electromagnetic radiation; for this explanation of the photoelectric effect, Einstein received the 1921 Nobel Prize in physics.[41]

Since the Maxwell theory of light allows for all possible energies of electromagnetic radiation, most physicists assumed initially that the energy quantization resulted from some unknown constraint on the matter that absorbs or emits the radiation. In 1905, Einstein was the first to propose that energy quantization was a property of electromagnetic radiation itself.[5] Although he accepted the validity of Maxwell's theory, Einstein pointed out that many anomalous experiments could be explained if the *energy* of a Maxwellian light wave were localized into point-like quanta that move independently of one another, even if the wave itself is spread continuously over space.[5] In 1909[40] and 1916,[42] Einstein showed that, if Planck's law of black-body radiation is accepted, the energy quanta must also carry momentum $p = h/\lambda$, making them full-fledged particles. This photon momentum was observed experimentally[43] by Arthur Compton, for which he received the Nobel Prize in 1927. The pivotal question was then: how to unify Maxwell's wave theory of light with its experimentally observed particle nature? The answer to this question occupied Albert Einstein for the rest of his life,[44] and was solved in quantum electrodynamics and its successor, the Standard Model (see Second quantization and The photon as a gauge boson, below).

17.4   Einstein's light quantum

Unlike Planck, Einstein entertained the possibility that there might be actual physical quanta of light—what we now call photons. He noticed that a light quantum with energy proportional to its frequency would explain a number of troubling puzzles and paradoxes, including an unpublished law by Stokes, the ultraviolet catastrophe, and of course the photoelectric effect. Stokes's law said simply that the frequency of fluorescent light cannot be greater than the frequency of the light (usually ultraviolet) inducing it. Einstein eliminated the ultraviolet catastrophe by imagining a gas of photons behaving like a gas of electrons that he had previously considered. He was advised by a colleague to be careful how he wrote up this paper, in order to not challenge Planck too directly, as he was a powerful figure, and indeed the warning was justified, as Planck never forgave him for writing it.[45]

17.5   Early objections

Einstein's 1905 predictions were verified experimentally in several ways in the first two decades of the 20th century, as recounted in Robert Millikan's Nobel lecture.[46] However, before Compton's experiment[43] showing that photons carried momentum proportional to their wave number (or frequency) (1922), most physicists were reluctant to believe that electromagnetic radiation itself might be particulate. (See, for example, the Nobel lectures of Wien,[38] Planck[39] and Millikan.[46]) Instead, there was a widespread belief that energy quantization resulted from some unknown constraint on the matter that absorbs or emits radiation. Attitudes changed over time. In part, the change can be traced to experiments such as Compton scattering, where it was much more difficult not to ascribe quantization to light itself to explain the observed results.[47]

Even after Compton's experiment, Niels Bohr, Hendrik Kramers and John Slater made one last attempt to preserve the Maxwellian continuous electromagnetic field model of light, the so-called BKS model.[48] To account for the data then available, two drastic hypotheses had to be made:

1. **Energy and momentum are conserved only on the average in interactions between matter and radiation, not in elementary processes such as absorption and emission.** This allows one to reconcile the discontinuously changing energy of the atom (jump between energy states) with the continuous release of energy into radiation.

2. **Causality is abandoned**. For example, spontaneous emissions are merely emissions induced by a "virtual" electromagnetic field.

However, refined Compton experiments showed that energy–momentum is conserved extraordinarily well in elementary processes; and also that the jolting of the electron and the generation of a new photon in Compton scattering obey causality to within 10 ps. Accordingly, Bohr and his co-workers gave their model "as honorable a funeral as possible".[44] Nevertheless, the failures of the BKS model inspired Werner Heisenberg in his development of matrix mechanics.[49]

A few physicists persisted[50] in developing semiclassical models in which electromagnetic radiation is not quantized, but matter appears to obey the laws of quantum mechanics. Although the evidence for photons from chemical and physical experiments was overwhelming by the 1970s, this evidence could not be considered as *absolutely* definitive; since it relied on the interaction of light with matter, a sufficiently complicated theory of matter could in principle account for the evidence. Nevertheless, *all* semiclassical theories were refuted definitively in the 1970s and 1980s by photon-correlation experiments.[Note 5] Hence, Einstein's hypothesis that quantization is a property of light itself is considered to be proven.

17.6   Wave–particle duality and uncertainty principles

See also: Wave–particle duality, Squeezed coherent state, Uncertainty principle and De Broglie–Bohm theory

Photons, like all quantum objects, exhibit wave-like and particle-like properties. Their dual wave–particle nature can be difficult to visualize. The photon displays clearly wave-like phenomena such as diffraction and interference on the length scale of its wavelength. For example, a single photon passing through a double-slit experiment lands on the screen exhibiting interference phenomena but only if no measure was made on the actual slit being run across. To account for the particle interpretation that phenomenon is called probability distribution but behaves according to Maxwell's equations.[51] However, experiments confirm that the photon is *not* a short pulse of electromagnetic radiation; it does not spread out as it propagates, nor does it divide when it encounters a beam splitter.[52] Rather, the photon seems to be a point-like particle since it is absorbed or emitted *as a whole* by arbitrarily small systems, systems much smaller than its wavelength, such as an atomic nucleus ($\approx 10^{-15}$ m across) or even the point-like electron. Nevertheless, the photon is *not* a point-like particle whose trajectory is shaped probabilistically by the electromagnetic field, as conceived by Einstein and others; that hypothesis was also refuted by the photon-correlation experiments cited above. According to our present understanding, the electromagnetic field itself is produced by photons, which in turn result from a local gauge symmetry and the laws of quantum field theory (see the Second quantization and Gauge boson sections below).

A key element of quantum mechanics is Heisenberg's uncertainty principle, which forbids the simultaneous measurement of the position and momentum of a particle along the same direction. Remarkably, the uncertainty principle for charged, material particles *requires* the quantization of light into photons, and even the frequency dependence of the photon's energy and momentum. An elegant illustration is Heisenberg's thought experiment for locating an electron with an ideal microscope.[53] The position of the electron can be determined to within the resolving power of the microscope, which is given by a formula from classical optics

$$\Delta x \sim \frac{\lambda}{\sin \theta}$$

where θ is the aperture angle of the microscope. Thus, the position uncertainty Δx can be made arbitrarily small by reducing the wavelength λ. The momentum of the electron is uncertain, since it received a "kick" Δp from the light scattering from it into the microscope. If light were *not* quantized into photons, the uncertainty Δp could be made arbitrarily small by reducing the light's intensity. In that case, since the wavelength and intensity of light can be varied independently, one could simultaneously determine the position and momentum to arbitrarily high accuracy, violating the uncertainty principle. By contrast, Einstein's formula for photon momentum preserves the uncertainty principle; since the photon is scattered anywhere within the aperture, the uncertainty of momentum transferred equals

$$\Delta p \sim p_{\text{photon}} \sin \theta = \frac{h}{\lambda} \sin \theta$$

giving the product $\Delta x \Delta p \sim h$, which is Heisenberg's uncertainty principle. Thus, the entire world is quantized; both matter and fields must obey a consistent set of quantum laws, if either one is to be quantized.[54]

The analogous uncertainty principle for photons forbids the simultaneous measurement of the number n of photons (see Fock state and the Second quantization section below) in an electromagnetic wave and the phase ϕ of that wave

$$\Delta n \Delta \phi > 1$$

See coherent state and squeezed coherent state for more details.

Both (photons and material) particles such as electrons create analogous interference patterns when passing through a double-slit experiment. For photons, this corresponds to the interference of a Maxwell light wave whereas, for material particles, this corresponds to the interference of the Schrödinger wave equation. Although this similarity might suggest that Maxwell's equations are simply Schrödinger's equation for photons, most physicists do not agree.[55][56] For one thing, they are mathematically different; most obviously, Schrödinger's one equation solves for a complex field, whereas Maxwell's four equations solve for real fields. More generally, the normal concept of a Schrödinger probability wave function cannot be applied to photons.[57] Being massless, they cannot be localized without being destroyed; technically, photons cannot have a position eigenstate $|\mathbf{r}\rangle$, and, thus, the normal Heisenberg uncertainty principle $\Delta x \Delta p > h/2$ does not pertain to photons. A few substitute wave functions have been suggested for the photon,[58][59][60][61] but they have not come into general use. Instead, physicists generally accept the second-quantized theory of photons described below, quantum electrodynamics, in which photons are quantized excitations of electromagnetic modes.

Another interpretation, that avoids duality, is the De Broglie–Bohm theory: knowned also as the *pilot-wave model*, the photon in this theory is both, wave and particle.[62] *"This idea seems to me so natural and simple, to resolve the wave-particle dilemma in such a clear and ordinary way, that it is a great mystery to me that it was so generally ignored"*,[63] J.S.Bell.

17.7   Bose–Einstein model of a photon gas

Main articles: Bose gas, Bose–Einstein statistics, Spin-statistics theorem and Gas in a box

In 1924, Satyendra Nath Bose derived Planck's law of black-body radiation without using any electromagnetism, but rather a modification of coarse-grained counting of phase space.[64] Einstein showed that this modification is equivalent to assuming that photons are rigorously identical and that it implied a "mysterious non-local interaction",[65][66] now understood as the requirement for a symmetric quantum mechanical state. This work led to the concept of coherent states and the development of the laser. In the same papers, Einstein extended Bose's formalism to material particles (bosons) and predicted that they would condense into their lowest quantum state at low enough temperatures; this Bose–Einstein condensation was observed experimentally in 1995.[67] It was later used by Lene Hau to slow, and then completely stop, light in 1999[68] and 2001.[69]

The modern view on this is that photons are, by virtue of their integer spin, bosons (as opposed to fermions with half-integer spin). By the spin-statistics theorem, all bosons obey Bose–Einstein statistics (whereas all fermions obey Fermi–Dirac statistics).[70]

17.8   Stimulated and spontaneous emission

Main articles: Stimulated emission and Laser
 In 1916, Einstein showed that Planck's radiation law could be derived from a semi-classical, statistical treatment of photons and atoms, which implies a relation between the rates at which atoms emit and absorb photons. The condition follows from the assumption that light is emitted and absorbed by atoms independently, and that the thermal equilibrium is preserved by interaction with atoms. Consider a cavity in thermal equilibrium and filled with electromagnetic radiation and atoms that can emit and absorb that radiation. Thermal equilibrium requires that the energy density $\rho(\nu)$ of photons with frequency ν (which is proportional to their number density) is, on average, constant in time; hence, the rate at which photons of any particular frequency are *emitted* must equal the rate of *absorbing* them.[71]

Einstein began by postulating simple proportionality relations for the different reaction rates involved. In his model, the rate R_{ji} for a system to *absorb* a photon of frequency ν and transition from a lower energy E_j to a higher energy E_i is proportional to the number N_j of atoms with energy E_j and to the energy density $\rho(\nu)$ of ambient photons with that frequency,

$$R_{ji} = N_j B_{ji} \rho(\nu)$$

where B_{ji} is the rate constant for absorption. For the reverse process, there are two possibilities: spontaneous emission of a photon, and a return to the lower-energy state that is initiated by the interaction with a passing photon. Following Einstein's approach, the corresponding rate R_{ij} for the emission of photons of frequency ν and transition from a higher energy E_i to a lower energy E_j is

$$R_{ij} = N_i A_{ij} + N_i B_{ij} \rho(\nu)$$

where A_{ij} is the rate constant for emitting a photon spontaneously, and B_{ij} is the rate constant for emitting it in response to ambient photons (induced or stimulated emission). In thermodynamic equilibrium, the number of atoms in state i and that of atoms in state j must, on average, be constant; hence, the rates R_{ji} and R_{ij} must be equal. Also, by arguments analogous to the derivation of Boltzmann statistics, the ratio of N_i and N_j is $g_i/g_j \exp{(E_j - E_i)/kT}$, where $g_{i,j}$ are the degeneracy of the state i and that of j, respectively, $E_{i,j}$ their energies, k the Boltzmann constant and T the system's temperature. From this, it is readily derived that $g_i B_{ij} = g_j B_{ji}$ and

$$A_{ij} = \frac{8\pi h \nu^3}{c^3} B_{ij}.$$

The A and Bs are collectively known as the *Einstein coefficients*.[72]

Einstein could not fully justify his rate equations, but claimed that it should be possible to calculate the coefficients A_{ij}, B_{ji} and B_{ij} once physicists had obtained "mechanics and electrodynamics modified to accommodate the quantum hypothesis".[73] In fact, in 1926, Paul Dirac derived the B_{ij} rate constants in using a semiclassical approach,[74] and, in 1927, succeeded in deriving *all* the rate constants from first principles within the framework of quantum theory.[75][76] Dirac's work was the foundation of quantum electrodynamics, i.e., the quantization of the electromagnetic field itself.

Dirac's approach is also called *second quantization* or quantum field theory;[77][78][79] earlier quantum mechanical treatments only treat material particles as quantum mechanical, not the electromagnetic field.

Einstein was troubled by the fact that his theory seemed incomplete, since it did not determine the *direction* of a spontaneously emitted photon. A probabilistic nature of light-particle motion was first considered by Newton in his treatment of birefringence and, more generally, of the splitting of light beams at interfaces into a transmitted beam and a reflected beam. Newton hypothesized that hidden variables in the light particle determined which path it would follow.[33] Similarly, Einstein hoped for a more complete theory that would leave nothing to chance, beginning his separation[44] from quantum mechanics. Ironically, Max Born's probabilistic interpretation of the wave function[80][81] was inspired by Einstein's later work searching for a more complete theory.[82]

17.9   Second quantization and high energy photon interactions

Main article: Quantum field theory

In 1910, Peter Debye derived Planck's law of black-body radiation from a relatively simple assumption.[83] He correctly decomposed the electromagnetic field in a cavity into its Fourier modes, and assumed that the energy in any mode was an integer multiple of $h\nu$, where ν is the frequency of the electromagnetic mode. Planck's law of black-body radiation follows immediately as a geometric sum. However, Debye's approach failed to give the correct formula for the energy fluctuations of blackbody radiation, which were derived by Einstein in 1909.[40]

In 1925, Born, Heisenberg and Jordan reinterpreted Debye's concept in a key way.[84] As may be shown classically, the Fourier modes of the electromagnetic field—a complete set of electromagnetic plane waves indexed by their wave vector **k** and polarization state—are equivalent to a set of uncoupled simple harmonic oscillators. Treated quantum mechanically, the energy levels of such oscillators are known to be $E = nh\nu$, where ν is the oscillator frequency. The key new step was to identify an electromagnetic mode with energy $E = nh\nu$ as a state with n photons, each of energy $h\nu$. This approach gives the correct energy fluctuation formula.

Dirac took this one step further.[75][76] He treated the interaction between a charge and an electromagnetic field as a small perturbation that induces transitions in the photon states, changing the numbers of photons in the modes, while conserving energy and momentum overall. Dirac was able to derive Einstein's A_{ij} and B_{ij} coefficients from first principles, and showed that the Bose–Einstein statistics of photons is a natural consequence of quantizing the electromagnetic field correctly (Bose's reasoning went in the opposite direction; he derived Planck's law of black-body radiation by *assuming* B–E statistics). In Dirac's time, it was not yet known that all bosons, including photons, must obey Bose–Einstein statistics.

Dirac's second-order perturbation theory can involve virtual photons, transient intermediate states of the electromagnetic field; the static electric and magnetic interactions are mediated by such virtual photons. In such quantum field theories, the probability amplitude of observable events is calculated by summing over *all* possible intermediate steps, even ones that are unphysical; hence, virtual photons are not constrained to satisfy $E = pc$, and may have extra polarization states; depending on the gauge used, virtual photons may have three or four polarization states, instead of the two states of real photons. Although these transient virtual photons can never be observed, they contribute measurably to the probabilities of observable events. Indeed, such second-order and higher-order perturbation calculations can give apparently infinite contributions to the sum. Such unphysical results are corrected for using the technique of renormalization.

Other virtual particles may contribute to the summation as well; for example, two photons may interact indirectly through virtual electron–positron pairs.[85] In fact, such photon-photon scattering (see two-photon physics), as well as electron-photon scattering, is meant to be one of the modes of operations of the planned particle accelerator, the International Linear Collider.[86]

In modern physics notation, the quantum state of the electromagnetic field is written as a Fock state, a tensor product of the states for each electromagnetic mode

$$|n_{k_0}\rangle \otimes |n_{k_1}\rangle \otimes \cdots \otimes |n_{k_n}\rangle \ldots$$

where $|n_{k_i}\rangle$ represents the state in which n_{k_i} photons are in the mode k_i. In this notation, the creation of a new photon in mode k_i (e.g., emitted from an atomic transition) is written as $|n_{k_i}\rangle \to |n_{k_i} + 1\rangle$. This notation merely expresses the concept of Born, Heisenberg and Jordan described above, and does not add any physics.

17.10 The hadronic properties of the photon

Measurements of the interaction between energetic photons and hadrons show that the interaction is much more intense than expected by the interaction of merely photons with the hadron's electric charge. Furthermore, the interaction of energetic photons with protons is similar to the interaction of photons with neutrons[87] in spite of the fact that the electric charge structures of protons and neutrons are substantially different.

A theory called Vector Meson Dominance (VMD) was developed to explain this effect. According to VMD, the photon is a superposition of the pure electromagnetic photon (which interacts only with electric charges) and vector meson.[88]

However, if experimentally probed at very short distances, the intrinsic structure of the photon is recognized as a flux of quark and gluon components, quasi-free according to asymptotic freedom in QCD and described by the photon structure function.[89][90] A comprehensive comparison of data with theoretical predictions is presented in a recent review.[91]

17.11 The photon as a gauge boson

Main article: Gauge theory

The electromagnetic field can be understood as a gauge field, i.e., as a field that results from requiring that a gauge symmetry holds independently at every position in spacetime.[92] For the electromagnetic field, this gauge symmetry is the Abelian U(1) symmetry of a complex number, which reflects the ability to vary the phase of a complex number without affecting observables or real valued functions made from it, such as the energy or the Lagrangian.

The quanta of an Abelian gauge field must be massless, uncharged bosons, as long as the symmetry is not broken; hence, the photon is predicted to be massless, and to have zero electric charge and integer spin. The particular form of the electromagnetic interaction specifies that the photon must have spin ± 1; thus, its helicity must be $\pm \hbar$. These two spin components correspond to the classical concepts of right-handed and left-handed circularly polarized light. However, the transient virtual photons of quantum electrodynamics may also adopt unphysical polarization states.[92]

In the prevailing Standard Model of physics, the photon is one of four gauge bosons in the electroweak interaction; the other three are denoted W^+, W^- and Z^0 and are responsible for the weak interaction. Unlike the photon, these gauge bosons have mass, owing to a mechanism that breaks their SU(2) gauge symmetry. The unification of the photon with W and Z gauge bosons in the electroweak interaction was accomplished by Sheldon Glashow, Abdus Salam and Steven Weinberg, for which they were awarded the 1979 Nobel Prize in physics.[93][94][95] Physicists continue to hypothesize grand unified theories that connect these four gauge bosons with the eight gluon gauge bosons of quantum chromodynamics; however, key predictions of these theories, such as proton decay, have not been observed experimentally.[96]

17.12 Contributions to the mass of a system

See also: Mass in special relativity and General relativity

The energy of a system that emits a photon is *decreased* by the energy E of the photon as measured in the rest frame of the emitting system, which may result in a reduction in mass in the amount E/c^2. Similarly, the mass of a system that absorbs a photon is *increased* by a corresponding amount. As an application, the energy balance of nuclear reactions involving photons is commonly written in terms of the masses of the nuclei involved, and terms of the form E/c^2 for the gamma photons (and for other relevant energies, such as the recoil energy of nuclei).[97]

This concept is applied in key predictions of quantum electrodynamics (QED, see above). In that theory, the mass of electrons (or, more generally, leptons) is modified by including the mass contributions of virtual photons, in a technique known as renormalization. Such "radiative corrections" contribute to a number of predictions of QED, such as the magnetic dipole moment of leptons, the Lamb shift, and the hyperfine structure of bound lepton pairs, such as muonium and positronium.[98]

Since photons contribute to the stress–energy tensor, they exert a gravitational attraction on other objects, according to the theory of general relativity. Conversely, photons are themselves affected by gravity; their normally straight trajectories may be bent by warped spacetime, as in gravitational lensing, and their frequencies may be lowered by moving to a higher gravitational potential, as in the Pound–Rebka experiment. However, these effects are not specific to photons; exactly the same effects would be predicted for classical electromagnetic waves.[99]

17.13 Photons in matter

See also: Group velocity and Photochemistry

Any 'explanation' of how photons travel through matter has to explain why different arrangements of matter are transparent or opaque at different wavelengths (light through carbon as diamond or not, as graphite) and why individual photons behave in the same way as large groups. Explanations that invoke 'absorption' and 're-emission' have to provide an explanation for the directionality of the photons (diffraction, reflection) and further explain how entangled photon pairs can travel through matter without their quantum state collapsing.

The simplest explanation is that light that travels through transparent matter does so at a lower speed than c, the speed of light in a vacuum. In addition, light can also undergo scattering and absorption. There are circumstances in which heat transfer through a material is mostly radiative, involving emission and absorption of photons within it. An example would be in the core of the Sun. Energy can take about a million years to reach the surface.[100] However, this phenomenon is distinct from scattered radiation passing diffusely through matter, as it involves local equilibrium between the radiation and the temperature. Thus, the time is how long it takes the *energy* to be transferred, not the *photons* themselves. Once in open space, a photon from the Sun takes only 8.3 minutes to reach Earth. The factor by which the speed of light is decreased in a material is called the refractive index of the material. In a classical wave picture, the slowing can be explained by the light inducing electric polarization in the matter, the polarized matter radiating new light, and the new light interfering with the original light wave to form a delayed wave. In a particle picture, the slowing can instead be described as a blending of the photon with quantum excitation of the matter (quasi-particles such as phonons and excitons) to form a polariton; this polariton has a nonzero effective mass, which means that it cannot travel at c.

Alternatively, photons may be viewed as *always* traveling at c, even in matter, but they have their phase shifted (delayed or advanced) upon interaction with atomic scatters: this modifies their wavelength and momentum, but not speed.[101] A light wave made up of these photons does travel slower than the speed of light. In this view the photons are "bare", and are scattered and phase shifted, while in the view of the preceding paragraph the photons are "dressed" by their interaction with matter, and move without scattering or phase shifting, but at a lower speed.

Light of different frequencies may travel through matter at different speeds; this is called dispersion. In some cases, it can result in extremely slow speeds of light in matter. The effects of photon interactions with other quasi-particles may be observed directly in Raman scattering and Brillouin scattering.[102]

Photons can also be absorbed by nuclei, atoms or molecules, provoking transitions between their energy levels. A classic example is the molecular transition of retinal $C_{20}H_{28}O$, which is responsible for vision, as discovered in 1958 by Nobel laureate biochemist George Wald and co-workers. The absorption provokes a cis-trans isomerization that, in combination with other such transitions, is transduced into nerve impulses. The absorption of photons can even break chemical bonds, as in the photodissociation of chlorine; this is the subject of photochemistry.[103][104] Analogously, gamma rays can in some circumstances dissociate atomic nuclei in a process called photodisintegration.

17.14 Technological applications

Photons have many applications in technology. These examples are chosen to illustrate applications of photons *per se*, rather than general optical devices such as lenses, etc. that could operate under a classical theory of light. The laser is an extremely important application and is discussed above under stimulated emission.

Individual photons can be detected by several methods. The classic photomultiplier tube exploits the photoelectric effect: a photon landing on a metal plate ejects an electron, initiating an ever-amplifying avalanche of electrons. Charge-coupled device chips use a similar effect in semiconductors: an incident photon generates a charge on a microscopic capacitor that can be detected. Other detectors such as Geiger counters use the ability of photons to ionize gas molecules, causing a detectable change in conductivity.[105]

Planck's energy formula $E = h\nu$ is often used by engineers and chemists in design, both to compute the change in energy resulting from a photon absorption and to predict the frequency of the light emitted for a given energy transition. For example, the emission spectrum of a fluorescent light bulb can be designed using gas molecules with different electronic energy levels and adjusting the typical energy with which an electron hits the gas molecules within the bulb.[Note 6]

Under some conditions, an energy transition can be excited by "two" photons that individually would be insufficient. This allows for higher resolution microscopy, because the sample absorbs energy only in the region where two beams of different colors overlap significantly, which can be made much smaller than the excitation volume of a single beam (see two-photon excitation microscopy). Moreover, these photons cause less damage to the sample, since they are of lower energy.[106]

In some cases, two energy transitions can be coupled so that, as one system absorbs a photon, another nearby system "steals" its energy and re-emits a photon of a different frequency. This is the basis of fluorescence resonance energy transfer, a technique that is used in molecular biology to study the interaction of suitable proteins.[107]

Several different kinds of hardware random number generator involve the detection of single photons. In one example, for each bit in the random sequence that is to be produced, a photon is sent to a beam-splitter. In such a situation, there are two possible outcomes of equal probability. The actual outcome is used to determine whether the next bit in the sequence is "0" or "1".[108][109]

17.15   Recent research

See also: Quantum optics

Much research has been devoted to applications of photons in the field of quantum optics. Photons seem well-suited to be elements of an extremely fast quantum computer, and the quantum entanglement of photons is a focus of research. Nonlinear optical processes are another active research area, with topics such as two-photon absorption, self-phase modulation, modulational instability and optical parametric oscillators. However, such processes generally do not require the assumption of photons *per se*; they may often be modeled by treating atoms as nonlinear oscillators. The nonlinear process of spontaneous parametric down conversion is often used to produce single-photon states. Finally, photons are essential in some aspects of optical communication, especially for quantum cryptography.[Note 7]

17.16   See also

- Advanced Photon Source at Argonne National Laboratory

- Ballistic photon

- Doppler shift

- Electromagnetic radiation

- HEXITEC

- Laser

- Light

- Luminiferous aether

- Medipix

- Phonons

- Photon counting

- Photon energy

- Photon polarization

- Photonic molecule

- Photography

- Photonics

- Quantum optics

- Single photon sources

- Static forces and virtual-particle exchange

- Two-photon physics

- EPR paradox

- Dirac equation

17.17   Notes

[1] Although the 1967 Elsevier translation of Planck's Nobel Lecture interprets Planck's *Lichtquant* as "photon", the more literal 1922 translation by Hans Thacher Clarke and Ludwik Silberstein *The origin and development of the quantum theory*, The Clarendon Press, 1922 (here) uses "light-quantum". No evidence is known that Planck himself used the term "photon" by 1926 (see also this note).

[2] Isaac Asimov credits Arthur Compton with defining quanta of energy as photons in 1923. Asimov, I. (1966). *The Neutrino, Ghost Particle of the Atom*. Garden City (NY): Doubleday. ISBN 0-380-00483-6. LCCN 66017073. and Asimov, I. (1966). *The Universe From Flat Earth To Quasar*. New York (NY): Walker. ISBN 0-8027-0316-X. LCCN 66022515.

[3] The mass of the photon is believed to be exactly zero, based on experiment and theoretical considerations described in the article. Some sources also refer to the *relativistic mass* concept, which is just the energy scaled to units of mass. For a photon with wavelength λ or energy E, this is $h/\lambda c$ or E/c^2. This usage for the term "mass" is no longer common in scientific literature. Further info: What is the mass of a photon? http://math.ucr.edu/home/baez/physics/ParticleAndNuclear/photon_mass.html

[4] The phrase "no matter how intense" refers to intensities below approximately 10^{13} W/cm^2 at which point perturbation theory begins to break down. In contrast, in the intense regime, which for visible light is above approximately 10^{14} W/cm^2, the classical wave description correctly predicts the energy acquired by electrons, called ponderomotive energy. (See also: Boreham *et al.* (1996). "Photon density and the correspondence principle of electromagnetic interaction".) By comparison, sunlight is only about 0.1 W/cm^2.

[5] These experiments produce results that cannot be explained by any classical theory of light, since they involve anticorrelations that result from the quantum measurement process. In 1974, the first such experiment was carried out by Clauser, who reported a violation of a classical Cauchy–Schwarz inequality. In 1977, Kimble *et al.* demonstrated an analogous anti-bunching effect of photons interacting with a beam splitter; this approach was simplified and sources of error eliminated in the photon-anticorrelation experiment of Grangier *et al.* (1986). This work is reviewed and simplified further in Thorn *et al.* (2004). (These references are listed below under #Additional references.)

[6] An example is US Patent Nr. 5212709.

[7] Introductory-level material on the various sub-fields of quantum optics can be found in Fox, M. (2006). *Quantum Optics: An Introduction*. Oxford University Press. ISBN 0-19-856673-5.

17.18 References

[1] Amsler, C. (Particle Data Group); Amsler; Doser; Antonelli; Asner; Babu; Baer; Band; Barnett; Bergren; Beringer; Bernardi; Bertl; Bichsel; Biebel; Bloch; Blucher; Blusk; Cahn; Carena; Caso; Ceccucci; Chakraborty; Chen; Chivukula; Cowan; Dahl; d'Ambrosio; Damour et al. (2008). "Review of Particle Physics: Gauge and Higgs bosons" (PDF). *Physics Letters B* **667**: 1. Bibcode:2008PhLB..667....1P. doi:10.1016/j.physletb.2008.07.018.

[2] Kimble, H.J.; Dagenais, M.; Mandel, L.; Dagenais; Mandel (1977). "Photon Anti-bunching in Resonance Fluorescence". *Physical Review Letters* **39** (11): 691–695. Bibcode:1977PhRvL..39..691K. doi:10.1103/PhysRevLett.39.691.

[3] Grangier, P.; Roger, G.; Aspect, A.; Roger; Aspect (1986). "Experimental Evidence for a Photon Anticorrelation Effect on a Beam Splitter: A New Light on Single-Photon Interferences". *Europhysics Letters* **1** (4): 173–179. Bibcode:1986EL......1..173G. doi:10.1209/0295-5075/1/4/004.

[4] Planck, M. (1901). "On the Law of Distribution of Energy in the Normal Spectrum". *Annalen der Physik* **4** (3): 553–563. Bibcode:1901AnP...309..553P. doi:10.1002/andp.19013090310. Archived from the original on 2008-04-18.

[5] Einstein, A. (1905). "Über einen die Erzeugung und Verwandlung des Lichtes betreffenden heuristischen Gesichtspunkt" (PDF). *Annalen der Physik* (in German) **17** (6): 132–148. Bibcode:1905AnP...322..132E. doi:10.1002/andp.19053220607.. An English translation is available from Wikisource.

[6] "Discordances entre l'expérience et la théorie électromagnétique du rayonnement." In Électrons et Photons. Rapports et Discussions de Cinquième Conseil de Physique, edited by Institut International de Physique Solvay. Paris: Gauthier-Villars, pp. 55-85.

[7] Helge Kragh: *Photon: New light on an old name.* Arxiv, 2014-2-28

[8] Villard, P. (1900). "Sur la réflexion et la réfraction des rayons cathodiques et des rayons déviables du radium". *Comptes Rendus des Séances de l'Académie des Sciences* (in French) **130**: 1010–1012.

[9] Villard, P. (1900). "Sur le rayonnement du radium". *Comptes Rendus des Séances de l'Académie des Sciences* (in French) **130**: 1178 1179.

[10] Rutherford, E.; Andrade, E.N.C. (1914). "The Wavelength of the Soft Gamma Rays from Radium B". *Philosophical Magazine* **27** (161): 854–868. doi:10.1080/14786440508635156.

[11] Kobychev, V.V.; Popov, S.B. (2005). "Constraints on the photon charge from observations of extragalactic sources". *Astronomy Letters* **31** (3): 147–151. arXiv:hep-ph/0411398. Bibcode:2005AstL...31..147K. doi:10.1134/1.1883345.

[12] Role as gauge boson and polarization section 5.1 in Aitchison, I.J.R.; Hey, A.J.G. (1993). *Gauge Theories in Particle Physics.* IOP Publishing. ISBN 0-85274-328-9.

[13] See p.31 in Amsler, C. et al. (2008). "Review of Particle Physics". *Physics Letters B* **667**: 1–1340. Bibcode:2008PhLB..667....1P. doi:10.1016/j.physletb.2008.07.018.

[14] Halliday, David; Resnick, Robert; Walker, Jerl (2005), *Fundamental of Physics* (7th ed.), USA: John Wiley and Sons, Inc., ISBN 0-471-23231-9

[15] See section 1.6 in Alonso, M.; Finn, E.J. (1968). *Fundamental University Physics Volume III: Quantum and Statistical Physics.* Addison-Wesley. ISBN 0-201-00262-0.

[16] Davison E. Soper, Electromagnetic radiation is made of photons, Institute of Theoretical Science, University of Oregon

[17] This property was experimentally verified by Raman and Bhagavantam in 1931: Raman, C.V.; Bhagavantam, S. (1931). "Experimental proof of the spin of the photon" (PDF). *Indian Journal of Physics* **6**: 353.

[18] Burgess, C.; Moore, G. (2007). "1.3.3.2". *The Standard Model. A Primer.* Cambridge University Press. ISBN 0-521-86036-9.

[19] Griffiths, David J. (2008), *Introduction to Elementary Particles* (2nd revised ed.), WILEY-VCH, ISBN 978-3-527-40601-2

[20] E.g., section 9.3 in Alonso, M.; Finn, E.J. (1968). *Fundamental University Physics Volume III: Quantum and Statistical Physics.* Addison-Wesley.

[21] E.g., Appendix XXXII in Born, M. (1962). *Atomic Physics.* Blackie & Son. ISBN 0-486-65984-4.

[22] Mermin, David (February 1984). "Relativity without light". *American Journal of Physics* **52** (2): 119–124. Bibcode:1984AmJPh..52..119M. doi:10.1119/1.13917.

[23] Plimpton, S.; Lawton, W. (1936). "A Very Accurate Test of Coulomb's Law of Force Between Charges". *Physical Review* **50** (11): 1066. Bibcode:1936PhRv...50.1066P. doi:10.1103/PhysRev.50.1066.

[24] Williams, E.; Faller, J.; Hill, H. (1971). "New Experimental Test of Coulomb's Law: A Laboratory Upper Limit on the Photon Rest Mass". *Physical Review Letters* **26** (12): 721. Bibcode:1971PhRvL..26..721W. doi:10.1103/PhysRevLett.26.721.

[25] Chibisov, G V (1976). "Astrophysical upper limits on the photon rest mass". *Soviet Physics Uspekhi* **19** (7): 624. Bibcode:1976SvPhU..19..624C. doi:10.1070/PU1976v019n07ABEH005277.

[26] Lakes, Roderic (1998). "Experimental Limits on the Photon Mass and Cosmic Magnetic Vector Potential". *Physical Review Letters* **80** (9): 1826. Bibcode:1998PhRvL..80.1826L. doi:10.1103/PhysRevLett.80.1826.

[27] Amsler, C; Doser, M; Antonelli, M; Asner, D; Babu, K; Baer, H; Band, H; Barnett, R et al. (2008). "Review of Particle Physics∗". *Physics Letters B* **667**: 1. Bibcode:2008PhLB..667....1P. doi:10.1016/j.physletb.2008.07.018. Summary Table

[28] Adelberger, Eric; Dvali, Gia; Gruzinov, Andrei (2007). "Photon-Mass Bound Destroyed by Vortices". *Physical Review Letters* **98** (1): 010402. arXiv:hep-ph/0306245. Bibcode:2007PhRvL..98a0402A. doi:10.1103/PhysRevLett.98.010402. PMID 17358459. preprint

[29] Wilczek, Frank (2010). *The Lightness of Being: Mass, Ether, and the Unification of Forces*. Basic Books. p. 212. ISBN 978-0-465-01895-6.

[30] Descartes, R. (1637). *Discours de la méthode (Discourse on Method)* (in French). Imprimerie de Ian Maire. ISBN 0-268-00870-1.

[31] Hooke, R. (1667). *Micrographia: or some physiological descriptions of minute bodies made by magnifying glasses with observations and inquiries thereupon ...* London (UK): Royal Society of London. ISBN 0-486-49564-7.

[32] Huygens, C. (1678). *Traité de la lumière* (in French).. An English translation is available from Project Gutenberg

[33] Newton, I. (1952) [1730]. *Opticks* (4th ed.). Dover (NY): Dover Publications. Book II, Part III, Propositions XII–XX; Queries 25–29. ISBN 0-486-60205-2.

[34] Buchwald, J.Z. (1989). *The Rise of the Wave Theory of Light: Optical Theory and Experiment in the Early Nineteenth Century*. University of Chicago Press. ISBN 0-226-07886-8. OCLC 18069573.

[35] Maxwell, J.C. (1865). "A Dynamical Theory of the Electromagnetic Field". *Philosophical Transactions of the Royal Society* **155**: 459–512. Bibcode:1865RSPT..155..459C. doi:10.1098/rstl.1865.0008. This article followed a presentation by Maxwell on 8 December 1864 to the Royal Society.

[36] Hertz, H. (1888). "Über Strahlen elektrischer Kraft". *Sitzungsberichte der Preussischen Akademie der Wissenschaften (Berlin)* (in German) **1888**: 1297–1307.

[37] Frequency-dependence of luminiscence p. 276f., photoelectric effect section 1.4 in Alonso, M.; Finn, E.J. (1968). *Fundamental University Physics Volume III: Quantum and Statistical Physics*. Addison-Wesley. ISBN 0-201-00262-0.

[38] Wien, W. (1911). "Wilhelm Wien Nobel Lecture".

[39] Planck, M. (1920). "Max Planck's Nobel Lecture".

[40] Einstein, A. (1909). "Über die Entwicklung unserer Anschauungen über das Wesen und die Konstitution der Strahlung" (PDF). *Physikalische Zeitschrift* (in German) **10**: 817–825.. An English translation is available from Wikisource.

[41] Presentation speech by Svante Arrhenius for the 1921 Nobel Prize in Physics, December 10, 1922. Online text from [nobelprize.org], The Nobel Foundation 2008. Access date 2008-12-05.

[42] Einstein, A. (1916). "Zur Quantentheorie der Strahlung". *Mitteilungen der Physikalischen Gesellschaft zu Zürich* **16**: 47. Also *Physikalische Zeitschrift*, **18**, 121–128 (1917). (German)

[43] Compton, A. (1923). "A Quantum Theory of the Scattering of X-rays by Light Elements". *Physical Review* **21** (5): 483–502. Bibcode:1923PhRv...21..483C. doi:10.1103/PhysRev.21.483.

[44] Pais, A. (1982). *Subtle is the Lord: The Science and the Life of Albert Einstein.* Oxford University Press. ISBN 0-19-853907-X.

[45] *Einstein and the Quantum: The Quest of the Valiant Swabian*, A. Douglas Stone, Princeton University Press, 2013.

[46] Millikan, R.A (1924). "Robert A. Millikan's Nobel Lecture".

[47] Hendry, J. (1980). "The development of attitudes to the wave-particle duality of light and quantum theory, 1900–1920". *Annals of Science* **37** (1): 59–79. doi:10.1080/00033798000200121.

[48] Bohr, N.; Kramers, H.A.; Slater, J.C. (1924). "The Quantum Theory of Radiation". *Philosophical Magazine* **47**: 785–802. doi:10.1080/14786442408565262. Also *Zeitschrift für Physik*, **24**, 69 (1924).

[49] Heisenberg, W. (1933). "Heisenberg Nobel lecture".

[50] Mandel, L. (1976). E. Wolf, ed. "The case for and against semiclassical radiation theory". *Progress in Optics.* Progress in Optics (North-Holland) **13**: 27–69. doi:10.1016/S0079-6638(08)70018-0. ISBN 978-0-444-10806-7.

[51] Taylor, G.I. (1909). *Interference fringes with feeble light. Proceedings of the Cambridge Philosophical Society* **15**: 114–115.

[52] Saleh, B. E. A. and Teich, M. C. (2007). *Fundamentals of Photonics.* Wiley. ISBN 0-471-35832-0.

[53] Heisenberg, W. (1927). "Über den anschaulichen Inhalt der quantentheoretischen Kinematik und Mechanik". *Zeitschrift für Physik* (in German) **43** (3–4): 172–198. Bibcode:1927ZPhy...43..172H. doi:10.1007/BF01397280.

[54] E.g., p. 10f. in Schiff, L.I. (1968). *Quantum Mechanics* (3rd ed.). McGraw-Hill. ASIN B001B3MINM. ISBN 0-07-055287-8.

[55] Kramers, H.A. (1958). *Quantum Mechanics.* Amsterdam: North-Holland. ASIN B0006AUW5C. ISBN 0-486-49533-7.

[56] Bohm, D. (1989) [1954]. *Quantum Theory.* Dover Publications. ISBN 0-486-65969-0.

[57] Newton, T.D.; Wigner, E.P. (1949). "Localized states for elementary particles". *Reviews of Modern Physics* **21** (3): 400–406. Bibcode:1949RvMP...21..400N. doi:10.1103/RevModPhys.21.400.

[58] Bialynicki-Birula, I. (1994). "On the wave function of the photon" (PDF). *Acta Physica Polonica A* **86**: 97–116.

[59] Sipe, J.E. (1995). "Photon wave functions". *Physical Review A* **52** (3): 1875–1883. Bibcode:1995PhRvA..52.1875S. doi:10.1103/PhysRevA.5

[60] Bialynicki-Birula, I. (1996). "Photon wave function". *Progress in Optics.* Progress in Optics **36**: 245–294. doi:10.1016/S0079-6638(08)70316-0. ISBN 978-0-444-82530-8.

[61] Scully, M.O.; Zubairy, M.S. (1997). *Quantum Optics.* Cambridge (UK): Cambridge University Press. ISBN 0-521-43595-1.

[62] The best illustration is the Couder experiment, demonstrating the behaviour of a mechanical analog, see https://www.youtube.com/watch?v=W9yWv5dqSKk

[63] Bell, J. S., "Speakable and Unspeakable in Quantum Mechanics", Cambridge: Cambridge University Press, 1987.

[64] Bose, S.N. (1924). "Plancks Gesetz und Lichtquantenhypothese". *Zeitschrift für Physik* (in German) **26**: 178–181. Bibcode:1924ZPhy...26..1 doi:10.1007/BF01327326.

[65] Einstein, A. (1924). "Quantentheorie des einatomigen idealen Gases". *Sitzungsberichte der Preussischen Akademie der Wissenschaften (Berlin), Physikalisch-mathematische Klasse* (in German) **1924**: 261–267.

[66] Einstein, A. (1925). "Quantentheorie des einatomigen idealen Gases, Zweite Abhandlung". *Sitzungsberichte der Preussischen Akademie der Wissenschaften (Berlin), Physikalisch-mathematische Klasse* (in German) **1925**: 3–14. doi:10.1002/3527608958.ch28. ISBN 978-3-527-60895-9.

[67] Anderson, M.H.; Ensher, J.R.; Matthews, M.R.; Wieman, C.E.; Cornell, E.A. (1995). "Observation of Bose–Einstein Condensation in a Dilute Atomic Vapor". *Science* **269** (5221): 198–201. Bibcode:1995Sci...269..198A. doi:10.1126/science.269.5221.198. JSTOR 2888436. PMID 17789847.

[68] "Physicists Slow Speed of Light". News.harvard.edu (1999-02-18). Retrieved on 2015-05-11.

[69] "Light Changed to Matter, Then Stopped and Moved". photonics.com (February 2007). Retrieved on 2015-05-11.

[70] Streater, R.F.; Wightman, A.S. (1989). *PCT, Spin and Statistics, and All That.* Addison-Wesley. ISBN 0-201-09410-X.

[71] Einstein, A. (1916). "Strahlungs-emission und -absorption nach der Quantentheorie". *Verhandlungen der Deutschen Physikalischen Gesellschaft* (in German) **18**: 318–323. Bibcode:1916DPhyG..18..318E.

[72] Section 1.4 in Wilson, J.; Hawkes, F.J.B. (1987). *Lasers: Principles and Applications*. New York: Prentice Hall. ISBN 0-13-523705-X.

[73] P. 322 in Einstein, A. (1916). "Strahlungs-emission und -absorption nach der Quantentheorie". *Verhandlungen der Deutschen Physikalischen Gesellschaft* (in German) **18**: 318–323. Bibcode:1916DPhyG..18..318E.:

> Die Konstanten A_m^n and B_m^n würden sich direkt berechnen lassen, wenn wir im Besitz einer im Sinne der Quantenhypothese modifizierten Elektrodynamik und Mechanik wären."

[74] Dirac, P.A.M. (1926). "On the Theory of Quantum Mechanics". *Proceedings of the Royal Society A* **112** (762): 661–677. Bibcode:1926RSPSA.112..661D. doi:10.1098/rspa.1926.0133.

[75] Dirac, P.A.M. (1927). "The Quantum Theory of the Emission and Absorption of Radiation" (PDF). *Proceedings of the Royal Society A* **114** (767): 243–265. Bibcode:1927RSPSA.114..243D. doi:10.1098/rspa.1927.0039.

[76] Dirac, P.A.M. (1927b). *The Quantum Theory of Dispersion. Proceedings of the Royal Society A* **114**: 710–728. doi:10.1098/rspa.1927.0071.

[77] Heisenberg, W.; Pauli, W. (1929). "Zur Quantentheorie der Wellenfelder". *Zeitschrift für Physik* (in German) **56**: 1. Bibcode:1929ZPhy...56....1H. doi:10.1007/BF01340129.

[78] Heisenberg, W.; Pauli, W. (1930). "Zur Quantentheorie der Wellenfelder". *Zeitschrift für Physik* (in German) **59** (3–4): 139. Bibcode:1930ZPhy...59..168H. doi:10.1007/BF01341423.

[79] Fermi, E. (1932). "Quantum Theory of Radiation" (PDF). *Reviews of Modern Physics* **4**: 87. Bibcode:1932RvMP....4...87F. doi:10.1103/RevModPhys.4.87.

[80] Born, M. (1926). "Zur Quantenmechanik der Stossvorgänge" (PDF). *Zeitschrift für Physik* (in German) **37** (12): 863–867. Bibcode:1926ZPhy...37..863B. doi:10.1007/BF01397477.

[81] Born, M. (1926). "Quantenmechanik der Stossvorgänge". *Zeitschrift für Physik* (in German) **38** (11–12): 803. Bibcode:1926ZPhy...38..803B. doi:10.1007/BF01397184.

[82] Pais, A. (1986). *Inward Bound: Of Matter and Forces in the Physical World*. Oxford University Press. p. 260. ISBN 0-19-851997-4. Specifically, Born claimed to have been inspired by Einstein's never-published attempts to develop a "ghost-field" theory, in which point-like photons are guided probabilistically by ghost fields that follow Maxwell's equations.

[83] Debye, P. (1910). "Der Wahrscheinlichkeitsbegriff in der Theorie der Strahlung". *Annalen der Physik* (in German) **33** (16): 1427–1434. Bibcode:1910AnP...338.1427D. doi:10.1002/andp.19103381617.

[84] Born, M.; Heisenberg, W.; Jordan, P. (1925). "Quantenmechanik II". *Zeitschrift für Physik* (in German) **35** (8–9): 557–615. Bibcode:1926ZPhy...35..557B. doi:10.1007/BF01379806.

[85] Photon-photon-scattering section 7-3-1, renormalization chapter 8-2 in Itzykson, C.; Zuber, J.-B. (1980). *Quantum Field Theory*. McGraw-Hill. ISBN 0-07-032071-3.

[86] Weiglein, G. (2008). "Electroweak Physics at the ILC". *Journal of Physics: Conference Series* **110** (4): 042033. arXiv:0711.3003. Bibcode:2008JPhCS.110d2033W. doi:10.1088/1742-6596/110/4/042033.

[87] Bauer, T. H.; Spital, R. D.; Yennie, D. R.; Pipkin, F. M. (1978). "The hadronic properties of the photon in high-energy interactions". *Reviews of Modern Physics* **50** (2): 261. Bibcode:1978RvMP...50..261B. doi:10.1103/RevModPhys.50.261.

[88] Sakurai, J. J. (1960). "Theory of strong interactions". *Annals of Physics* **11**: 1. Bibcode:1960AnPhy..11....1S. doi:10.1016/0003-4916(60)90126-3.

[89] Walsh, T. F.; Zerwas, P. (1973). "Two-photon processes in the parton model". *Physics Letters B* **44** (2): 195. Bibcode:1973PhLB...44..195W. doi:10.1016/0370-2693(73)90520-0.

[90] Witten, E. (1977). "Anomalous cross section for photon-photon scattering in gauge theories". *Nuclear Physics B* **120** (2): 189. Bibcode:1977NuPhB.120..189W. doi:10.1016/0550-3213(77)90038-4.

[91] Nisius, R. (2000). "The photon structure from deep inelastic electron–photon scattering". *Physics Reports* **332** (4–6): 165. Bibcode:2000PhR...332..165N. doi:10.1016/S0370-1573(99)00115-5.

[92] Ryder, L.H. (1996). *Quantum field theory* (2nd ed.). Cambridge University Press. ISBN 0-521-47814-6.

[93] Sheldon Glashow Nobel lecture, delivered 8 December 1979.

[94] Abdus Salam Nobel lecture, delivered 8 December 1979.

[95] Steven Weinberg Nobel lecture, delivered 8 December 1979.

[96] E.g., chapter 14 in Hughes, I. S. (1985). *Elementary particles* (2nd ed.). Cambridge University Press. ISBN 0-521-26092-2.

[97] E.g., section 10.1 in Dunlap, R.A. (2004). *An Introduction to the Physics of Nuclei and Particles.* Brooks/Cole. ISBN 0-534-39294-6.

[98] Radiative correction to electron mass section 7-1-2, anomalous magnetic moments section 7-2-1, Lamb shift section 7-3-2 and hyperfine splitting in positronium section 10-3 in Itzykson, C.; Zuber, J.-B. (1980). *Quantum Field Theory.* McGraw-Hill. ISBN 0-07-032071-3.

[99] E. g. sections 9.1 (gravitational contribution of photons) and 10.5 (influence of gravity on light) in Stephani, H.; Stewart, J. (1990). *General Relativity: An Introduction to the Theory of Gravitational Field.* Cambridge University Press. pp. 86 ff, 108 ff. ISBN 0-521-37941-5.

[100] Naeye, R. (1998). *Through the Eyes of Hubble: Birth, Life and Violent Death of Stars.* CRC Press. ISBN 0-7503-0484-7. OCLC 40180195.

[101] Ch 4 in Hecht, Eugene (2001). *Optics.* Addison Wesley. ISBN 978-0-8053-8566-3.

[102] Polaritons section 10.10.1, Raman and Brillouin scattering section 10.11.3 in Patterson, J.D.; Bailey, B.C. (2007). *Solid-State Physics: Introduction to the Theory.* Springer. pp. 569 ff, 580 ff. ISBN 3-540-24115-9.

[103] E.g., section 11-5 C in Pine, S.H.; Hendrickson, J.B.; Cram, D.J.; Hammond, G.S. (1980). *Organic Chemistry* (4th ed.). McGraw-Hill. ISBN 0-07-050115-7.

[104] Nobel lecture given by G. Wald on December 12, 1967, online at nobelprize.org: The Molecular Basis of Visual Excitation.

[105] Photomultiplier section 1.1.10, CCDs section 1.1.8, Geiger counters section 1.3.2.1 in Kitchin, C.R. (2008). *Astrophysical Techniques.* Boca Raton (FL): CRC Press. ISBN 1-4200-8243-4.

[106] Denk, W.; Svoboda, K. (1997). "Photon upmanship: Why multiphoton imaging is more than a gimmick". *Neuron* **18** (3): 351–357. doi:10.1016/S0896-6273(00)81237-4. PMID 9115730.

[107] Lakowicz, J.R. (2006). *Principles of Fluorescence Spectroscopy.* Springer. pp. 529 ff. ISBN 0-387-31278-1.

[108] Jennewein, T.; Achleitner, U.; Weihs, G.; Weinfurter, H.; Zeilinger, A. (2000). "A fast and compact quantum random number generator". *Review of Scientific Instruments* **71** (4): 1675–1680. arXiv:quant-ph/9912118. Bibcode:2000RScI...71.1675J. doi:10.1063/1.1150518.

[109] Stefanov, A.; Gisin, N.; Guinnard, O.; Guinnard, L.; Zbiden, H. (2000). "Optical quantum random number generator". *Journal of Modern Optics* **47** (4): 595–598. doi:10.1080/095003400147908.

17.19 Additional references

By date of publication:

- Clauser, J.F. (1974). "Experimental distinction between the quantum and classical field-theoretic predictions for the photoelectric effect". *Physical Review D* **9** (4): 853–860. Bibcode:1974PhRvD...9..853C. doi:10.1103/PhysRevD.9.853.

- Kimble, H.J.; Dagenais, M.; Mandel, L. (1977). "Photon Anti-bunching in Resonance Fluorescence". *Physical Review Letters* **39** (11): 691–695. Bibcode:1977PhRvL..39..691K. doi:10.1103/PhysRevLett.39.691.

- Pais, A. (1982). *Subtle is the Lord: The Science and the Life of Albert Einstein.* Oxford University Press.

- Feynman, Richard (1985). *QED: The Strange Theory of Light and Matter.* Princeton University Press. ISBN 978-0-691-12575-6.

- Grangier, P.; Roger, G.; Aspect, A. (1986). "Experimental Evidence for a Photon Anticorrelation Effect on a Beam Splitter: A New Light on Single-Photon Interferences". *Europhysics Letters* **1** (4): 173–179. Bibcode:1986EL......1..173G. doi:10.1209/0295-5075/1/4/004.

- Lamb, W.E. (1995). "Anti-photon". *Applied Physics B* **60** (2–3): 77–84. Bibcode:1995ApPhB..60...77L. doi:10.1007/BF01135846.

- Special supplemental issue of *Optics and Photonics News* (vol. 14, October 2003) article web link

  - Roychoudhuri, C.; Rajarshi, R. (2003). "The nature of light: what is a photon?". *Optics and Photonics News* **14**: S1 (Supplement).

  - Zajonc, A. "Light reconsidered". *Optics and Photonics News* **14**: S2–S5 (Supplement).

  - Loudon, R. "What is a photon?". *Optics and Photonics News* **14**: S6–S11 (Supplement).

  - Finkelstein, D. "What is a photon?". *Optics and Photonics News* **14**: S12–S17 (Supplement).

  - Muthukrishnan, A.; Scully, M.O.; Zubairy, M.S. "The concept of the photon—revisited". *Optics and Photonics News* **14**: S18–S27 (Supplement).

  - Mack, H.; Schleich, W.P.. "A photon viewed from Wigner phase space". *Optics and Photonics News* **14**: S28–S35 (Supplement).

- Glauber, R. (2005). "One Hundred Years of Light Quanta" (PDF). *2005 Physics Nobel Prize Lecture.*

- Hentschel, K. (2007). "Light quanta: The maturing of a concept by the stepwise accretion of meaning". *Physics and Philosophy* **1** (2): 1–20.

Education with single photons:

- Thorn, J.J.; Neel, M.S.; Donato, V.W.; Bergreen, G.S.; Davies, R.E.; Beck, M. (2004). "Observing the quantum behavior of light in an undergraduate laboratory" (PDF). *American Journal of Physics* **72** (9): 1210–1219. Bibcode:2004AmJPh..72.1210T. doi:10.1119/1.1737397.

- Bronner, P.; Strunz, Andreas; Silberhorn, Christine; Meyn, Jan-Peter (2009). "Interactive screen experiments with single photons". *European Journal of Physics* **30** (2): 345–353. Bibcode:2009EJPh...30..345B. doi:10.1088/0143-0807/30/2/014.

17.20   External links

- The dictionary definition of photon at Wiktionary

- Media related to Photon at Wikimedia Commons

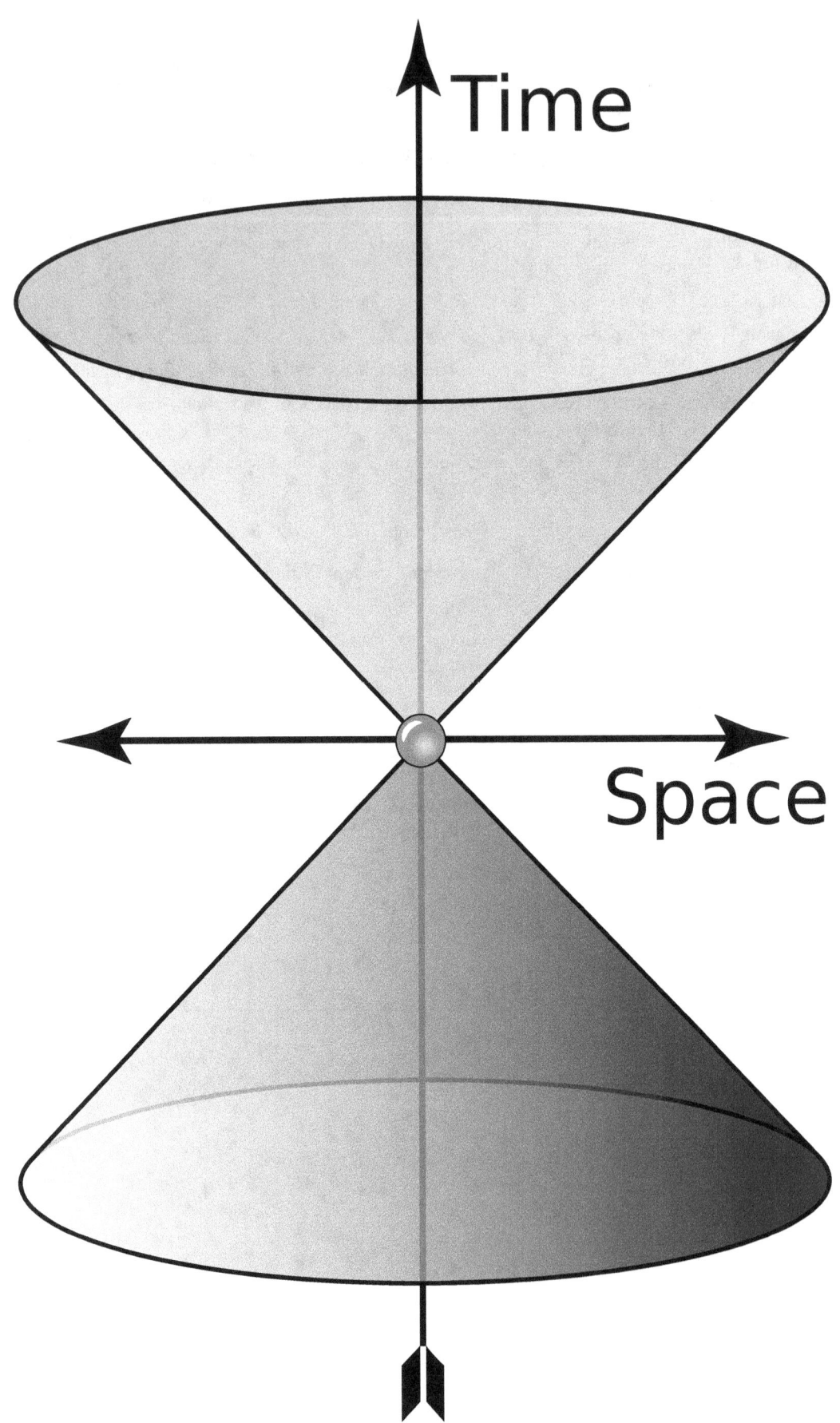

The cone shows possible values of wave 4-vector of a photon. The "time" axis gives the angular frequency (rad☐s⁻¹) and the "space" axes represent the angular wavenumber (rad☐m⁻¹). Green and indigo represent left and right polarization

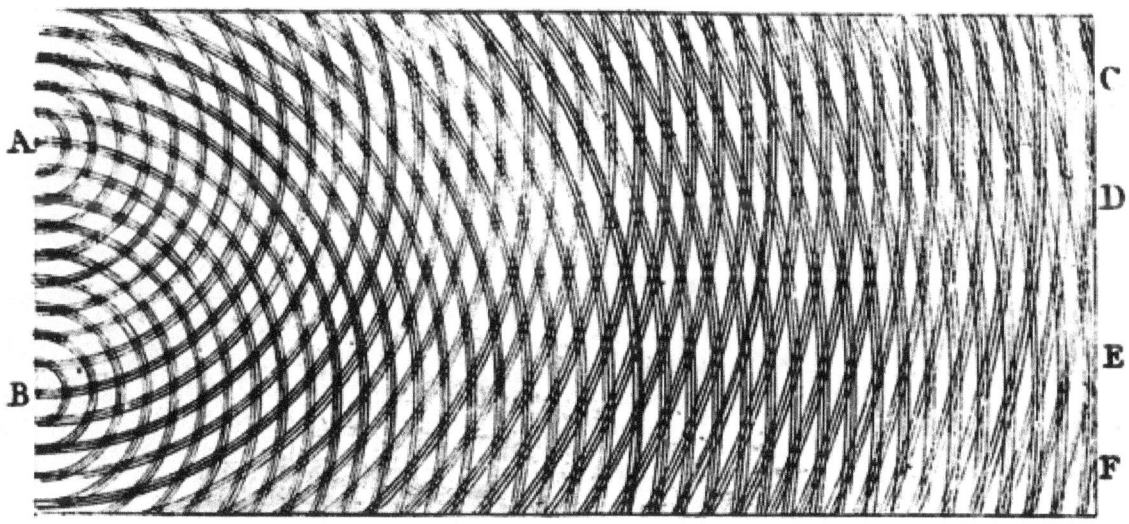

Thomas Young's double-slit experiment in 1801 showed that light can act as a wave, helping to invalidate early particle theories of light.[14]:964

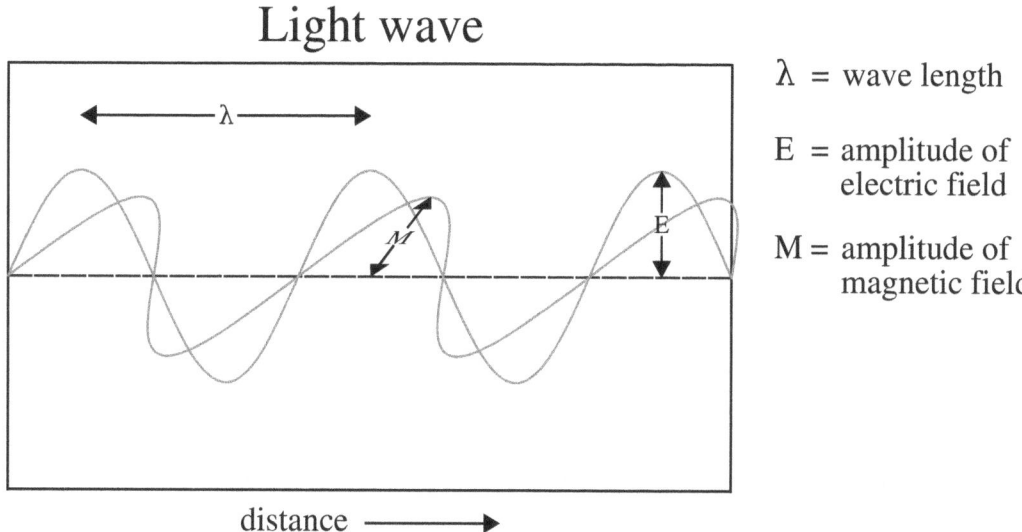

In 1900, Maxwell's theoretical model of light as oscillating electric and magnetic fields seemed complete. However, several observations could not be explained by any wave model of electromagnetic radiation, leading to the idea that light-energy was packaged into quanta described by E=hν. Later experiments showed that these light-quanta also carry momentum and, thus, can be considered particles: the photon concept was born, leading to a deeper understanding of the electric and magnetic fields themselves.

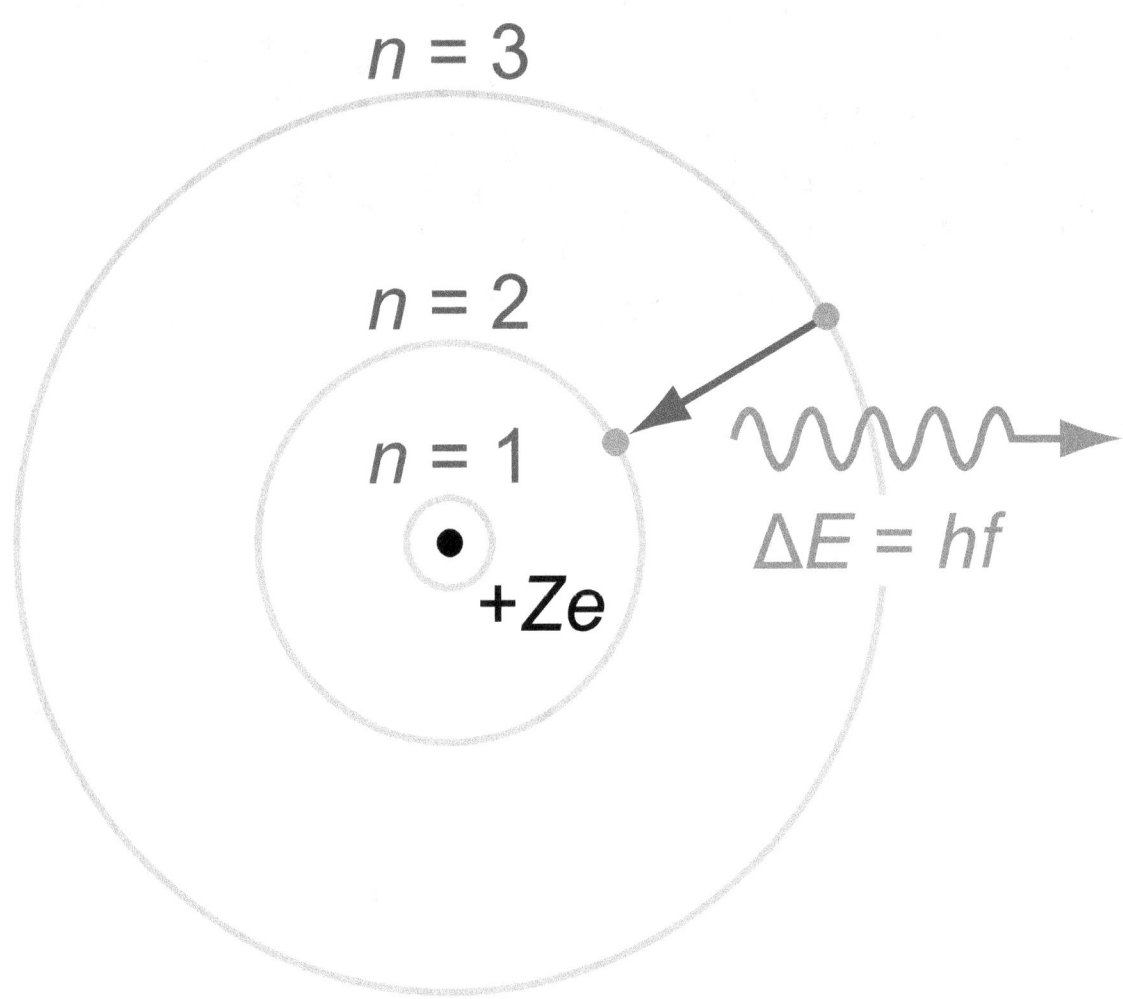

Up to 1923, most physicists were reluctant to accept that light itself was quantized. Instead, they tried to explain photon behavior by quantizing only matter, *as in the Bohr model of the hydrogen atom (shown here). Even though these semiclassical models were only a first approximation, they were accurate for simple systems and they led to quantum mechanics.*

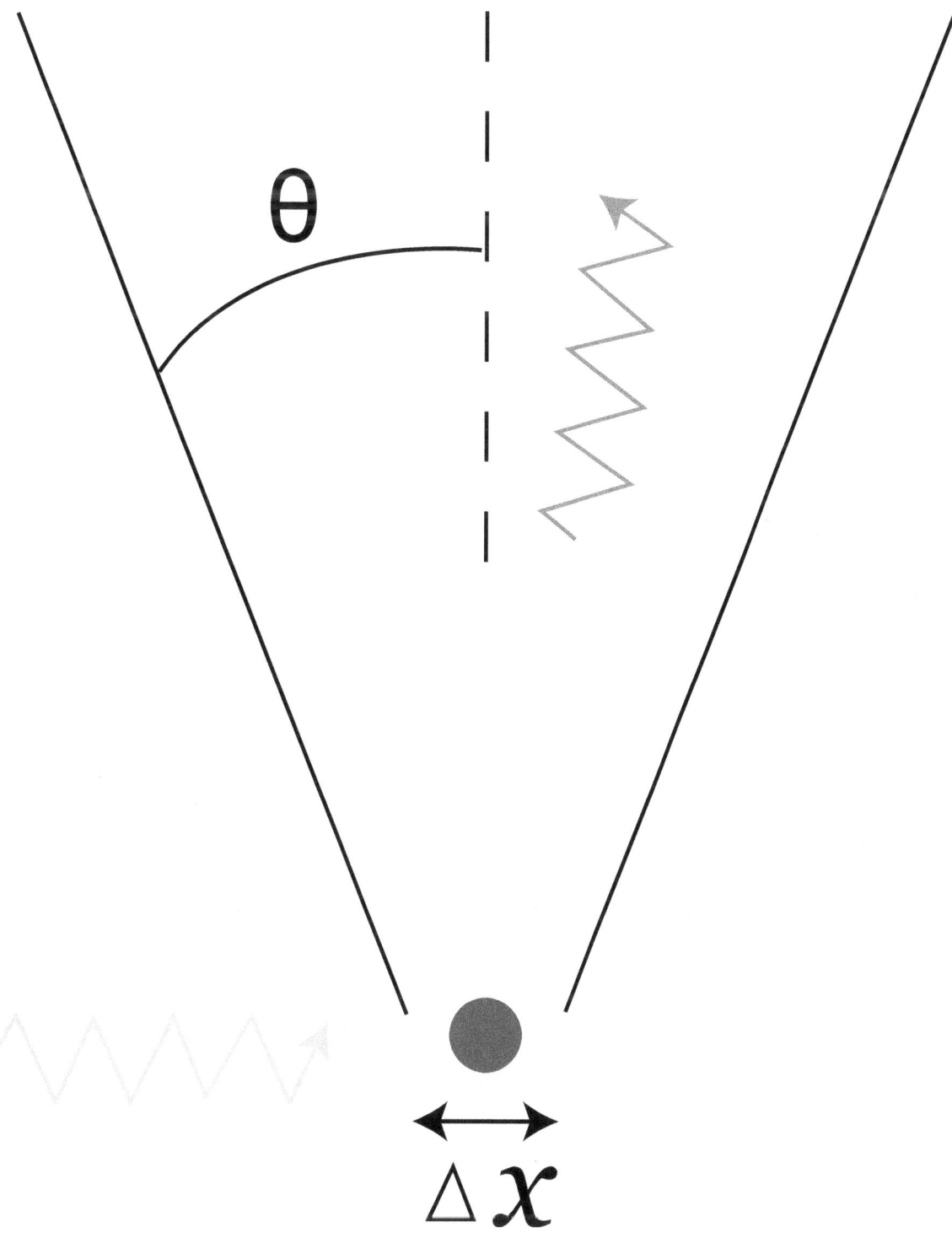

Heisenberg's thought experiment for locating an electron (shown in blue) with a high-resolution gamma-ray microscope. The incoming gamma ray (shown in green) is scattered by the electron up into the microscope's aperture angle θ. The scattered gamma ray is shown in red. Classical optics shows that the electron position can be resolved only up to an uncertainty Δx that depends on θ and the wavelength λ of the incoming light.

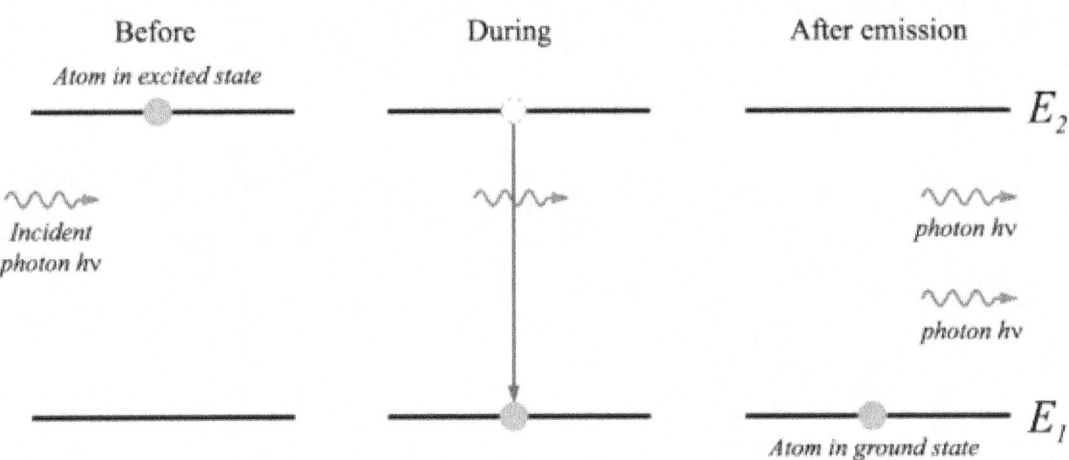

Stimulated emission (in which photons "clone" themselves) was predicted by Einstein in his kinetic analysis, and led to the development of the laser. Einstein's derivation inspired further developments in the quantum treatment of light, which led to the statistical interpretation of quantum mechanics.

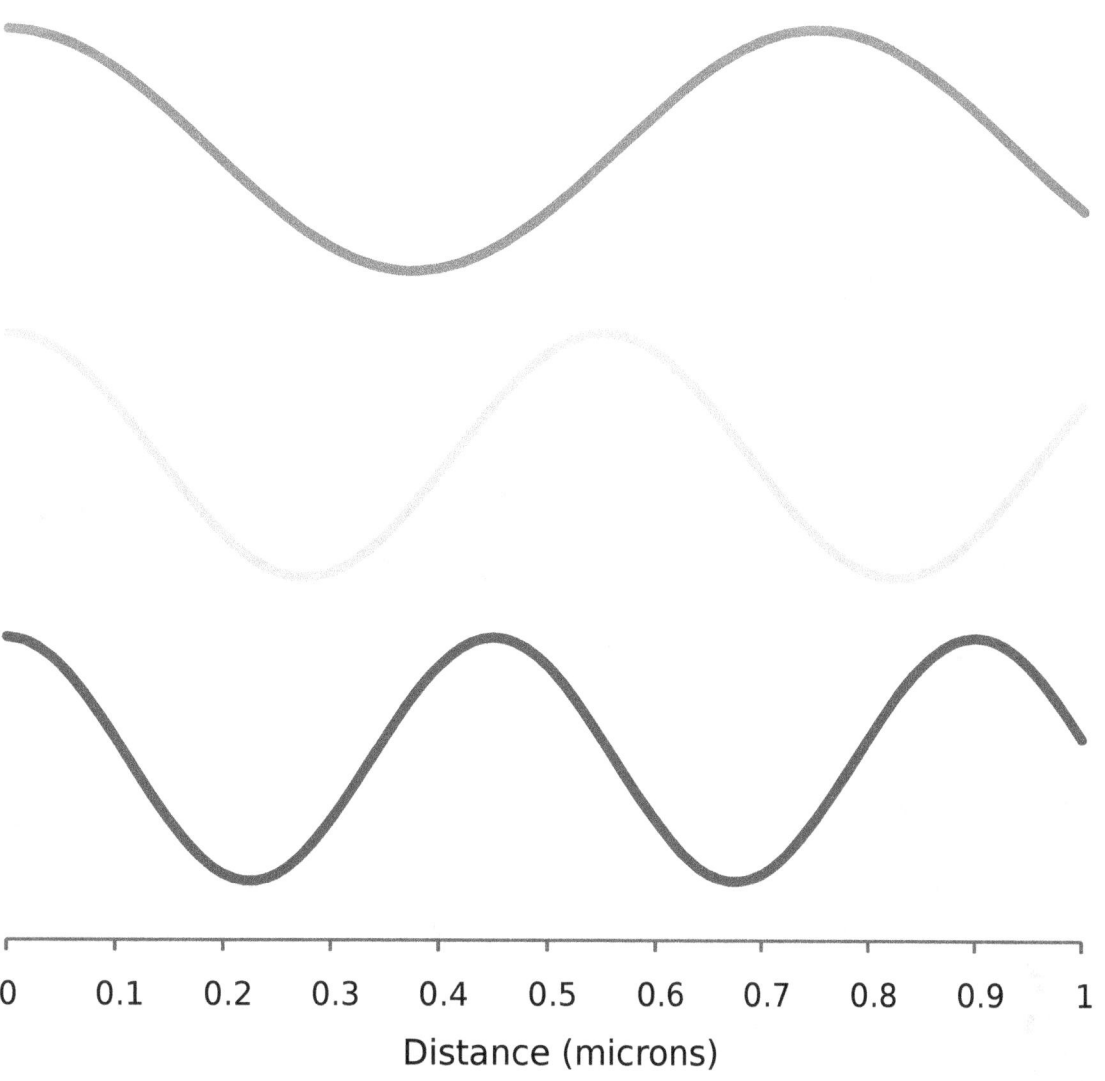

Different electromagnetic modes *(such as those depicted here) can be treated as independent simple harmonic oscillators. A photon corresponds to a unit of energy E=hν in its electromagnetic mode.*

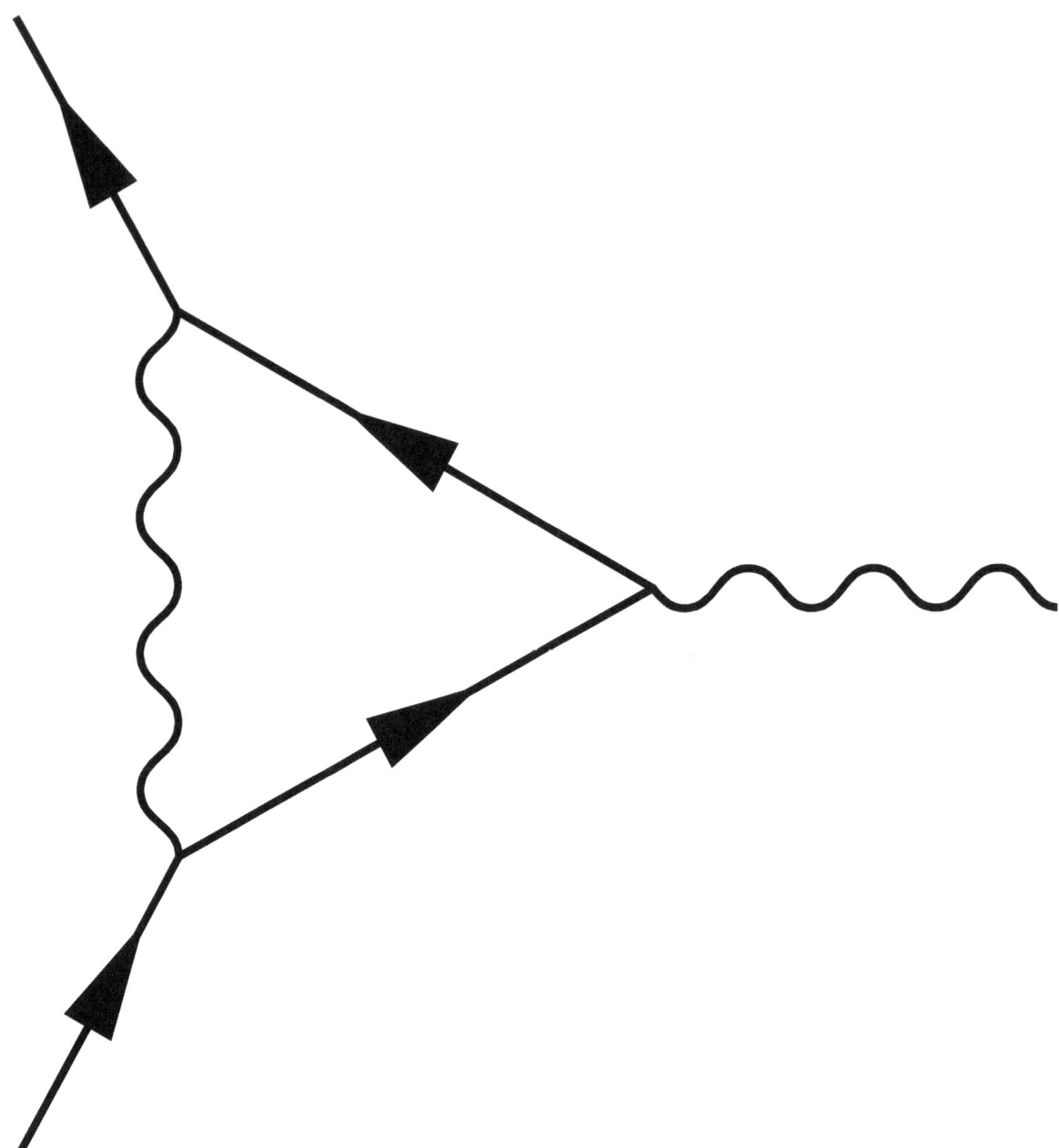

In quantum field theory, the probability of an event is computed by summing the probability amplitude (a complex number) for all possible ways in which the event can occur, as in the Feynman diagram shown here; the probability equals the square of the modulus of the total amplitude.

Chapter 18

W and Z bosons

The **W and Z bosons** (together known as the **weak bosons** or, less specifically, the **intermediate vector bosons**) are the elementary particles that mediate the weak interaction; their symbols are W+, W−, and Z. The W bosons have a positive and negative electric charge of 1 elementary charge respectively and are each other's antiparticles. The Z boson is electrically neutral and is its own antiparticle. The three particles have a spin of 1, and the W bosons have a magnetic moment, while the Z has none. All three of these particles are very short-lived, with a half-life of about 3×10^{-25} s. Their discovery was a major success for what is now called the Standard Model of particle physics.

The W bosons are named after the *w*eak force. The physicist Steven Weinberg named the additional particle the "Z particle",[3] later giving the explanation that it was the last additional particle needed by the model – the W bosons had already been named – and that it has *z*ero electric charge.[4]

The two **W bosons** are best known as mediators of neutrino absorption and emission, where their charge is associated with electron or positron emission or absorption, always causing nuclear transmutation. The Z boson is not involved in the absorption or emission of electrons and positrons.

The **Z boson** mediates the transfer of momentum, spin, and energy when neutrinos scatter *elastically* from matter, something that must happen without the production or absorption of new, charged particles. Such behaviour (which is almost as common as inelastic neutrino interactions) is seen in bubble chambers irradiated with neutrino beams. Whenever an electron simply "appears" in such a chamber as a new free particle suddenly moving with kinetic energy, and moves in the direction of the neutrinos as the apparent result of a new impulse, and this behavior happens more often when the neutrino beam is present, it is inferred to be a result of a neutrino interacting directly with the electron. Here, the neutrino simply strikes the electron and scatters away from it, transferring some of the neutrino's momentum to the electron. Since (i) neither neutrinos nor electrons are affected by the strong force, (ii) neutrinos are electrically neutral (therefore don't interact electromagnetically), and (iii) the incredibly small masses of these particles make any gravitational force between them negligible, such an interaction can only happen via the weak force. Since such an electron is not created from a nucleon, and is unchanged except for the new force impulse imparted by the neutrino, this weak force interaction between the neutrino and the electron must be mediated by a weak-force boson particle with no charge. Thus, this interaction requires a Z boson.

18.1   Basic properties

These bosons are among the heavyweights of the elementary particles. With masses of 80.4 GeV/c^2 and 91.2 GeV/c^2, respectively, the W and Z bosons are almost 100 times as massive as the proton – heavier, even, than entire atoms of iron. The masses of these bosons are significant because they act as the force carriers of a quite short-range fundamental force: their high masses thus limit the range of the weak nuclear force. By way of contrast, the electromagnetic force has an infinite range, because its force carrier, the photon, has zero mass, and the same is supposed of the hypothetical graviton.

All three bosons have particle spin $s = 1$. The emission of a W+ or W− boson either raises or lowers the electric charge of the emitting particle by one unit, and also alters the spin by one unit. At the same time, the emission or absorption of a W boson can change the type of the particle – for example changing a strange quark into an up quark. The neutral Z boson cannot change the electric charge of any particle, nor can it change any other of the so-called "charges" (such as strangeness, baryon number, charm, etc.). The emission or absorption of a Z boson can only change the spin, momentum, and energy of the other particle. (See also *weak neutral current*.)

18.2 Weak nuclear force

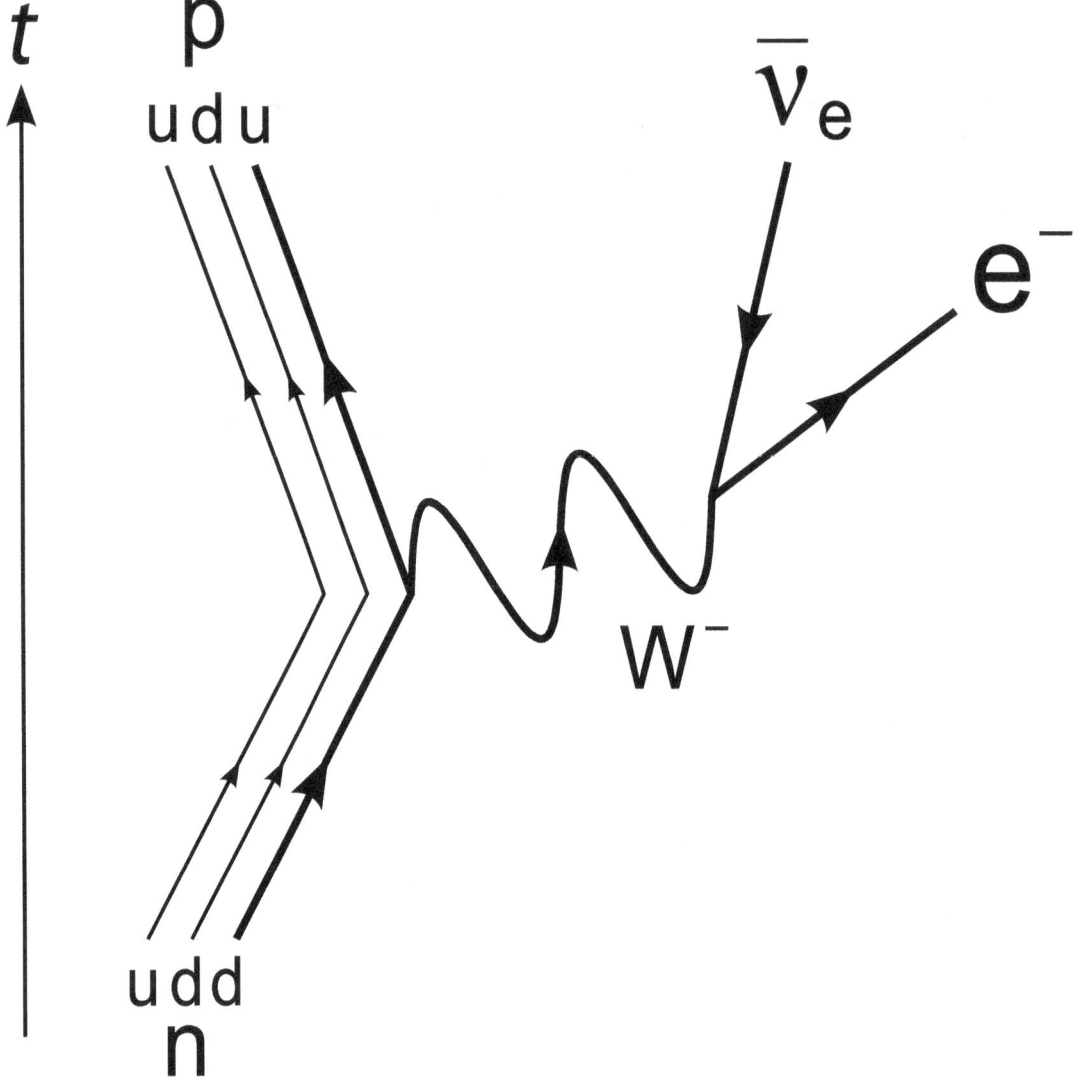

The Feynman diagram for beta decay of a neutron into a proton, electron, and electron antineutrino via an intermediate heavy W boson

The W and Z bosons are carrier particles that mediate the weak nuclear force, much as the photon is the carrier particle for the electromagnetic force.

18.2.1 W bosons

The W bosons are best known for their role in nuclear decay. Consider, for example, the beta decay of cobalt-60.

60
27Co → 60
28Ni$^+$ + e− + ν
e

This reaction does not involve the whole cobalt-60 nucleus, but affects only one of its 33 neutrons. The neutron is converted into a proton while also emitting an electron (called a beta particle in this context) and an electron antineutrino:

n0 → p+ + e− + ν
e

Again, the neutron is not an elementary particle but a composite of an up quark and two down quarks (udd). It is in fact one of the down quarks that interacts in beta decay, turning into an up quark to form a proton (uud). At the most fundamental level, then, the weak force changes the flavour of a single quark:

d → u + W−

which is immediately followed by decay of the W− itself:

W− → e− + ν
e

18.2.2 Z boson

The Z boson is its own antiparticle. Thus, all of its flavour quantum numbers and charges are zero. The exchange of a Z boson between particles, called a neutral current interaction, therefore leaves the interacting particles unaffected, except for a transfer of momentum. Z boson interactions involving neutrinos have distinctive signatures: They provide the only known mechanism for elastic scattering of neutrinos in matter; neutrinos are almost as likely to scatter elastically (via Z boson exchange) as inelastically (via W boson exchange). The first prediction of Z bosons was made by Brazilian physicist José Leite Lopes in 1958,[5] by devising an equation which showed the analogy of the weak nuclear interactions with electromagnetism. Steve Weinberg, Sheldon Glashow and Abdus Salam used later these results to develop the electroweak unification,[6] in 1973. Weak neutral currents via Z boson exchange were confirmed shortly thereafter in 1974, in a neutrino experiment in the Gargamelle bubble chamber at CERN.

18.3   Predicting the W and Z

Following the spectacular success of quantum electrodynamics in the 1950s, attempts were undertaken to formulate a similar theory of the weak nuclear force. This culminated around 1968 in a unified theory of electromagnetism and weak interactions by Sheldon Glashow, Steven Weinberg, and Abdus Salam, for which they shared the 1979 Nobel Prize in Physics.[7] Their electroweak theory postulated not only the W bosons necessary to explain beta decay, but also a new Z boson that had never been observed.

The fact that the W and Z bosons have mass while photons are massless was a major obstacle in developing electroweak theory. These particles are accurately described by an SU(2) gauge theory, but the bosons in a gauge theory must be massless. As a case in point, the photon is massless because electromagnetism is described by a U(1) gauge theory. Some mechanism is required to break the SU(2) symmetry, giving mass to the W and Z in the process. One explanation, the Higgs mechanism, was forwarded by the 1964 PRL symmetry breaking papers. It predicts the existence of yet another

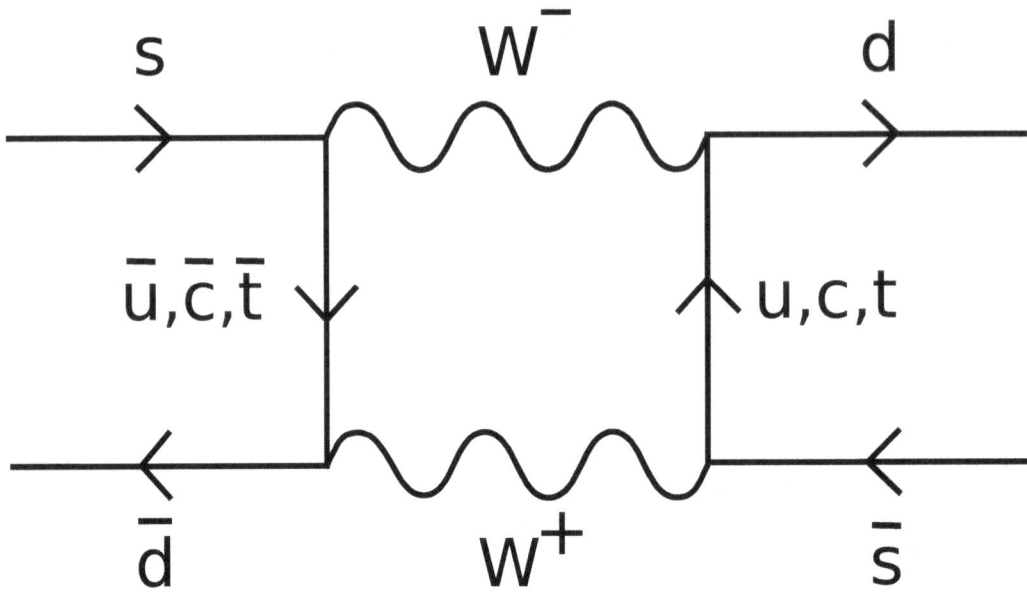

A Feynman diagram showing the exchange of a pair of W bosons. This is one of the leading terms contributing to neutral Kaon oscillation.

new particle; the Higgs boson. Of the four components of a Goldstone boson created by the Higgs field, three are "eaten" by the W^+, Z^0, and W^- bosons to form their longitudinal components and the remainder appears as the spin 0 Higgs boson.

The combination of the SU(2) gauge theory of the weak interaction, the electromagnetic interaction, and the Higgs mechanism is known as the Glashow-Weinberg-Salam model. These days it is widely accepted as one of the pillars of the Standard Model of particle physics. As of 13 December 2011, intensive search for the Higgs boson carried out at CERN has indicated that if the particle is to be found, it seems likely to be found around 125 GeV. On 4 July 2012, the CMS and the ATLAS experimental collaborations at CERN announced the discovery of a new particle with a mass of 125.3 $\pm$ 0.6 GeV that appears consistent with a Higgs boson.

18.4 Discovery

Unlike beta decay, the observation of neutral current interactions that involve particles *other than neutrinos* requires huge investments in particle accelerators and detectors, such as are available in only a few high-energy physics laboratories in the world (and then only after 1983). This is because Z-bosons behave in somewhat the same manner as photons, but do not become important until the energy of the interaction is comparable with the relatively huge mass of the Z boson.

The discovery of the W and Z bosons was considered a major success for CERN. First, in 1973, came the observation of neutral current interactions as predicted by electroweak theory. The huge Gargamelle bubble chamber photographed the tracks of a few electrons suddenly starting to move, seemingly of their own accord. This is interpreted as a neutrino interacting with the electron by the exchange of an unseen Z boson. The neutrino is otherwise undetectable, so the only observable effect is the momentum imparted to the electron by the interaction.

The discovery of the W and Z bosons themselves had to wait for the construction of a particle accelerator powerful enough to produce them. The first such machine that became available was the Super Proton Synchrotron, where unambiguous signals of W bosons were seen in January 1983 during a series of experiments made possible by Carlo Rubbia and Simon van der Meer. The actual experiments were called UA1 (led by Rubbia) and UA2 (led by Pierre Darriulat),[8] and were the collaborative effort of many people. Van der Meer was the driving force on the accelerator end (stochastic cooling). UA1 and UA2 found the Z boson a few months later, in May 1983. Rubbia and van der Meer were promptly awarded

The Gargamelle bubble chamber, now exhibited at CERN

the 1984 Nobel Prize in Physics, a most unusual step for the conservative Nobel Foundation.[9]

The W+, W−, and Z0 bosons, together with the photon (γ), comprise the four gauge bosons of the electroweak interaction.

18.5   Decay

The W and Z bosons decay to fermion–antifermion pairs but neither the W nor the Z bosons can decay into the higher-mass top quark. Neglecting phase space effects and higher order corrections, simple estimates of their branching fractions can be calculated from the coupling constants.

18.5.1   W bosons

W bosons can decay to a lepton and neutrino or to an up-type quark and a down-type quark. The decay width of the W boson to a quark–antiquark pair is proportional to the corresponding squared CKM matrix element and the number of quark colours, $NC = 3$. The decay widths for the W bosons are then proportional to:

Here, e+, μ+, τ+ denote the three flavours of leptons (more exactly, the positive charged antileptons). ν
e, ν
μ, ν
τ denote the three flavours of neutrinos. The other particles, starting with u and d, all denote quarks and antiquarks (factor NC is applied). The various V_{ij} denote the corresponding CKM matrix coefficients.

Unitarity of the CKM matrix implies that $|V_{ud}|^2 + |V_{us}|^2 + |V_{ub}|^2 = |V_{cd}|^2 + |V_{cs}|^2 + |V_{cb}|^2 = 1$. Therefore the leptonic branching ratios of the W boson are approximately B(e+ν
e) = B(μ+ν

μ) = $B(\tau + \nu$

τ) = $^1/_9$. The hadronic branching ratio is dominated by the CKM-favored ud and cs final states. The sum of the hadronic branching ratios has been measured experimentally to be 67.60±0.27%, with $B(l^+\nu_l)$ = 10.80±0.09%.[1]

18.5.2 Z bosons

Z bosons decay into a fermion and its antiparticle. As the Z-boson is a mixture of the pre-symmetry-breaking W^0 and B^0 bosons (see weak mixing angle), each vertex factor includes a factor $T_3 - Qsin^2\theta W$, where T_3 is the third component of the weak isospin of the fermion, Q is the electric charge of the fermion (in units of the elementary charge), and θW is the weak mixing angle. Because the weak isospin is different for fermions of different chirality, either left-handed or right-handed, the coupling is different as well.

The **relative** strengths of each coupling can be estimated by considering that the decay rates include the square of these factors, and all possible diagrams (e.g. sum over quark families, and left and right contributions). This is just an estimate, as we are considering only tree-level diagrams in the Fermi theory.

Here, L and R denote the left- and right-handed chiralities of the fermions respectively. (The right-handed neutrinos do not exist in the standard model. However, in some extensions beyond the standard model they do.) The notation $x = sin^2\theta W$ is used.

18.6 See also

- Bose–Einstein statistics
- Boson
- List of particles
- Standard Model (mathematical formulation)
- W' and Z' bosons
- X and Y bosons: analogous pair of bosons predicted by the Grand Unified Theory

18.7 References

[1] J. Beringer et al. (2012). "2012 Review of Particle Physics - Gauge and Higgs Bosons" (PDF). *Physical Review D* **86**: 1. Bibcode:2012PhRvD..86a0001B. doi:10.1103/PhysRevD.86.010001.

[2] (PDF) http://pdg.lbl.gov/2013/reviews/rpp2013-rev-w-mass.pdf. Missing or empty |title= (help)

[3] Steven Weinberg, A Model of Leptons, Phys. Rev. Lett. 19, 1264–1266 (1967) – the electroweak unification paper.

[4] Weinberg, Steven (1993). *Dreams of a Final Theory: the search for the fundamental laws of nature*. Vintage Press. p. 94. ISBN 0-09-922391-0.

[5] "Forty years of the first attempt at the electroweak unification and of the prediction of the weak neutral boson".

[6] "The Nobel Prize in Physics 1979". Nobel Foundation. Retrieved 2008-09-10.

[7] Nobel Prize in Physics for 1979 (see also Nobel Prize in Physics on Wikipedia)

[8] The UA2 Collaboration collection

[9] 1984 Nobel Prize in physics

[10] C. Amsler et al. (Particle Data Group), PL B667, 1 (2008) and 2009 partial update for the 2010 edition

18.8   External links

- The Review of Particle Physics, the ultimate source of information on particle properties.

- The W and Z particles: a personal recollection by Pierre Darriulat

- When CERN saw the end of the alphabet by Daniel Denegri

- W and Z particles at Hyperphysics

Chapter 19

Gluon

Gluons /'gluːɒnz/ are elementary particles that act as the exchange particles (or gauge bosons) for the strong force between quarks, analogous to the exchange of photons in the electromagnetic force between two charged particles.[6]

In technical terms, gluons are vector gauge bosons that mediate strong interactions of quarks in quantum chromodynamics (QCD). Gluons themselves carry the color charge of the strong interaction. This is unlike the photon, which mediates the electromagnetic interaction but lacks an electric charge. Gluons therefore participate in the strong interaction in addition to mediating it, making QCD significantly harder to analyze than QED (quantum electrodynamics).

19.1 Properties

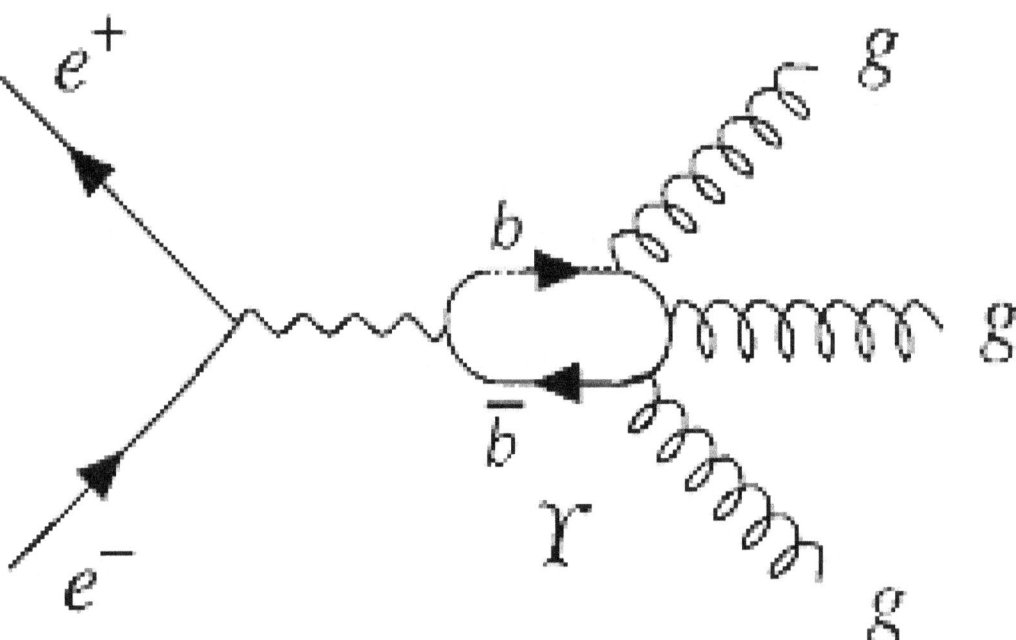

Diagram 2: $e^+e^- \to \Upsilon(9.46) \to 3g$

The gluon is a vector boson; like the photon, it has a spin of 1. While massive spin-1 particles have three polarization states,

143

massless gauge bosons like the gluon have only two polarization states because gauge invariance requires the polarization to be transverse. In quantum field theory, unbroken gauge invariance requires that gauge bosons have zero mass (experiment limits the gluon's rest mass to less than a few meV/c^2). The gluon has negative intrinsic parity.

19.2 Numerology of gluons

Unlike the single photon of QED or the three W and Z bosons of the weak interaction, there are eight independent types of gluon in QCD.

This may be difficult to understand intuitively. Quarks carry three types of color charge; antiquarks carry three types of anticolor. Gluons may be thought of as carrying both color and anticolor, but to correctly understand how they are combined, it is necessary to consider the mathematics of color charge in more detail.

19.2.1 Color charge and superposition

In quantum mechanics, the states of particles may be added according to the principle of superposition; that is, they may be in a "combined state" with a *probability*, if some particular quantity is measured, of giving several different outcomes. A relevant illustration in the case at hand would be a gluon with a color state described by:

$$(r\bar{b} + b\bar{r})/\sqrt{2}.$$

This is read as "red–antiblue plus blue–antired". (The factor of the square root of two is required for normalization, a detail that is not crucial to understand in this discussion.) If one were somehow able to make a direct measurement of the color of a gluon in this state, there would be a 50% chance of it having red-antiblue color charge and a 50% chance of blue-antired color charge.

19.2.2 Color singlet states

It is often said that the stable strongly interacting particles (such as the proton and the neutron, i.e. hadrons) observed in nature are "colorless", but more precisely they are in a "color singlet" state, which is mathematically analogous to a *spin* singlet state.[7] Such states allow interaction with other color singlets, but not with other color states; because long-range gluon interactions do not exist, this illustrates that gluons in the singlet state do not exist either.[7]

The color singlet state is:[7]

$$(r\bar{r} + b\bar{b} + g\bar{g})/\sqrt{3}.$$

In words, if one could measure the color of the state, there would be equal probabilities of it being red-antired, blue-antiblue, or green-antigreen.

19.2.3 Eight gluon colors

There are eight remaining independent color states, which correspond to the "eight types" or "eight colors" of gluons. Because states can be mixed together as discussed above, there are many ways of presenting these states, which are known as the "color octet". One commonly used list is:[7]

These are equivalent to the Gell-Mann matrices; the translation between the two is that red-antired is the upper-left matrix entry, red-antiblue is the upper middle entry, blue-antigreen is the middle right entry, and so on. The critical feature of these particular eight states is that they are linearly independent, and also independent of the singlet state; there is no way to add any combination of states to produce any other. (It is also impossible to add them to make rr, gg, or bb[8] otherwise the forbidden singlet state could also be made.) There are many other possible choices, but all are mathematically equivalent, at least equally complex, and give the same physical results.

19.2.4 Group theory details

Technically, QCD is a gauge theory with SU(3) gauge symmetry. Quarks are introduced as spinor fields in N_f flavors, each in the fundamental representation (triplet, denoted **3**) of the color gauge group, SU(3). The gluons are vector fields in the adjoint representation (octets, denoted **8**) of color SU(3). For a general gauge group, the number of force-carriers (like photons or gluons) is always equal to the dimension of the adjoint representation. For the simple case of SU(N), the dimension of this representation is $N^2 - 1$.

In terms of group theory, the assertion that there are no color singlet gluons is simply the statement that quantum chromodynamics has an SU(3) rather than a U(3) symmetry. There is no known *a priori* reason for one group to be preferred over the other, but as discussed above, the experimental evidence supports SU(3).[7].The U(1) group for electromagnetic field combines with a slightly more complicated group known as SU(2),S stands for "special".

19.3 Confinement

Main article: Color confinement

Since gluons themselves carry color charge, they participate in strong interactions. These gluon-gluon interactions constrain color fields to string-like objects called "flux tubes", which exert constant force when stretched. Due to this force, quarks are confined within composite particles called hadrons. This effectively limits the range of the strong interaction to 1×10^{-15} meters, roughly the size of an atomic nucleus. Beyond a certain distance, the energy of the flux tube binding two quarks increases linearly. At a large enough distance, it becomes energetically more favorable to pull a quark-antiquark pair out of the vacuum rather than increase the length of the flux tube.

Gluons also share this property of being confined within hadrons. One consequence is that gluons are not directly involved in the nuclear forces between hadrons. The force mediators for these are other hadrons called mesons.

Although in the normal phase of QCD single gluons may not travel freely, it is predicted that there exist hadrons that are formed entirely of gluons — called glueballs. There are also conjectures about other exotic hadrons in which real gluons (as opposed to virtual ones found in ordinary hadrons) would be primary constituents. Beyond the normal phase of QCD (at extreme temperatures and pressures), quark–gluon plasma forms. In such a plasma there are no hadrons; quarks and gluons become free particles.

19.4 Experimental observations

Quarks and gluons (colored) manifest themselves by fragmenting into more quarks and gluons, which in turn hadronize into normal (colorless) particles, correlated in jets. As shown in 1978 summer conferences[2] the PLUTO detector at the electron-positron collider DORIS (DESY) produced the first evidence that the hadronic decays of the very narrow resonance $\Upsilon(9.46)$ could be interpreted as three-jet event topologies produced by three gluons. Later published analyses by the same experiment confirmed this interpretation and also the spin 1 nature of the gluon[9][10] (see also the recollection[2] and PLUTO experiments).

In summer 1979 at higher energies at the electron-positron collider PETRA (DESY) again three-jet topologies were observed, now interpreted as qq gluon bremsstrahlung, now clearly visible, by TASSO,[11] MARK-J[12] and PLUTO experiments[13] (later in 1980 also by JADE[14]). The spin 1 of the gluon was confirmed in 1980 by TASSO[15] and PLUTO experiments[16] (see also the review[3]). In 1991 a subsequent experiment at the LEP storage ring at CERN again confirmed this result.[17]

The gluons play an important role in the elementary strong interactions between quarks and gluons, described by QCD and studied particularly at the electron-proton collider HERA at DESY. The number and momentum distribution of the gluons in the proton (gluon density) have been measured by two experiments, H1 and ZEUS,[18] in the years 1996 till today (2012). The gluon contribution to the proton spin has been studied by the HERMES experiment at HERA.[19] The gluon density in the photon (when behaving hadronically) also has been measured.[20]

Color confinement is verified by the failure of free quark searches (searches of fractional charges). Quarks are normally produced in pairs (quark + antiquark) to compensate the quantum color and flavor numbers; however at Fermilab single production of top quarks has been shown (technically this still involves a pair production, but quark and antiquark are of different flavor).[21] No glueball has been demonstrated.

Deconfinement was claimed in 2000 at CERN SPS[22] in heavy-ion collisions, and it implies a new state of matter: quark–gluon plasma, less interacting than in the nucleus, almost as in a liquid. It was found at the Relativistic Heavy Ion Collider (RHIC) at Brookhaven in the years 2004–2010 by four contemporaneous experiments.[23] A quark–gluon plasma state has been confirmed at the CERN Large Hadron Collider (LHC) by the three experiments ALICE, ATLAS and CMS in 2010.[24]

19.5   See also

- Quark

- Hadron

- Meson

- Gauge boson

- Quark model

- Quantum chromodynamics

- Quark–gluon plasma

- Color confinement

- Glueball

- Gluon field

- Gluon field strength tensor

- Exotic hadrons

- Standard Model

- Three-jet events

- Deep inelastic scattering

19.6   References

[1] M. Gell-Mann (1962). "Symmetries of Baryons and Mesons". *Physical Review* **125** (3): 1067–1084. Bibcode:1962PhRv..125.1067G. doi:10.1103/PhysRev.125.1067.

[2] B.R. Stella and H.-J. Meyer (2011). "ϒ(9.46 GeV) and the gluon discovery (a critical recollection of PLUTO results)". *European Physical Journal H* **36** (2): 203–243. arXiv:1008.1869v3. Bibcode:2011EPJH...36..203S. doi:10.1140/epjh/e2011-10029-3.

[3] P. Söding (2010). "On the discovery of the gluon". *European Physical Journal H* **35** (1): 3–28. Bibcode:2010EPJH...35....3S. doi:10.1140/epjh/e2010-00002-5.

[4] W.-M. Yao et al. (2006). "Review of Particle Physics" (PDF). *Journal of Physics G* **33**: 1. arXiv:astro-ph/0601168. Bibcode:2006JPhG...33....1Y. doi:10.1088/0954-3899/33/1/001.

[5] F. Yndurain (1995). "Limits on the mass of the gluon". *Physics Letters B* **345** (4): 524. Bibcode:1995PhLB..345..524Y. doi:10.1016/0370-2693(94)01677-5.

[6] C.R. Nave. "The Color Force". *HyperPhysics*. Georgia State University, Department of Physics. Retrieved 2012-04-02.

[7] David Griffiths (1987). *Introduction to Elementary Particles*. John Wiley & Sons. pp. 280–281. ISBN 0-471-60386-4.

[8] J. Baez. "Why are there eight gluons and not nine?". Retrieved 2009-09-13.

[9] Ch. Berger *et al.* (PLUTO Collaboration) (1979). "Jet analysis of the $\Upsilon(9.46)$ decay into charged hadrons". *Physics Letters B* **82** (3–4): 449. Bibcode:1979PhLB...82..449B. doi:10.1016/0370-2693(79)90265-X.

[10] Ch. Berger *et al.* (PLUTO Collaboration) (1981). "Topology of the Υ decay". *Zeitschrift für Physik C* **8** (2): 101. Bibcode:1981ZPhyC...8..101 doi:10.1007/BF01547873.

[11] R. Brandelik *et al.* (TASSO collaboration) (1979). "Evidence for Planar Events in e^+e^- Annihilation at High Energies". *Physics Letters B* **86** (2): 243–249. Bibcode:1979PhLB...86..243B. doi:10.1016/0370-2693(79)90830-X.

[12] D.P. Barber *et al.* (MARK-J collaboration) (1979). "Discovery of Three-Jet Events and a Test of Quantum Chromodynamics at PETRA". *Physical Review Letters* **43** (12): 830. Bibcode:1979PhRvL..43..830B. doi:10.1103/PhysRevLett.43.830.

[13] Ch. Berger *et al.* (PLUTO Collaboration) (1979). "Evidence for Gluon Bremsstrahlung in e^+e^- Annihilations at High Energies". *Physics Letters B* **86** (3–4): 418. Bibcode:1979PhLB...86..418B. doi:10.1016/0370-2693(79)90869-4.

[14] W. Bartel *et al.* (JADE Collaboration) (1980). "Observation of planar three-jet events in e^+e^- annihilation and evidence for gluon bremsstrahlung". *Physics Letters B* **91**: 142. Bibcode:1980PhLB...91..142B. doi:10.1016/0370-2693(80)90680-2.

[15] R. Brandelik *et al.* (TASSO Collaboration) (1980). "Evidence for a spin-1 gluon in three-jet events". *Physics Letters B* **97** (3–4): 453. Bibcode:1980PhLB...97..453B. doi:10.1016/0370-2693(80)90639-5.

[16] Ch. Berger *et al.* (PLUTO Collaboration) (1980). "A study of multi-jet events in e^+e^- annihilation". *Physics Letters B* **97** (3–4): 459. Bibcode:1980PhLB...97..459B. doi:10.1016/0370-2693(80)90640-1.

[17] G. Alexander *et al.* (OPAL Collaboration) (1991). "Measurement of Three-Jet Distributions Sensitive to the Gluon Spin in e^+e^- Annihilations at $\sqrt{s} = 91$ GeV". *Zeitschrift für Physik C* **52** (4): 543. Bibcode:1991ZPhyC..52..543A. doi:10.1007/BF01562326.

[18] L. Lindeman (H1 and ZEUS collaborations) (1997). "Proton structure functions and gluon density at HERA". *Nuclear Physics B Proceedings Supplements* **64**: 179–183. Bibcode:1998NuPhS..64..179L. doi:10.1016/S0920-5632(97)01057-8.

[19] http://www-hermes.desy.de

[20] C. Adloff *et al.* (H1 collaboration) (1999). "Charged particle cross sections in the photoproduction and extraction of the gluon density in the photon". *European Physical Journal C* **10**: 363–372. arXiv:hep-ex/9810020. Bibcode:1999EPJC...10..363H. doi:10.1007/s100520050761.

[21] M. Chalmers (6 March 2009). "Top result for Tevatron". *Physics World*. Retrieved 2012-04-02.

[22] M.C. Abreu et al. (2000). "Evidence for deconfinement of quark and antiquark from the J/Ψ suppression pattern measured in Pb-Pb collisions at the CERN SpS". *Physics Letters B* **477**: 28–36. Bibcode:2000PhLB..477...28A. doi:10.1016/S0370-2693(00)00237-9.

[23] D. Overbye (15 February 2010). "In Brookhaven Collider, Scientists Briefly Break a Law of Nature". *New York Times*. Retrieved 2012-04-02.

[24] "LHC experiments bring new insight into primordial universe" (Press release). CERN. 26 November 2010. Retrieved 2012-04-02.

19.7 Further reading

- A. Ali and G. Kramer (2011). "JETS and QCD: A historical review of the discovery of the quark and gluon jets and its impact on QCD". *European Physical Journal H* **36** (2): 245–326. arXiv:1012.2288. Bibcode:2011EPJH...36..245A. doi:10.1140/epjh/e2011-10047-1.

Chapter 20

Unitary group

In mathematics, the **unitary group** of degree n, denoted U(n), is the group of $n \times n$ unitary matrices, with the group operation that of matrix multiplication. The unitary group is a subgroup of the general linear group GL(n, **C**). **Hyperorthogonal group** is an archaic name for the unitary group, especially over finite fields.

In the simple case $n = 1$, the group U(1) corresponds to the circle group, consisting of all complex numbers with absolute value 1 under multiplication. All the unitary groups contain copies of this group.

The unitary group U(n) is a real Lie group of dimension n^2. The Lie algebra of U(n) consists of $n \times n$ skew-Hermitian matrices, with the Lie bracket given by the commutator.

The **general unitary group** (also called the **group of unitary similitudes**) consists of all matrices A such that $A*A$ is a nonzero multiple of the identity matrix, and is just the product of the unitary group with the group of all positive multiples of the identity matrix.

20.1   Properties

Since the determinant of a unitary matrix is a complex number with norm 1 , the determinant gives a group homomorphism

$$\det : \mathrm{U}(n) \to \mathrm{U}(1).$$

The kernel of this homomorphism is the set of unitary matrices with determinant 1 . This subgroup is called the **special unitary group**, denoted SU(n) . We then have a short exact sequence of Lie groups:

$$1 \to \mathrm{SU}(n) \to \mathrm{U}(n) \to \mathrm{U}(1) \to 1.$$

This short exact sequence splits so that U(n) may be written as a semidirect product of SU(n) by U(1) . Here the U(1) subgroup of U(n) can be taken to consist of matrices, which are diagonal, have $e^{i\theta}$ in the upper left corner and 1 on the rest of the diagonal.

The unitary group U(n) is nonabelian for $n > 1$. The center of U(n) is the set of scalar matrices λI with $\lambda \in$ U(1) . This follows from Schur's lemma. The center is then isomorphic to U(1) . Since the center of U(n) is a 1 -dimensional abelian normal subgroup of U(n) , the unitary group is not semisimple.

20.2   Topology

The unitary group U(n) is endowed with the relative topology as a subset of M(n, **C**), the set of all $n \times n$ complex matrices, which is itself homeomorphic to a $2n^2$-dimensional Euclidean space.

As a topological space, U(n) is both compact and connected. The compactness of U(n) follows from the Heine–Borel theorem and the fact that it is a closed and bounded subset of M(n, $\mathbf{C}$). To show that U(n) is connected, recall that any unitary matrix A can be diagonalized by another unitary matrix S. Any diagonal unitary matrix must have complex numbers of absolute value 1 on the main diagonal. We can therefore write

$$A = S \operatorname{diag}(e^{i\theta_1}, \ldots, e^{i\theta_n}) S^{-1}.$$

A path in U(n) from the identity to A is then given by

$$t \mapsto S \operatorname{diag}(e^{it\theta_1}, \ldots, e^{it\theta_n}) S^{-1}.$$

The unitary group is not simply connected; the fundamental group of U(n) is infinite cyclic for all n:

$$\pi_1(U(n)) \cong \mathbf{Z}.$$

To see this, note that the above splitting of U(n) as a semidirect product of SU(n) and U(1) induces a topological product structure on U(n), so that

$$\pi_1(U(n)) \cong \pi_1(SU(n)) \times \pi_1(U(1)).$$

Now the first unitary group U(1) is topologically a circle, which is well known to have a fundamental group isomorphic to $\mathbf{Z}$, and the inclusion map U(n) $\to$ U(n+1) is an isomorphism on π_1. (It has quotient the Stiefel manifold.)

The determinant map det: U(n) $\to$ U(1) induces an isomorphism of fundamental groups, with the splitting U(1) $\to$ U(n) inducing the inverse.

The Weyl group of U(n) is the symmetric group Sn, acting on the diagonal torus by permuting the entries:

$$\operatorname{diag}(e^{i\theta_1}, \ldots, e^{i\theta_n}) \mapsto \operatorname{diag}(e^{i\theta_{\sigma(1)}}, \ldots, e^{i\theta_{\sigma(n)}})$$

20.3 Related groups

20.3.1 2-out-of-3 property

The unitary group is the 3-fold intersection of the orthogonal, symplectic, and complex groups:

$$U(n) = O(2n) \cap Sp(2n, \mathbf{R}) \cap GL(n, \mathbf{C}).$$

Thus a unitary structure can be seen as an orthogonal structure, a complex structure, and a symplectic structure, which are required to be *compatible* (meaning that one uses the same J in the complex structure and the symplectic form, and that this J is orthogonal; writing all the groups as matrix groups fixes a J (which is orthogonal) and ensures compatibility).

In fact, it is the intersection of any *two* of these three; thus a compatible orthogonal and complex structure induce a symplectic structure, and so forth.[1][2]

At the level of equations, this can be seen as follows:

Any two of these equations implies the third.

At the level of forms, this can be seen by decomposing a Hermitian form into its real and imaginary parts: the real part is symmetric (orthogonal), and the imaginary part is skew-symmetric (symplectic)—and these are related by the complex structure (which is the compatibility). On an almost Kähler manifold, one can write this decomposition as $h = g + i\omega$, where h is the Hermitian form, g is the Riemannian metric, i is the almost complex structure, and ω is the almost symplectic structure.

From the point of view of Lie groups, this can partly be explained as follows: $O(2n)$ is the maximal compact subgroup of $GL(2n, \mathbf{R})$, and $U(n)$ is the maximal compact subgroup of both $GL(n, \mathbf{C})$ and $Sp(2n)$. Thus the intersection $O(2n) \cap GL(n, \mathbf{C})$ or $O(2n) \cap Sp(2n)$ is the maximal compact subgroup of both of these, so $U(n)$. From this perspective, what is unexpected is the intersection $GL(n, \mathbf{C}) \cap Sp(2n) = U(n)$.

20.3.2   Special unitary and projective unitary groups

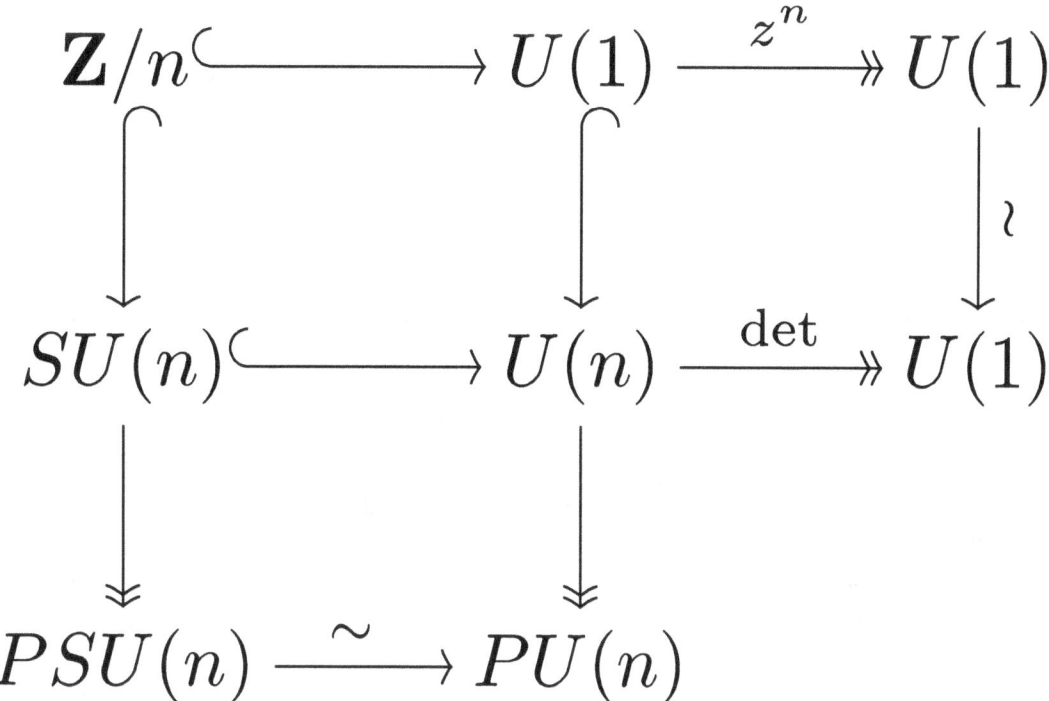

Main article: Projective unitary group

Just as the orthogonal group has the special orthogonal group $SO(n)$ as subgroup and the projective orthogonal group $PO(n)$ as quotient, and the projective special orthogonal group $PSO(n)$ as subquotient, the unitary group has associated to it the special unitary group $SU(n)$, the projective unitary group $PU(n)$, and the projective special unitary group $PSU(n)$. These are related as by the commutative diagram at right; notably, both projective groups are equal: $PSU(n) = PU(n)$.

The above is for the classical unitary group (over the complex numbers) – for unitary groups over finite fields, one similarly obtains special unitary and projective unitary groups, but in general $PSU(n, q^2) \neq PU(n, q^2)$.

20.4 G-structure: almost Hermitian

In the language of G-structures, a manifold with a U(n)-structure is an almost Hermitian manifold.

20.5 Generalizations

From the point of view of Lie theory, the classical unitary group is a real form of the Steinberg group ^2A_n , which is an algebraic group that arises from the combination of the *diagram automorphism* of the general linear group (reversing the Dynkin diagram An, which corresponds to transpose inverse) and the *field automorphism* of the extension **C/R** (namely complex conjugation). Both these automorphisms are automorphisms of the algebraic group, have order 2, and commute, and the unitary group is the fixed points of the product automorphism, as an algebraic group. The classical unitary group is a real form of this group, corresponding to the standard Hermitian form Ψ, which is positive definite.

This can be generalized in a number of ways:

- generalizing to other Hermitian forms yields indefinite unitary groups U(p,q);

- the field extension can be replaced by any degree 2 separable algebra, most notably a degree 2 extension of a finite field;

- generalizing to other diagrams yields other groups of Lie type, namely the other Steinberg groups $^2D_n, {}^2E_6, {}^3D_4$, (in addition to ^2A_n) and Suzuki-Ree groups

$$^2B_2\left(2^{2n+1}\right), {}^2F_4\left(2^{2n+1}\right), {}^2G_2\left(3^{2n+1}\right);$$

- considering a generalized unitary group as an algebraic group, one can take its points over various algebras.

20.5.1 Indefinite forms

Analogous to the indefinite orthogonal groups, one can define an **indefinite unitary group**, by considering the transforms that preserve a given Hermitian form, not necessarily positive definite (but generally taken to be non-degenerate). Here one is working with a vector space over the complex numbers.

Given a Hermitian form Ψ on a complex vector space V, the unitary group U(Ψ) is the group of transforms that preserve the form: the transform M such that $\Psi(Mv,Mw) = \Psi(v,w)$ for all $v,w \in V$. In terms of matrices, representing the form by a matrix denoted Φ, this says that $M^*\Phi M = \Phi$.

Just as for symmetric forms over the reals, Hermitian forms are determined by signature, and are all unitarily congruent to a diagonal form with p entries of 1 on the diagonal and q entries of –1. The non-degenerate assumption is equivalent to $p+q = n$. In a standard basis, this is represented as a quadratic form as:

$$\|z\|_\Psi^2 = \|z_1\|^2 + \cdots + \|z_p\|^2 - \|z_{p+1}\|^2 - \cdots - \|z_n\|^2$$

and as a symmetric form as:

$$\Psi(w, z) = \bar{w}_1 z_1 + \cdots + \bar{w}_p z_p - \bar{w}_{p+1} z_{p+1} - \cdots - \bar{w}_n z_n.$$

The resulting group is denoted U(p,q).

20.5.2   Finite fields

Over the finite field with $q = p^r$ elements, $\mathbf{F}q$, there is a unique quadratic extension field, $\mathbf{F}q^2$, with order 2 automorphism $\alpha \colon x \mapsto x^q$ (the rth power of the Frobenius automorphism). This allows one to define a Hermitian form on an $\mathbf{F}q^2$ vector space V, as an $\mathbf{F}q$-bilinear map $\Psi \colon V \times V \to K$ such that $\Psi(w,v) = \alpha\,(\Psi(v,w))$ and $\Psi(w,cv) = c\Psi(w,v)$ for $c \in \mathbf{F}q^2$. *Further, all non-degenerate Hermitian forms on a vector space over a finite field are unitarily congruent to the standard one, represented by the identity matrix, that is, any Hermitian form is unitarily equivalent to*

$$\Psi(w,v) = w^\alpha \cdot v = \sum_{i=1}^{n} w_i^q v_i$$

where w_i, v_i represent the coordinates of $w,v \in V$ in some particular $\mathbf{F}q^2$-basis of the n-dimensional space V (Grove 2002, Thm. 10.3).

Thus one can define a (unique) unitary group of dimension n for the extension $\mathbf{F}q^2/\mathbf{F}_q$, denoted either as U($n$, q) or U(n, q^2) depending on the author. The subgroup of the unitary group consisting of matrices of determinant 1 is called the **special unitary group** and denoted SU(n, q) or SU(n, q^2). For convenience, this article will use the U(n, q^2) convention. The center of U(n, q^2) has order $q+1$ and consists of the scalar matrices which are unitary, that is those matrices cIV with $c^{q+1} = 1$. The center of the special unitary group has order gcd(n, $q+1$) and consists of those unitary scalars which also have order dividing n. The quotient of the unitary group by its center is called the **projective unitary group**, PU(n, q^2), and the quotient of the special unitary group by its center is the **projective special unitary group** PSU(n, q^2). In most cases ($n > 1$ and (n, q^2) $\notin$ {(2,2^2), (2,3^2), (3,2^2)}), SU(n, q^2) is a perfect group and PSU(n, q^2) is a finite simple group, (Grove 2002, Thm. 11.22 and 11.26).

20.5.3   Degree-2 separable algebras

More generally, given a field k and a degree-2 separable k-algebra K (which may be a field extension but need not be), one can define unitary groups with respect to this extension.

First, there is a unique k-automorphism of K $a \mapsto \bar{a}$ which is an involution and fixes exactly k ($a = \bar{a}$ if and only if $a \in k$).[3] This generalizes complex conjugation and the conjugation of degree 2 finite field extensions, and allows one to define Hermitian forms and unitary groups as above.

20.5.4   Algebraic groups

The equations defining a unitary group are polynomial equations over k (but not over K): for the standard form $\Phi = I$ the equations are given in matrices as $A^*A = I$, where $A^* = \bar{A}^\mathrm{T}$ is the conjugate transpose. Given a different form, they are $A^*\Phi A = \Phi$. The unitary group is thus an algebraic group, whose points over a k-algebra R are given by:

$$U(n, K/k, \Phi)(R) := \{ A \in \mathrm{GL}(n, K \otimes_k R) : A^*\Phi A = \Phi \}.$$

For the field extension $\mathbf{C}/\mathbf{R}$ and the standard (positive definite) Hermitian form, these yield an algebraic group with real and complex points given by:

$$U(n, \mathbf{C}/\mathbf{R})(\mathbf{R}) = U(n)$$

$$U(n, \mathbf{C}/\mathbf{R})(\mathbf{C}) = \mathrm{GL}(n, \mathbf{C}).$$

In fact, the unitary group is a linear algebraic group.

Unitary group of a quadratic module

The unitary group of a quadratic module is a generalisation of the linear algebraic group U just defined, which incorporates as special cases many different classical algebraic groups. The definition goes back to Anthony Bak's thesis.[4]

To define it, one has to define quadratic modules first:

Let R be a ring with anti-automorphism J, $\varepsilon \in R^\times$ such that $r^{J^2} = \varepsilon r \varepsilon^{-1}$ for all r in R and $\varepsilon^J = \varepsilon^{-1}$. Define

$$\Lambda_{min} := \{r \in R \ : \ r - r^J \varepsilon\},$$

$$\Lambda_{max} := \{r \in R \ : \ r^J \varepsilon = -r\}.$$

Let $\Lambda \subsetneq R$ be an additive subgroup of R, then Λ is called *form parameter* if $\Lambda_{min} \subseteq \Lambda \subseteq \Lambda_{max}$ and $r^J \Lambda r \subseteq \Lambda$. A pair (R, Λ) such that R is a ring and Λ a form parameter is called *form ring*.

Let M be an R-module and f a J-sesquilinear form on M (i.e. $f(xr, ys) = r^J f(x, y)s$ for any $x, y \in M$ and $r, s \in R$). Define $h(x, y) := f(x, y) + f(y, x)^J \varepsilon \in R$ and $q(x) := f(x, x) \in R/\Lambda$, then f is said to *define* the Λ-*quadratic form* (h, q) on M. A *quadratic module* over (R, Λ) is a triple (M, h, q) such that M is an R-module and (h, q) is a Λ-quadratic form.

To any quadratic module *(M,h,q)* defined by a J-sesquilinear form f on M over a form ring (R, Λ) one can associate the *unitary group*

$$U(M) := \{\sigma \in GL(M) \ : \ \forall x, y \in M, h(\sigma x, \sigma y) = h(x, y) \text{ and } q(\sigma x) = q(x)\}.$$

The special case where $\Lambda = \Lambda_{\max}$, with J any non-trivial involution (i.e. $J \neq id_R$, $J^2 = id_R$ and $\varepsilon = -1$ gives back the "classical" unitary group (as an algebraic group).

20.6 Polynomial invariants

The unitary groups are the automorphisms of two polynomials in real non-commutative variables:

$$C_1 = (u^2 + v^2) + (w^2 + x^2) + (y^2 + z^2) + \dots$$

$$C_2 = (uv - vu) + (wx - xw) + (yz - zy) + \dots$$

These are easily seen to be the real and imaginary parts of the complex form $Z\overline{Z}$. The two invariants separately are invariants of O(2n) and Sp(2n, R). Combined they make the invariants of U(n) which is a subgroup of both these groups. The variables must be non-commutative in these invariants otherwise the second polynomial is identically zero.

20.7 Classifying space

The classifying space for U(n) is described in the article classifying space for U(n).

20.8 See also

- projective unitary group
- orthogonal group
- symplectic group

20.9   Notes

[1] Arnold, V.I. (1989). *Mathematical Methods of Classical Mechanics* (Second ed.). Springer. p. 225.

[2] Baez, John. "Symplectic, Quaternionic, Fermionic". Retrieved 1 February 2012.

[3] Milne, Algebraic Groups and Arithmetic Groups, p. 103

[4] Bak, Anthony - On modules with quadratic forms, pp. 55-66 in Lecture Notes in Mathematics, Vol. 108, Springer, Berlin Heidelberg New York, 1969

20.10   References

- Grove, Larry C. (2002), *Classical groups and geometric algebra*, Graduate Studies in Mathematics **39**, Providence, R.I.: American Mathematical Society, ISBN 978-0-8218-2019-3, MR 1859189

Chapter 21

Special unitary group

"SU(5)" redirects here. For the specific grand unification theory, see Georgi–Glashow model.

The **special unitary group** of degree n, denoted SU(n), is the group of $n \times n$ unitary matrices with determinant 1 (i.e., real-valued determinant, not complex as for general unitary matrices). The group operation is that of matrix multiplication. The special unitary group is a subgroup of the unitary group U(n), consisting of all $n \times n$ unitary matrices. As a compact classical group, U(n) is the group that preserves the standard inner product on $\mathbf{C}^n$.[nb 1] It is itself a subgroup of the general linear group, SU(n) $\subset$ U(n) $\subset$ GL(n, $\mathbf{C}$).

The SU(n) groups find wide application in the Standard Model of particle physics, especially SU(2) in the electroweak interaction and SU(3) in quantum chromodynamics.[1]

The simplest case, SU(1), is the trivial group, having only a single element. The group SU(2) is isomorphic to the group of quaternions of norm 1, and is thus diffeomorphic to the 3-sphere. Since unit quaternions can be used to represent rotations in 3-dimensional space (up to sign), there is a surjective homomorphism from SU(2) to the rotation group SO(3) whose kernel is $\{+I, -I\}$.[nb 2] SU(2) is also identical to one of the symmetry groups of spinors, Spin(3), that enables a spinor presentation of rotations.

21.1 Properties

The special unitary group SU(n) is a real Lie group (though not a complex Lie group). Its dimension as a real manifold is $n^2 - 1$. Topologically, it is compact and simply connected. Algebraically, it is a simple Lie group (meaning its Lie algebra is simple; see below). The center of SU(n) is isomorphic to the cyclic group Zn, and is composed of the diagonal matrices ζI for ζ an nth root of unity and I the $n \times n$ identity matrix. Its outer automorphism group, for $n \geq 3$, is Z$_2$, while the outer automorphism group of SU(2) is the trivial group.

A maximal torus, of rank $n - 1$, is given by the set of diagonal matrices with determinant 1. The Weyl group is the symmetric group Sn, which is represented by signed permutation matrices (the signs being necessary to ensure the determinant is 1).

The Lie algebra of SU(n), denoted by **su**(n), can be identified with the set of traceless antihermitian $n \times n$ complex matrices, with the regular commutator as Lie bracket. Particle physicists often use a different, equivalent representation: the set of traceless hermitian $n \times n$ complex matrices with Lie bracket given by $-i$ times the commutator.

21.2 Infinitesimal generators

The Lie algebra **su**(n) can be generated by n^2 operators $\hat{O}_{ij}$, $i, j = 1, 2, ..., n$, which satisfy the commutator relationships

$$\left[\hat{O}_{ij}, \hat{O}_{k\ell}\right] = \delta_{jk}\hat{O}_{i\ell} - \delta_{i\ell}\hat{O}_{kj}$$

for $i, j, k, \ell = 1, 2, ..., n$, where δ_{jk} denotes the Kronecker delta. Additionally, the operator

$$\hat{N} = \sum_{i=1}^{n} \hat{O}_{ii}$$

satisfies

$$\left[\hat{N}, \hat{O}_{ij}\right] = 0,$$

which implies that the number of *independent* generators of the Lie algebra is $n^2 - 1$.[2]

21.2.1   Fundamental representation

In the defining, or fundamental, representation of **su**(n) the generators Ta are represented by traceless hermitian matrices complex $n \times n$ matrices, where:

$$T_a T_b = \frac{1}{2n}\delta_{ab}I_n + \frac{1}{2}\sum_{c=1}^{n^2-1}(if_{abc} + d_{abc})T_c$$

where the f are the structure constants and are antisymmetric in all indices, while the d-coefficients are symmetric in all indices. As a consequence:

$$[T_a, T_b]_+ = \frac{1}{n}\delta_{ab}I_n + \sum_{c=1}^{n^2-1} d_{abc}T_c$$

$$[T_a, T_b]_- = i\sum_{c=1}^{n^2-1} f_{abc}T_c \,.$$

We also take

$$\sum_{c,e=1}^{n^2-1} d_{ace}d_{bce} = \frac{n^2 - 4}{n}\delta_{ab}$$

as a normalization convention.

21.2.2   Adjoint representation

In the $(n^2 - 1)$ -dimensional adjoint representation, the generators are represented by $(n^2 - 1) \times (n^2 - 1)$ matrices, whose elements are defined by the structure constants themselves:

$$(T_a)_{jk} = -if_{ajk}.$$

21.3   *n = 2*

See also: Versor

SU(2) is the following group:

$$\text{SU}(2) = \left\{ \begin{pmatrix} \alpha & -\overline{\beta} \\ \beta & \overline{\alpha} \end{pmatrix} : \ \alpha, \beta \in \mathbf{C}, |\alpha|^2 + |\beta|^2 = 1 \right\},$$

where the overline denotes complex conjugation. Now consider the following map:

$$\varphi : \mathbf{C}^2 \to M(2, \mathbf{C})$$
$$\varphi(\alpha, \beta) = \begin{pmatrix} \alpha & -\overline{\beta} \\ \beta & \overline{\alpha} \end{pmatrix},$$

where $M(2, \mathbf{C})$ denotes the set of 2 by 2 complex matrices. By considering $\mathbf{C}^2$ diffeomorphic to $\mathbf{R}^4$ and $M(2, \mathbf{C})$ diffeomorphic to $\mathbf{R}^8$ we can see that φ is an injective real linear map and hence an embedding. Now, considering the restriction of φ to the 3-sphere (since modulus is 1), denoted S^3, we can see that this is an embedding of the 3-sphere onto a compact submanifold of $M(2, \mathbf{C})$. However it is also clear that $\varphi(S^3) = \text{SU}(2)$. Therefore as a manifold S^3 is diffeomorphic to SU(2) and so SU(2) is a compact, connected Lie group.

The Lie algebra of SU(2) is:

$$\mathfrak{su}(2) = \left\{ \begin{pmatrix} ia & -\overline{z} \\ z & -ia \end{pmatrix} : \ a \in \mathbf{R}, z \in \mathbf{C} \right\}$$

It is easily verified that matrices of this form have trace zero and are antihermitian. The Lie algebra is then generated by the following matrices

$$u_1 = \begin{pmatrix} 0 & i \\ i & 0 \end{pmatrix} \qquad u_2 = \begin{pmatrix} 0 & -1 \\ 1 & 0 \end{pmatrix} \qquad u_3 = \begin{pmatrix} i & 0 \\ 0 & -i \end{pmatrix},$$

which are easily seen to have the form of the general element specified above. These satisfy $u_3 u_2 = -u_2 u_3 = -u_1$ and $u_2 u_1 = -u_1 u_2 = -u_3$. The commutator bracket is therefore specified by

$$[u_3, u_1] = 2u_2, \qquad [u_1, u_2] = 2u_3, \qquad [u_2, u_3] = 2u_1.$$

The above generators are related to the Pauli matrices by $u_1 = i\,\sigma_1, u_2 = -i\,\sigma_2$ and $u_3 = i\,\sigma_3$. This representation is often used in quantum mechanics to represent the spin of fundamental particles such as electrons. They also serve as unit vectors for the description of our 3 spatial dimensions in loop quantum gravity.

The Lie algebra is used to work out the representations of SU(2).

21.4   *n = 3*

The generators of **su**(3), T, in the defining representation, are:

$$T_a = \frac{\lambda_a}{2}.$$

where λ the Gell-Mann matrices, are the SU(3) analog of the Pauli matrices for SU(2):

These λ_a span all traceless Hermitian matrices H of the Lie algebra, as required.

They obey the relations

$$[T_a, T_b] = i \sum_{c=1}^{8} f_{abc} T_c$$

$$\{T_a, T_b\} = \frac{1}{3}\delta_{ab} + \sum_{c=1}^{8} d_{abc} T_c$$

$$\{\lambda_a, \lambda_b\} = \frac{4}{3}\delta_{ab} + 2\sum_{c=1}^{8} d_{abc}\lambda_c$$

The f are the structure constants of the Lie algebra, given by:

$$f_{123} = 1$$

$$f_{147} = -f_{156} = f_{246} = f_{257} = f_{345} = -f_{367} = \frac{1}{2}$$

$$f_{458} = f_{678} = \frac{\sqrt{3}}{2},$$

while all other f_{abc} not related to these by permutation are zero.

The symmetric coefficients d take the values:

$$d_{118} = d_{228} = d_{338} = -d_{888} = \frac{1}{\sqrt{3}}$$

$$d_{448} = d_{558} = d_{668} = d_{778} = -\frac{1}{2\sqrt{3}}$$

$$d_{146} = d_{157} = -d_{247} = d_{256} = d_{344} = d_{355} = -d_{366} = -d_{377} = \frac{1}{2}.$$

As a topological space, *SU(3)* is a direct product of a 3-sphere and a 5-sphere, $S^3 \boxtimes S^5$.

A generic *SU(3)* group element generated by a traceless 3×3 hermitian matrix H, normalized as tr$(H^2) = 2$, is given by[3]

$$\exp(i\theta H) =$$

$$\left[-\frac{1}{3} I \sin(\phi + 2\pi/3)\sin(\phi - 2\pi/3) - \frac{1}{2\sqrt{3}} H \sin(\phi) - \frac{1}{4} H^2\right] \frac{\exp\left(\frac{2}{\sqrt{3}} i\theta \sin\phi\right)}{\cos(\phi + 2\pi/3)\cos(\phi - 2\pi/3)}$$

$$+ \left[-\frac{1}{3} I \sin(\phi)\sin(\phi - 2\pi/3) - \frac{1}{2\sqrt{3}} H \sin(\phi + 2\pi/3) - \frac{1}{4} H^2\right] \frac{\exp\left(\frac{2}{\sqrt{3}} i\theta \sin(\phi + 2\pi/3)\right)}{\cos(\phi)\cos(\phi - 2\pi/3)}$$

$$+ \left[-\frac{1}{3} I \sin(\phi)\sin(\phi + 2\pi/3) - \frac{1}{2\sqrt{3}} H \sin(\phi - 2\pi/3) - \frac{1}{4} H^2\right] \frac{\exp\left(\frac{2}{\sqrt{3}} i\theta \sin(\phi - 2\pi/3)\right)}{\cos(\phi)\cos(\phi + 2\pi/3)}$$

where

$$\phi \equiv \frac{1}{3}\left(\arccos\left(\frac{3}{2}\sqrt{3}\det H\right) - \frac{\pi}{2}\right)$$

21.5 Lie algebra structure

The above representation bases generalize to $n > 3$, using generalized Pauli matrices.

If we choose an (arbitrary) particular basis, then the subspace of traceless diagonal $n \times n$ matrices with imaginary entries forms an $(n-1)$-dimensional Cartan subalgebra.

Complexify the Lie algebra, so that any traceless $n \times n$ matrix is now allowed. The weight eigenvectors are the Cartan subalgebra itself, as well as the matrices with only one nonzero entry which is off diagonal. Even though the Cartan subalgebra **h** is only $(n-1)$-dimensional, to simplify calculations, it is often convenient to introduce an auxiliary element, the unit matrix which commutes with everything else (which is not an element of the Lie algebra!) for the purpose of computing weights—and that only. So, we have a basis where the i-th basis vector is the matrix with 1 on the i-th diagonal entry and zero elsewhere. Weights would then be given by n coordinates and the sum over all n coordinates has to be zero (because the unit matrix is only auxiliary).

So, SU(n) is of rank $n-1$ and its Dynkin diagram is given by An_{-1}, a chain of $n-1$ vertices, o–o–o–o---o. Its root system consists of $n(n-1)$ roots spanning a $n-1$ Euclidean space. Here, we use n redundant coordinates instead of $n-1$ to emphasize the symmetries of the root system (the n coordinates have to add up to zero).

In other words, we are embedding this $n-1$ dimensional vector space in an n-dimensional one. Thus, the roots consists of all the $n(n-1)$ permutations of $(1, -1, 0, ..., 0)$. The construction given above explains why. A choice of simple roots is

$$(1, -1, 0, \ldots, 0),$$
$$(0, 1, -1, \ldots, 0),$$
$$\ldots$$
$$(0, 0, 0, \ldots, 1, -1).$$

Its Cartan matrix is

$$\begin{pmatrix} 2 & -1 & 0 & \ldots & 0 \\ -1 & 2 & -1 & \ldots & 0 \\ 0 & -1 & 2 & \ldots & 0 \\ \vdots & \vdots & \vdots & \ddots & \vdots \\ 0 & 0 & 0 & \ldots & 2 \end{pmatrix}.$$

Its Weyl group or Coxeter group is the symmetric group Sn, the symmetry group of the $(n-1)$-simplex.

21.6 Generalized special unitary group

For a field F, the **generalized special unitary group over F**, SU(p, q; F), is the group of all linear transformations of determinant 1 of a vector space of rank $n = p + q$ over F which leave invariant a nondegenerate, Hermitian form of signature (p, q). This group is often referred to as the **special unitary group of signature p q over F**. The field F can be replaced by a commutative ring, in which case the vector space is replaced by a free module.

Specifically, fix a Hermitian matrix A of signature p q in GL(n, **R**), then all

$$M \in \text{SU}(p, q, R)$$

satisfy

$$M^* A M = A$$
$$\det M = 1.$$

Often one will see the notation SU(p, q) without reference to a ring or field; in this case, the ring or field being referred to is **C** and this gives one of the classical Lie groups. The standard choice for A when $F = \mathbf{C}$ is

$$A = \begin{bmatrix} 0 & 0 & i \\ 0 & I_{n-2} & 0 \\ -i & 0 & 0 \end{bmatrix}.$$

However there may be better choices for A for certain dimensions which exhibit more behaviour under restriction to subrings of **C**.

21.6.1 Example

An important example of this type of group is the Picard modular group SU(2, 1; **Z**[i]) which acts (projectively) on complex hyperbolic space of degree two, in the same way that SL(2,9;**Z**) acts (projectively) on real hyperbolic space of dimension two. In 2005 Gábor Francsics and Peter Lax computed an explicit fundamental domain for the action of this group on HC^2.[4]

A further example is SU(1, 1; **C**), which is isomorphic to SL(2,**R**).

21.7 Important subgroups

In physics the special unitary group is used to represent bosonic symmetries. In theories of symmetry breaking it is important to be able to find the subgroups of the special unitary group. Subgroups of SU(n) that are important in GUT physics are, for $p > 1$, $n - p > 1$:

SU(n) $\supset$ SU(p) $\times$ SU($n - p$) $\times$ U(1)

where $\times$ denotes the direct product and U(1), known as the circle group, is the multiplicative group of all complex numbers with absolute value 1.

For completeness there are also the orthogonal and symplectic subgroups:

SU(n) $\supset$ SO(n),
SU($2n$) $\supset$ Sp(n).

Since the rank of SU(n) is $n - 1$ and of U(1) is 1, a useful check is that the sum of the ranks of the subgroups is less than or equal to the rank of the original group. SU(n) is a subgroup of various other Lie groups:

SO($2n$) $\supset$ SU(n)
Sp(n) $\supset$ SU(n)
Spin(4) = SU(2) $\times$ SU(2)
E$_6$ $\supset$ SU(6)
E$_7$ $\supset$ SU(8)
G$_2$ $\supset$ SU(3)

See spin group, and simple Lie groups for E$_6$, E$_7$, and G$_2$.

There are also the identities SU(4) = Spin(6) , SU(2) = Spin(3) = Sp(1) ,[nb 3] and U(1) = Spin(2) = SO(2) .

One should finally mention that SU(2) is the double covering group of SO(3), a relation that plays an important role in the theory of rotations of 2-spinors in non-relativistic quantum mechanics.

21.8 See also

- Projective special unitary group, PSU(n)

- Generalizations of Pauli matrices

21.9 Remarks

[1] For a characterization of U(n) and hence SU(n) in terms of preservation of the standard inner product on $\mathbb{C}^n$, see Classical group.

[2] For an explicit description of the homomorphism SU(2) $\to$ SO(3), see Connection between SO(3) and SU(2).

[3] Sp(n) is the compact real form of Sp($2n$, $\mathbf{C}$). It is sometimes denoted USp($2n$). The dimension of the Sp(n)-matrices is $2n \times 2n$.

21.10 References

[1] Halzen, Francis; Martin, Alan (1984). *Quarks & Leptons: An Introductory Course in Modern Particle Physics*. John Wiley & Sons. ISBN 0-471-88741-2.

[2] R.R. Puri, *Mathematical Methods of Quantum Optics*, Springer, 2001.

[3] Rosen, S P (1971). "Finite Transformations in Various Representations of SU(3)". *Journal of Mathematical Physics* **12** (4): 673. doi:10.1063/1.1665634. ISSN 0022-2488.; Curtright, T L; Zachos, C K (2015). "Elementary results for the fundamental representation of SU(3)". *Researchgate*. doi:10.13140/RG.2.1.1743.2163.

[4] Francsics, Gabor; Lax, Peter D. "An Explicit Fundamental Domain For The Picard Modular Group In Two Complex Dimensions". arXiv:math/0509708v1.

Chapter 22

Lanczos tensor

The **Lanczos tensor** or **Lanczos potential** is a rank 3 tensor in general relativity that generates the Weyl tensor.[1] It was first introduced by Cornelius Lanczos in 1949.[2] The theoretical importance of the Lanczos tensor is that it serves as the gauge field for the gravitational field in the same way that, by analogy, the electromagnetic four-potential generates the electromagnetic field.[3][4]

22.1   Definition

The Lanczos tensor can be defined in a few different ways. The most common modern definition is through the Weyl–Lanczos equations, which demonstrate the generation of the Weyl tensor from the Lanczos tensor.[4] These equations, presented below, were given by Takeno in 1964.[1] The way that Lanczos introduced the tensor originally was as a Lagrange multiplier[2][5] on constraint terms studied in the variational approach to general relativity.[6] Under any definition, the Lanczos tensor exhibits the following symmetries:

$$H_{abc} + H_{bac} = 0,$$

$$H_{abc} + H_{bca} + H_{cab} = 0.$$

The Lanczos tensor always exists in four dimensions[7] but does not generalize to higher dimensions.[8] This highlights the specialness of four dimensions.[3] Note further that the full Riemann tensor cannot in general be derived from derivatives of the Lanczos potential alone.[7][9] The Einstein field equations must provide the Ricci tensor to complete the components of the Ricci decomposition.

22.1.1   Weyl–Lanczos equations

The Weyl–Lanczos equations express the Weyl tensor entirely as derivatives of the Lanczos tensor:[10]

$$C_{abcd} = H_{abc;d} + H_{cda;b} + H_{bad;c} + H_{dcb;a} + (H^e{}_{(ac);e} + H_{(a|e|}{}^e{}_{;c)})g_{bd} + (H^e{}_{(bd);e} + H_{(b|e|}{}^e{}_{;d)})g_{ac} - (H^e{}_{(ad);e} + H_{(a|e|}{}^e{}_{;d)})g_{bc} - (H^e{}_{(bc);e}$$

where C_{abcd} is the Weyl tensor, the semicolon denotes the covariant derivative, and the subscripted parentheses indicate symmetrization. Although the above equations can be used to define the Lanczos tensor, they also show that it is not unique but rather has gauge freedom under an affine group.[11] If Φ^a is an arbitrary vector field, then the Weyl–Lanczos equations are invariant under the gauge transformation

$$H'_{abc} = H_{abc} + \Phi_{[a}g_{b]c}$$

where the subscripted brackets indicate antisymmetrization. An often convenient choice is the Lanczos algebraic gauge, $\Phi_a = -\frac{2}{3}H_{ab}{}^b$, which sets $H'_{ab}{}^b = 0$. The gauge can be further restricted through the Lanczos differential gauge $H_{ab}{}^c{}_{;c} = 0$. These gauge choices reduce the Weyl–Lanczos equations to the simpler form

$$C_{abcd} = H_{abc;d} + H_{cda;b} + H_{bad;c} + H_{dcb;a} + H^e{}_{ac;e}g_{bd} + H^e{}_{bd;e}g_{ac} - H^e{}_{ad;e}g_{bc} - H^e{}_{bc;e}g_{ad}.$$

22.2 Wave equation

The Lanczos potential tensor satisfies a wave equation[12]

$$\begin{aligned}\Box H_{abc} =&J_{abc}\\ &- 2R_c{}^d H_{abd} + R_a{}^d H_{bcd} + R_b{}^d H_{acd}\\ &+ (H_{dbe}g_{ac} - H_{dae}g_{bc})R^{de} + \frac{1}{2}RH_{abc},\end{aligned}$$

where $\Box$ is the d'Alembert operator and

$$J_{abc} = R_{ca;b} - R_{cb;a} - \frac{1}{6}(g_{ca}R_{;b} - g_{cb}R_{;a})$$

is known as the Cotton tensor. Since the Cotton tensor depends only on covariant derivatives of the Ricci tensor, it can perhaps be interpreted as a kind of matter current.[13] The additional self-coupling terms have no direct electromagnetic equivalent. These self-coupling terms, however, do not affect the vacuum solutions, where the Ricci tensor vanishes and the curvature is described entirely by the Weyl tensor. Thus in vacuum, the Einstein field equations are equivalent to the homogeneous wave equation $\Box H_{abc} = 0$, in perfect analogy to the vacuum wave equation $\Box A_a = 0$ of the electromagnetic four-potential. This shows a formal similarity between gravitational waves and electromagnetic waves, with the Lanczos tensor well-suited for studying gravitational waves.[14]

In the weak field approximation where $g_{ab} = \eta_{ab} + h_{ab}$, a convenient form for the Lanczos tensor in the Lanczos gauge is[13]

$$4H_{abc} \approx h_{ac,b} - h_{bc,a} - \frac{1}{6}(\eta_{ac}h^d{}_{d,b} - \eta_{bc}h^d{}_{d,a}).$$

22.3 Example

The most basic nontrivial case for expressing the Lanczos tensor is, of course, for the Schwarzschild metric.[4] The simplest, explicit component representation in natural units for the Lanczos tensor in this case is

$$H_{trt} = \frac{GM}{r^2}$$

with all other components vanishing up to symmetries. This form, however, is not in the Lanczos gauge. The nonvanishing terms of the Lanczos tensor in the Lanczos gauge are

$$H_{trt} = \frac{2GM}{3r^2}$$

$$H_{r\theta\theta} = \frac{-GM}{3(1 - 2GM/r)}$$

$$H_{r\phi\phi} = \frac{-GM \sin^2 \theta}{3(1 - 2GM/r)}$$

It is further possible to show, even in this simple case, that the Lanczos tensor cannot in general be reduced to a linear combination of the spin coefficients of the Newman–Penrose formalism, which attests to the Lanczos tensor's fundamental nature.[10] Similar calculations have been used to construct arbitrary Petrov type D solutions.[15]

22.4   See also

- Bach tensor

- Ricci calculus

- Schouten tensor

22.5   References

[1] Hyôitirô Takeno, "On the spintensor of Lanczos", *Tensor*, **15** (1964) pp. 103–119.

[2] Cornelius Lanczos, "Lagrangian Multiplier and Riemannian Spaces", *Rev. Mod. Phys.*, **21** (1949) pp. 497–502. doi:10.1103/RevModPhys.21.497

[3] P. O'Donnell and H. Pye, "A Brief Historical Review of the Important Developments in Lanczos Potential Theory", *EJTP*, **7** (2010) pp. 327–350. www.ejtp.com/articles/ejtpv7i24p327.pdf

[4] M. Novello and A. L. Velloso, "The Connection Between General Observers and Lanczos Potential", *General Relativity and Gravitation*, **19** (1987) pp. 1251-1265. doi:10.1007/BF00759104

[5] Cornelius Lanczos, "The Splitting of the Riemann Tensor", *Rev. Mod. Phys.*, **34** (1962) pp. 379–389. doi:10.1103/RevModPhys.34.379

[6] Cornelius Lanczos, "A Remarkable Property of the Riemann–Christoffel Tensor in Four Dimensions", *Annals of Mathematics*, **39** (1938) pp. 842-850. www.jstor.org/stable/1968467

[7] F. Bampi and G. Caviglia, "Third-order tensor potentials for the Riemann and Weyl tensors", *General Relativity and Gravitation*, **15** (1983) pp. 375-386. doi:10.1007/BF00759166

[8] S. B. Edgar, "Nonexistence of the Lanczos potential for the Riemann tensor in higher dimensions", *General Relativity and Gravitation*, **26** (1994) pp. 329-332. doi:10.1007/BF02108015

[9] E. Massa and E. Pagani, "Is the Riemann tensor derivable from a tensor potential?", *General Relativity and Gravitation*, **16** (1984) pp. 805-816. doi:10.1007/BF00762934

[10] P. O'Donnell, "A Solution of the Weyl–Lanczos Equations for the Schwarzschild Space-Time", *General Relativity and Gravitation*, **36** (2004) pp. 1415-1422. doi:10.1023/B:GERG.0000022577.11259.e0

[11] K. S. Hammon and L. K. Norris "The Affine Geometry of the Lanczos H-tensor Formalism", *General Relativity and Gravitation*, **25** (1993) pp. 55-80. doi:10.1007/BF00756929

[12] P. Dolan and C. W. Kim "The wave equation for the Lanczos potential", *Proc. R. Soc. Lond. A*, **447** (1994) pp. 557-575. doi:10.1098/rspa.1994.0155

[13] Mark D. Roberts, "The Physical Interpretation of the Lanczos Tensor." *Nuovo Cim. B* **110** (1996) 1165-1176. doi:10.1007/BF02724607 arXiv:gr-qc/9904006

[14] J. L. López-Bonilla, G. Ovando and J. J. Peña, "A Lanczos Potential for Plane Gravitational Waves." *Foundations of Physics Letters* **12** (1999) 401-405. doi:10.1023/A:1021656622094

[15] Zafar Ahsan and Mohd Bilal, "A Solution of Weyl-Lanczos Equations for Arbitrary Petrov Type D Vacuum Spacetimes." *Int J Theor Phys* **49** (2010) 2713-2722. doi:10.1007/s10773-010-0464-5

22.6 External links

- Peter O'Donnell, *Introduction To 2-Spinors In General Relativity*. World Scientific, 2003.

Chapter 23

General covariance

In theoretical physics, **general covariance** (also known as **diffeomorphism covariance** or **general invariance**) is the invariance of the *form* of physical laws under arbitrary differentiable coordinate transformations. The essential idea is that coordinates do not exist *a priori* in nature, but are only artifices used in describing nature, and hence should play no role in the formulation of fundamental physical laws.

A physical law expressed in a generally covariant fashion takes the same mathematical form in all coordinate systems,[1] and is usually expressed in terms of tensor fields. The classical (non-quantum) theory of electrodynamics is one theory that has such a formulation.

Albert Einstein proposed this principle for his special theory of relativity; however, that theory was limited to space-time coordinate systems related to each other by uniform relative motions only , the so-called "inertial frames." Einstein recognized that the general principle of relativity should also apply to accelerated relative motions, and he used the newly developed tool of tensor calculus to extend the special theory's global Lorentz covariance (applying only to inertial frames) to the more general local Lorentz covariance (which applies to all frames), eventually producing his general theory of relativity. The local reduction of the general metric tensor to the Minkowski metric corresponds to free-falling (geodesic) motion, in this theory, thus encompassing the phenomenon of gravitation.

Much of the work on classical unified field theories consisted of attempts to further extend the general theory of relativity to interpret additional physical phenomena, particularly electromagnetism, within the framework of general covariance, and more specifically as purely geometric objects in the space-time continuum.

23.1 Remarks

The relationship between general covariance and general relativity may be summarized by quoting a standard textbook:[2]

> Mathematics was not sufficiently refined in 1917 to cleave apart the demands for "no prior geometry" and for a geometric, coordinate-independent formulation of physics. Einstein described both demands by a single phrase, "general covariance." The "no prior geometry" demand actually fathered general relativity, but by doing so anonymously, disguised as "general covariance", it also fathered half a century of confusion.

A more modern interpretation of the physical content of the original principle of general covariance is that the Lie group GL4(**R**) is a fundamental "external" symmetry of the world. Other symmetries, including "internal" symmetries based on compact groups, now play a major role in fundamental physical theories.

23.2 See also

- Coordinate conditions

- Coordinate-free
- Covariance and contravariance
- Covariant derivative
- Diffeomorphism
- Fictitious force
- Galilean invariance
- Gauge covariant derivative
- General covariant transformations
- Harmonic coordinate condition
- Inertial frame of reference
- Lorentz covariance
- Principle of covariance
- Special relativity
- Symmetry in physics

23.3 Notes

[1] More precisely, only coordinate systems related through sufficiently differentiable transformations are considered.

[2] Charles W. Misner, Kip S. Thorne, and John Archibald Wheeler (1973). *Gravitation*. Freeman. p. 431. ISBN 0-7167-0344-0.

23.4 References

- O'Hanian, Hans C.; & Ruffini, Remo (1994). *Gravitation and Spacetime* (2nd edition ed.). New York: W. W. Norton. ISBN 0-393-96501-5. See *section 7.1*.

23.5 External links

- General covariance and the foundations of general relativity: eight decades of dispute, by J. D. Norton (file size: 4 MB)
  re-typeset version (file size: 460 KB)

Chapter 24

Graviton

This article is about the hypothetical particle. For other uses, see Graviton (disambiguation).

In physics, the **graviton** is a hypothetical elementary particle that mediates the force of gravitation in the framework of quantum field theory. If it exists, the graviton is expected to be massless (because the gravitational force appears to have unlimited range) and must be a spin-2 boson. The spin follows from the fact that the source of gravitation is the stress–energy tensor, a second-rank tensor (compared to electromagnetism's spin-1 photon, the source of which is the four-current, a first-rank tensor). Additionally, it can be shown that any massless spin-2 field would give rise to a force indistinguishable from gravitation, because a massless spin-2 field must couple to (interact with) the stress–energy tensor in the same way that the gravitational field does. Seeing as the graviton is hypothetical, its discovery would unite quantum theory with gravity.[4] This result suggests that, if a massless spin-2 particle is discovered, it must be the graviton, so that the only experimental verification needed for the graviton may simply be the discovery of a massless spin-2 particle.[5]

24.1 Theory

The four other known forces of nature are mediated by elementary particles: electromagnetism by the photon, the strong interaction by the gluons, the Higgs field by the Higgs Boson, and the weak interaction by the W and Z bosons. The hypothesis is that the gravitational interaction is likewise mediated by an – as yet undiscovered – elementary particle, dubbed as *the graviton*. In the classical limit, the theory would reduce to general relativity and conform to Newton's law of gravitation in the weak-field limit.[6][7][8]

24.1.1 Gravitons and renormalization

When describing graviton interactions, the classical theory (i.e., the tree diagrams) and semiclassical corrections (one-loop diagrams) behave normally, but Feynman diagrams with two (or more) loops lead to ultraviolet divergences; that is, infinite results that cannot be removed because the quantized general relativity is not renormalizable, unlike quantum electrodynamics. That is, the usual ways physicists calculate the probability that a particle will emit or absorb a graviton give nonsensical answers and the theory loses its predictive power. These problems, together with some conceptual puzzles, led many physicists to believe that a theory more complete than quantized general relativity must describe the behavior near the Planck scale.

24.1.2 Comparison with other forces

Unlike the force carriers of the other forces, gravitation plays a special role in general relativity in defining the spacetime in which events take place. In some descriptions, matter modifies the 'shape' of spacetime itself, and gravity is a result of this shape, an idea which at first glance may appear hard to match with the idea of a force acting between particles.[9]

Because the diffeomorphism invariance of the theory does not allow any particular space-time background to be singled out as the "true" space-time background, general relativity is said to be background independent. In contrast, the Standard Model is *not* background independent, with Minkowski space enjoying a special status as the fixed background space-time.[10] A theory of quantum gravity is needed in order to reconcile these differences.[11] Whether this theory should be background independent is an open question. The answer to this question will determine our understanding of what specific role gravitation plays in the fate of the universe.[12]

24.1.3 Gravitons in speculative theories

String theory predicts the existence of gravitons and their well-defined interactions. A graviton in perturbative string theory is a closed string in a very particular low-energy vibrational state. The scattering of gravitons in string theory can also be computed from the correlation functions in conformal field theory, as dictated by the AdS/CFT correspondence, or from matrix theory.

A feature of gravitons in string theory is that, as closed strings without endpoints, they would not be bound to branes and could move freely between them. If we live on a brane (as hypothesized by brane theories) this "leakage" of gravitons from the brane into higher-dimensional space could explain why gravitation is such a weak force, and gravitons from other branes adjacent to our own could provide a potential explanation for dark matter. However if gravitons were to move completely freely between branes this would dilute gravity too much, causing a violation of Newton's inverse square law. To combat this, Lisa Randall found that a three-brane (such as ours) would have a gravitational pull of its own, preventing gravitons from drifting freely, possibly resulting in the diluted gravity we observe while roughly maintaining Newton's inverse square law.[13] See brane cosmology.

A theory by Ahmed Farag Ali and Saurya Das adds quantum mechanical corrections (using Bohm trajectories) to general relativistic geodesics. If gravitons are given a small but non-zero mass, it could explain the cosmological constant without need for dark energy and solve the smallness problem.[14]

24.2 Experimental observation

Unambiguous detection of individual gravitons, though not prohibited by any fundamental law, is impossible with any physically reasonable detector.[15] The reason is the extremely low cross section for the interaction of gravitons with matter. For example, a detector with the mass of Jupiter and 100% efficiency, placed in close orbit around a neutron star, would only be expected to observe one graviton every 10 years, even under the most favorable conditions. It would be impossible to discriminate these events from the background of neutrinos, since the dimensions of the required neutrino shield would ensure collapse into a black hole.[15]

However, experiments to detect gravitational waves, which may be viewed as coherent states of many gravitons, are underway (e.g., LIGO and VIRGO). Although these experiments cannot detect individual gravitons, they might provide information about certain properties of the graviton.[16] For example, if gravitational waves were observed to propagate slower than c (the speed of light in a vacuum), that would imply that the graviton has mass (however, gravitational waves must propagate slower than "c" in a region with non-zero mass density if they are to be detectable).[17] Astronomical observations of the kinematics of galaxies, especially the galaxy rotation problem and modified Newtonian dynamics, might point toward gravitons having non-zero mass.[18]

24.3 Difficulties and outstanding issues

Most theories containing gravitons suffer from severe problems. Attempts to extend the Standard Model or other quantum field theories by adding gravitons run into serious theoretical difficulties at high energies (processes involving energies close to or above the Planck scale) because of infinities arising due to quantum effects (in technical terms, gravitation is nonrenormalizable). Since classical general relativity and quantum mechanics seem to be incompatible at such energies, from a theoretical point of view, this situation is not tenable. One possible solution is to replace particles with strings.

String theories are quantum theories of gravity in the sense that they reduce to classical general relativity plus field theory at low energies, but are fully quantum mechanical, contain a graviton, and are believed to be mathematically consistent.[19]

24.4   See also

- Gravitomagnetism
- Gravitational wave
- Planck mass
- Gravitation
- Static forces and virtual-particle exchange
- Multiverse
- Gravitino

24.5   References

[1] G is used to avoid confusion with gluons (symbol g)

[2] Rovelli, C. (2001). "Notes for a brief history of quantum gravity". arXiv:gr-qc/0006061 [gr-qc].

[3] Blokhintsev, D. I.; Gal'perin, F. M. (1934). "Gipoteza neitrino i zakon sokhraneniya energii" [Neutrino hypothesis and conservation of energy]. *Pod Znamenem Marxisma* (in Russian) **6**: 147–157.

[4] Lightman, A. P.; Press, W. H.; Price, R. H.; Teukolsky, S. A. (1975). "Problem 12.16". *Problem book in Relativity and Gravitation*. Princeton University Press. ISBN 0-691-08162-X.

[5] For a comparison of the geometric derivation and the (non-geometric) spin-2 field derivation of general relativity, refer to box 18.1 (and also 17.2.5) of Misner, C. W.; Thorne, K. S.; Wheeler, J. A. (1973). *Gravitation*. W. H. Freeman. ISBN 0-7167-0344-0.

[6] Feynman, R. P.; Morinigo, F. B.; Wagner, W. G.; Hatfield, B. (1995). *Feynman Lectures on Gravitation*. Addison-Wesley. ISBN 0-201-62734-5.

[7] Zee, A. (2003). *Quantum Field Theory in a Nutshell*. Princeton University Press. ISBN 0-691-01019-6.

[8] Randall, L. (2005). *Warped Passages: Unraveling the Universe's Hidden Dimensions*. Ecco Press. ISBN 0-06-053108-8.

[9] See the other articles on General relativity, Gravitational field, Gravitational wave, etc

[10] Colosi, D. et al. (2005). "Background independence in a nutshell: The dynamics of a tetrahedron". *Classical and Quantum Gravity* **22** (14): 2971. arXiv:gr-qc/0408079. Bibcode:2005CQGra..22.2971C. doi:10.1088/0264-9381/22/14/008.

[11] Witten, E. (1993). "Quantum Background Independence In String Theory". arXiv:hep-th/9306122 [hep-th].

[12] Smolin, L. (2005). "The case for background independence". arXiv:hep-th/0507235 [hep-th].

[13] Kaku, Michio (2006). *Parallel Worlds - The science of alternative universes and our future in the Cosmos*. pp. 218–221.

[14] Ali, Ahmed Farang (2014). "Cosmology from quantum potential". *Physical Letters B* **741**: 276–279. arXiv:1404.3093v3. doi:10.1016/j.physletb.2014.12.057.

[15] Rothman, T.; Boughn, S. (2006). "Can Gravitons be Detected?". *Foundations of Physics* **36** (12): 1801–1825. arXiv:gr-qc/0601043. Bibcode:2006FoPh...36.1801R. doi:10.1007/s10701-006-9081-9.

[16] Freeman Dyson (8 October 2013). "Is a graviton detectable?". *International Journal of Modern Physics A* **28** (25): 1330041-1–1330035–14. Bibcode:2013IJMPA..2830041D. doi:10.1142/S0217751X1330041X.

[17] Will, C. M. (1998). "Bounding the mass of the graviton using gravitational-wave observations of inspiralling compact binaries". *Physical Review D* **57** (4): 2061–2068. arXiv:gr-qc/9709011. Bibcode:1998PhRvD..57.2061W. doi:10.1103/PhysRevD.57.2061.

[18] Trippe, S. (2013), "A Simplified Treatment of Gravitational Interaction on Galactic Scales", J. Kor. Astron. Soc. **46**, 41. arXiv:1211.4692

[19] Sokal, A. (July 22, 1996). "Don't Pull the String Yet on Superstring Theory". *The New York Times.* Retrieved March 26, 2010.

24.6   External links

-

- Graviton on *In Our Time* at the BBC. (listen now)

Chapter 25

Diffeomorphism

In mathematics, a **diffeomorphism** is an isomorphism of smooth manifolds. It is an invertible function that maps one differentiable manifold to another such that both the function and its inverse are smooth.

25.1 Definition

Given two manifolds M and N, a differentiable map $f : M \to N$ is called a **diffeomorphism** if it is a bijection and its inverse $f^{-1} : N \to M$ is differentiable as well. If these functions are r times continuously differentiable, f is called a C^r-**diffeomorphism**.

Two manifolds M and N are **diffeomorphic** (symbol usually being $\simeq$) if there is a diffeomorphism f from M to N. They are C^r **diffeomorphic** if there is an r times continuously differentiable bijective map between them whose inverse is also r times continuously differentiable.

25.2 Diffeomorphisms of subsets of manifolds

Given a subset X of a manifold M and a subset Y of a manifold N, a function $f : X \to Y$ is said to be smooth if for all p in X there is a neighborhood $U \subset M$ of p and a smooth function $g : U \to N$ such that the restrictions agree $g_{|U \cap X} = f_{|U \cap X}$ (note that g is an extension of f). f is said to be a diffeomorphism if it is bijective, smooth and its inverse is smooth.

25.3 Local description

Model example

If U, V are connected open subsets of $\mathbf{R}^n$ such that V is simply connected, a differentiable map $f : U \to V$ is a **diffeomorphism** if it is proper and if the differential $Df_x : \mathbf{R}^n \to \mathbf{R}^n$ is bijective at each point x in U.

First remark

It is essential for V to be simply connected for the function f to be globally invertible (under the sole condition that its derivative is a bijective map at each point). For example, consider the "realification" of the complex square function

$$\begin{cases} f : \mathbf{R}^2 \setminus \{(0,0)\} \to \mathbf{R}^2 \setminus \{(0,0)\} \\ (x,y) \mapsto (x^2 - y^2, 2xy) \end{cases}$$

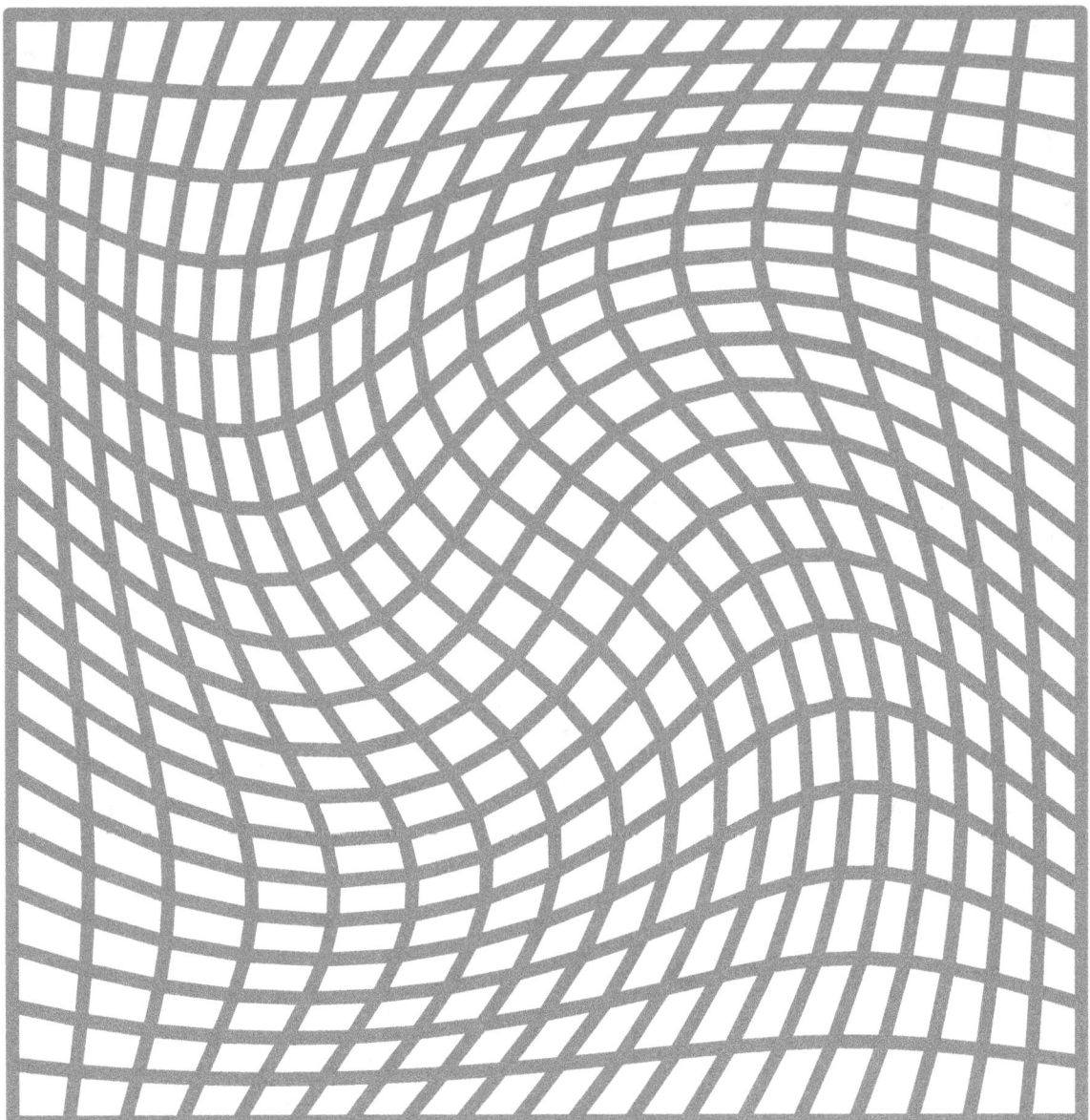

The image of a rectangular grid on a square under a diffeomorphism from the square onto itself.

Then f is surjective and it satisfies

$$\det Df_x = 4(x^2 + y^2) \neq 0$$

Thus, though *Dfx* is bijective at each point, f is not invertible because it fails to be injective (e.g. $f(1,0) = (1,0) = f(-1,0)$).

Second remark
Since the differential at a point (for a differentiable function)

$$Df_x : T_x U \to T_{f(x)} V$$

is a linear map, it has a well-defined inverse if and only if Dfx is a bijection. The matrix representation of Dfx is the $n \times n$ matrix of first-order partial derivatives whose entry in the i-th row and j-th column is $\partial f_i / \partial x_j$. This so-called Jacobian matrix is often used for explicit computations.

Third remark

Diffeomorphisms are necessarily between manifolds of the same dimension. Imagine f going from dimension n to dimension k. If $n < k$ then Dfx could never be surjective; and if $n > k$ then Dfx could never be injective. In both cases, therefore, Dfx fails to be a bijection.

Fourth remark

If Dfx is a bijection at x then f is said to be a local diffeomorphism (since, by continuity, Dfy will also be bijective for all y sufficiently close to x).

Fifth remark

Given a smooth map from dimension n to dimension k, if Df (or, locally, Dfx) is surjective, f is said to be a submersion (or, locally, a "local submersion"); and if Df (or, locally, Dfx) is injective, f is said to be an immersion (or, locally, a "local immersion").

Sixth remark

A differentiable bijection is *not* necessarily a diffeomorphism. $f(x) = x^3$, for example, is not a diffeomorphism from **R** to itself because its derivative vanishes at 0 (and hence its inverse is not differentiable at 0). This is an example of a homeomorphism that is not a diffeomorphism.

Seventh remark

When f is a map between *differentiable* manifolds, a diffeomorphic f is a stronger condition than a homeomorphic f. For a diffeomorphism, f and its inverse need to be differentiable; for a homeomorphism, f and its inverse need only be continuous. Every diffeomorphism is a homeomorphism, but not every homeomorphism is a diffeomorphism.

$f : M \to N$ is called a **diffeomorphism** if, in coordinate charts, it satisfies the definition above. More precisely: Pick any cover of M by compatible coordinate charts and do the same for N. Let φ and ψ be charts on, respectively, M and N, with U and V as, respectively, the images of φ and ψ. The map $\psi f \varphi^{-1} : U \to V$ is then a diffeomorphism as in the definition above, whenever $f(\varphi^{-1}(U)) \subset \psi^{-1}(V)$.

25.4 Examples

Since any manifold can be locally parametrised, we can consider some explicit maps from $\mathbf{R}^2$ into $\mathbf{R}^2$.

- Let

$$f(x, y) = \left(x^2 + y^3, x^2 - y^3 \right).$$

We can calculate the Jacobian matrix:

$$J_f = \begin{pmatrix} 2x & 3y^2 \\ 2x & -3y^2 \end{pmatrix}.$$

The Jacobian matrix has zero determinant if, and only if $xy = 0$. We see that f is a diffeomorphism away from the x-axis and the y-axis.

- Let

$$g(x, y) = (a_0 + a_{1,0}x + a_{0,1}y + \cdots, \; b_0 + b_{1,0}x + b_{0,1}y + \cdots)$$

where the $a_{i,j}$ and $b_{i,j}$ are arbitrary real numbers, and the omitted terms are of degree at least two in x and y. We can calculate the Jacobian matrix at $\mathbf{0}$:

$$J_g(0,0) = \begin{pmatrix} a_{1,0} & a_{0,1} \\ b_{1,0} & b_{0,1} \end{pmatrix}.$$

We see that g is a local diffeomorphism at $\mathbf{0}$ if, and only if,

$$a_{1,0}b_{0,1} - a_{0,1}b_{1,0} \neq 0,$$

i.e. the linear terms in the components of g are linearly independent as polynomials.

- Let

$$h(x, y) = \left(\sin(x^2 + y^2), \cos(x^2 + y^2) \right).$$

We can calculate the Jacobian matrix:

$$J_h = \begin{pmatrix} 2x\cos(x^2 + y^2) & 2y\cos(x^2 + y^2) \\ -2x\sin(x^2 + y^2) & -2y\sin(x^2 + y^2) \end{pmatrix}.$$

The Jacobian matrix has zero determinant everywhere! In fact we see that the image of h is the unit circle.

25.4.1 Surface deformations

In mechanics, a stress-induced transformation is called a deformation and may be described by a diffeomorphism. A diffeomorphism $f : U \to V$ between two surfaces U and V has a Jacobian matrix Df that is an invertible matrix. In fact, it is required that for p in U, there is a neighborhood of p in which the Jacobian Df stays non-singular. Since the Jacobian is a 2×2 real matrix, Df can be read as one of three types of complex number: ordinary complex, split complex number, or dual number. Suppose that in a chart of the surface, $f(x, y) = (u, v)$.

The total differential of u is

$$du = \frac{\partial u}{\partial x}dx + \frac{\partial u}{\partial y}dy, \;, \text{ and similarly for } v.$$

Then the image $(du, dv) = (dx, dy)Df$ is a linear transformation, fixing the origin, and expressible as the action of a complex number of a particular type. When (dx, dy) is also interpreted as that type of complex number, the action is of complex multiplication in the appropriate complex number plane. As such, there is a type of angle (Euclidean, hyperbolic, or slope) that is preserved in such a multiplication. Due to Df being invertible, the type of complex number is uniform over the surface.

Consequently, a surface deformation or diffeomorphism of surfaces has the **conformal property** of preserving (the appropriate type of) angles.

25.5 Diffeomorphism group

Let M be a differentiable manifold that is second-countable and Hausdorff. The **diffeomorphism group** of M is the group of all C^r diffeomorphisms of M to itself, denoted by $\mathrm{Diff}^r(M)$ or, when r is understood, $\mathrm{Diff}(M)$. This is a "large" group, in the sense that – provided M is not zero-dimensional – it is not locally compact.

25.5.1   Topology

The diffeomorphism group has two natural topologies: *weak* and *strong* (Hirsch 1997). When the manifold is compact, these two topologies agree. The weak topology is always metrizable. When the manifold is not compact, the strong topology captures the behavior of functions "at infinity" and is not metrizable. It is, however, still Baire.

Fixing a Riemannian metric on M, the weak topology is the topology induced by the family of metrics

$$d_K(f,g) = \sup_{x \in K} d(f(x), g(x)) + \sum_{1 \leq p \leq r} \sup_{x \in K} \| D^p f(x) - D^p g(x) \|$$

as K varies over compact subsets of M. Indeed, since M is σ-compact, there is a sequence of compact subsets Kn whose union is M. Then:

$$d(f,g) = \sum_n 2^{-n} \frac{d_{K_n}(f,g)}{1 + d_{K_n}(f,g)}.$$

The diffeomorphism group equipped with its weak topology is locally homeomorphic to the space of C^r vector fields (Leslie 1967). Over a compact subset of M, this follows by fixing a Riemannian metric on M and using the exponential map for that metric. If r is finite and the manifold is compact, the space of vector fields is a Banach space. Moreover, the transition maps from one chart of this atlas to another are smooth, making the diffeomorphism group into a Banach manifold with smooth right translations; left translations and inversion are only continuous. If $r = \infty$, the space of vector fields is a Fréchet space. Moreover, the transition maps are smooth, making the diffeomorphism group into a Fréchet manifold and even into a regular Fréchet Lie group.

If the manifold is σ-compact and not compact the full diffeomorphism group is not locally contractible for any of the two topologies. One has to restrict the group by controlling the deviation from the identity near infinity to obtain a diffeomorphism group which is a manifold; see (Michor & Mumford 2013).

25.5.2   Lie algebra

The Lie algebra of the diffeomorphism group of M consists of all vector fields on M equipped with the Lie bracket of vector fields. Somewhat formally, this is seen by making a small change to the coordinate x at each point in space:

$$x^\mu \to x^\mu + \varepsilon h^\mu(x)$$

so the infinitesimal generators are the vector fields

$$L_h = h^\mu(x) \frac{\partial}{\partial x_\mu}.$$

25.5.3   Examples

- When $M = G$ is a Lie group, there is a natural inclusion of G in its own diffeomorphism group via left-translation. Let Diff(G) denote the diffeomorphism group of G, then there is a splitting Diff(G) $\simeq G \times$ Diff(G, e), where Diff(G, e) is the subgroup of Diff(G) that fixes the identity element of the group.

- The diffeomorphism group of Euclidean space $\mathbf{R}^n$ consists of two components, consisting of the orientation preserving and orientation reversing diffeomorphisms. In fact, the general linear group is a deformation retract of subgroup Diff($\mathbf{R}^n$, 0) of diffeomorphisms fixing the origin under the map $f(x) \mapsto f(tx)/t$, $t \in$& (0,1]. In particular, the general linear group is also a deformation retract of the full diffeomorphism group.

- For a finite set of points, the diffeomorphism group is simply the symmetric group. Similarly, if M is any manifold there is a group extension $0 \to \mathrm{Diff}_0(M) \to \mathrm{Diff}(M) \to \Sigma(\pi_0(M))$. Here $\mathrm{Diff}_0(M)$is the subgroup of $\mathrm{Diff}(M)$ that preserves all the components of M, and $\Sigma(\pi_0(M))$ is the permutation group of the set $\pi_0(M)$ (the components of M). Moreover, the image of the map $\mathrm{Diff}(M) \to \Sigma(\pi_0(M))$ is the bijections of $\pi_0(M)$ that preserve diffeomorphism classes.

25.5.4 Transitivity

For a connected manifold M, the diffeomorphism group acts transitively on M. More generally, the diffeomorphism group acts transitively on the configuration space CkM. If M is at least two-dimensional, the diffeomorphism group acts transitively on the configuration space FkM and the action on M is multiply transitive (Banyaga 1997, p. 29).

25.5.5 Extensions of diffeomorphisms

In 1926, Tibor Radó asked whether the harmonic extension of any homeomorphism or diffeomorphism of the unit circle to the unit disc yields a diffeomorphism on the open disc. An elegant proof was provided shortly afterwards by Hellmuth Kneser. In 1945, Gustave Choquet, apparently unaware of this result, produced a completely different proof.

The (orientation-preserving) diffeomorphism group of the circle is pathwise connected. This can be seen by noting that any such diffeomorphism can be lifted to a diffeomorphism f of the reals satisfying $[f(x+1) = f(x) + 1]$; this space is convex and hence path-connected. A smooth, eventually constant path to the identity gives a second more elementary way of extending a diffeomorphism from the circle to the open unit disc (a special case of the Alexander trick). Moreover, the diffeomorphism group of the circle has the homotopy-type of the orthogonal group O(2).

The corresponding extension problem for diffeomorphisms of higher-dimensional spheres $\mathbf{S}^{n-1}$ was much studied in the 1950s and 1960s, with notable contributions from René Thom, John Milnor and Stephen Smale. An obstruction to such extensions is given by the finite Abelian group Γn, the "group of twisted spheres", defined as the quotient of the Abelian component group of the diffeomorphism group by the subgroup of classes extending to diffeomorphisms of the ball B^n.

25.5.6 Connectedness

For manifolds, the diffeomorphism group is usually not connected. Its component group is called the mapping class group. In dimension 2 (i.e. surfaces), the mapping class group is a finitely presented group generated by Dehn twists (Dehn, Lickorish, Hatcher). Max Dehn and Jakob Nielsen showed that it can be identified with the outer automorphism group of the fundamental group of the surface.

William Thurston refined this analysis by classifying elements of the mapping class group into three types: those equivalent to a periodic diffeomorphism; those equivalent to a diffeomorphism leaving a simple closed curve invariant; and those equivalent to pseudo-Anosov diffeomorphisms. In the case of the torus $\mathbf{S}^1 \times \mathbf{S}^1 = \mathbf{R}^2/\mathbf{Z}^2$, the mapping class group is simply the modular group SL(2, $\mathbf{Z}$) and the classification becomes classical in terms of elliptic, parabolic and hyperbolic matrices. Thurston accomplished his classification by observing that the mapping class group acted naturally on a compactification of Teichmüller space; as this enlarged space was homeomorphic to a closed ball, the Brouwer fixed-point theorem became applicable.

Smale conjectured that if M is an oriented smooth closed manifold, the identity component of the group of orientation-preserving diffeomorphisms is simple. This had first been proved for a product of circles by Michel Herman; it was proved in full generality by Thurston.

25.5.7 Homotopy types

- The diffeomorphism group of $\mathbf{S}^2$ has the homotopy-type of the subgroup O(3). This was proved by Steve Smale.[1]

- The diffeomorphism group of the torus has the homotopy-type of its linear automorphisms: $\mathbf{S}^1 \times \mathbf{S}^1 \times$ GL(2, $\mathbf{Z}$).

- The diffeomorphism groups of orientable surfaces of genus $g > 1$ have the homotopy-type of their mapping class groups (i.e. the components are contractible).

- The homotopy-type of the diffeomorphism groups of 3-manifolds are fairly well-understood via the work of Ivanov, Hatcher, Gabai and Rubinstein, although there are a few outstanding open cases (primarily 3-manifolds with finite fundamental groups).

- The homotopy-type of diffeomorphism groups of n-manifolds for $n > 3$ are poorly undersood. For example, it is an open problem whether or not Diff(S^4) has more than two components. Via Milnor, Kahn and Antonelli, however, it is known that provided $n > 6$, Diff(S^n) does not have the homotopy-type of a finite CW-complex.

25.6   Homeomorphism and diffeomorphism

Unlike non-diffeomorphic homeomorphisms, it is relatively difficult to find a pair of homeomorphic manifolds that are not diffeomorphic. In dimensions 1, 2, 3, any pair of homeomorphic smooth manifolds are diffeomorphic. In dimension 4 or greater, examples of homeomorphic but not diffeomorphic pairs have been found. The first such example was constructed by John Milnor in dimension 7. He constructed a smooth 7-dimensional manifold (called now Milnor's sphere) that is homeomorphic to the standard 7-sphere but not diffeomorphic to it. There are, in fact, 28 oriented diffeomorphism classes of manifolds homeomorphic to the 7-sphere (each of them is the total space of a fiber bundle over the 4-sphere with the 3-sphere as the fiber).

More unusual phenomena occur for 4-manifolds. In the early 1980s, a combination of results due to Simon Donaldson and Michael Freedman led to the discovery of exotic R4s: there are uncountably many pairwise non-diffeomorphic open subsets of $\mathbf{R}^4$ each of which is homeomorphic to $\mathbf{R}^4$, and also there are uncountably many pairwise non-diffeomorphic differentiable manifolds homeomorphic to $\mathbf{R}^4$ that do not embed smoothly in $\mathbf{R}^4$.

25.7   See also

- Étale morphism

- Large diffeomorphism

- Local diffeomorphism

- Superdiffeomorphism

25.8   Notes

[1] Smale, "Diffeomorphisms of the 2-sphere", *Proc. Amer. Math. Soc.* 10 (1959), pp. 621–626.

25.9   References

- Chaudhuri, Shyamoli, Hakuru Kawai and S.-II Henry Tye. "Path-integral formulation of closed strings", *Phys. Rev. D*, 36: 1148 (1987).

- Banyaga, Augustin (1997), *The structure of classical diffeomorphism groups*, Mathematics and its Applications, 400, Kluwer Academic, ISBN 0-7923-4475-8

- Duren, Peter L. (2004), *Harmonic Mappings in the Plane*, Cambridge Mathematical Tracts, 156, Cambridge University Press, ISBN 0-521-64121-7

- Hazewinkel, Michiel, ed. (2001), "Diffeomorphism", *Encyclopedia of Mathematics*, Springer, ISBN 978-1-55608-010-4

- Hirsch, Morris (1997), *Differential Topology*, Berlin, New York: Springer-Verlag, ISBN 978-0-387-90148-0

- Kriegl, Andreas; Michor, Peter (1997), *The convenient setting of global analysis*, Mathematical Surveys and Monographs, 53, American Mathematical Society, ISBN 0-8218-0780-3

- Leslie, J. A. (1967), "On a differential structure for the group of diffeomorphisms", *Topology* **6** (2): 263–271, doi:10.1016/0040-9383(67)90038-9, ISSN 0040-9383, MR 0210147

- Michor, Peter W.; Mumford, David (2013), "A zoo of diffeomorphism groups on $\mathbf{R}^n$.", *Annals of Global Analysis and Geometry* **44** (4): 529–540, doi:10.1007/s10455-013-9380-2 (arXiv:1211.5704)

- Milnor, John W. (2007), *Collected Works Vol. III, Differential Topology*, American Mathematical Society, ISBN 0-8218-4230-7

- Omori, Hideki (1997), *Infinite-dimensional Lie groups*, Translations of Mathematical Monographs, 158, American Mathematical Society, ISBN 0-8218-4575-6

- Kneser, Hellmuth (1926), "Lösung der Aufgabe 41.", *Jahresbericht der Deutschen Mathematiker-Vereinigung* (in German) **35** (2): 123

Chapter 26

Gauge theory gravity

Gauge theory gravity (GTG) is a theory of gravitation cast in the mathematical language of geometric algebra. To those familiar with general relativity, it is highly reminiscent of the tetrad formalism although there are significant conceptual differences. Most notably, the background in GTG is flat, Minkowski spacetime. The equivalence principle is not assumed, but instead follows from the fact that the gauge covariant derivative is minimally coupled. As in general relativity, equations structurally identical to the Einstein field equations are derivable from a variational principle. A spin tensor can also be supported in a manner similar to Einstein–Cartan–Sciama–Kibble theory. GTG was first proposed by Lasenby, Doran, and Gull in 1998[1] as a fulfillment of partial results presented in 1993.[2] The theory has not been widely adopted by the rest of the physics community, who have mostly opted for differential geometry approaches like that of the related gauge gravitation theory.

26.1   Mathematical foundation

The foundation of GTG comes from two principles. First, *position-gauge invariance* demands that arbitrary local displacements of fields not affect the physical content of the field equations. Second, *rotation-gauge invariance* demands that arbitrary local rotations of fields not affect the physical content of the field equations. These principles lead to the introduction of a new pair of linear functions, the position-gauge field and the rotation-gauge field. A displacement by some arbitrary function f

$$x \mapsto x' = f(x)$$

gives rise to the position-gauge field defined by the mapping on its adjoint,

$$\bar{\mathsf{h}}(a, x) \mapsto \bar{\mathsf{h}}'(a, x) = \bar{\mathsf{h}}(f^{-1}(a), f(x)),$$

which is linear in its first argument and a is a constant vector. Similarly, a rotation by some arbitrary rotor R gives rise to the rotation-gauge field

$$\bar{}(a, x) \mapsto \bar{}'(a, x) = R\bar{}(a, x)R^{\dagger} - 2a \cdot \nabla R R^{\dagger}.$$

We can define two different covariant directional derivatives

$$a \cdot D = a \cdot \bar{\mathsf{h}}(\nabla) + \frac{1}{2}\,(\mathsf{h}(a))$$

$$a \cdot \mathcal{D} = a \cdot \bar{\mathsf{h}}(\nabla) + (\mathsf{h}(a))$$

or with the specification of a coordinate system

$$D_\mu = \partial_\mu + \frac{1}{2}\Omega_\mu$$

$$\mathcal{D}_\mu = \partial_\mu + \Omega_\mu \times,$$

where $\times$ denotes the commutator product.

The first of these derivatives is better suited for dealing directly with spinors whereas the second is better suited for observables. The GTG analog of the Riemann tensor is built from the commutation rules of these derivatives.

$$[D_\mu, D_\nu]\psi = \frac{1}{2}\mathsf{R}_{\mu\nu}\psi$$

$$\mathcal{R}(a \wedge b) = \mathsf{R}(\mathsf{h}(a \wedge b))$$

26.2 Field equations

The field equations are derived by postulating the Einstein–Hilbert action governs the evolution of the gauge fields, i.e.

$$S = \int \left[\frac{1}{2\kappa} (\mathcal{R} - 2\Lambda) + \mathcal{L}_\mathrm{M} \right] (\det \mathsf{h})^{-1} \, \mathrm{d}^4 x.$$

Minimizing variation of the action with respect to the two gauge fields results in the field equations

$$\mathcal{G}(a) - \Lambda a = \kappa \mathcal{T}(a)$$

$$\mathcal{D} \wedge \bar{\mathsf{h}}(a) = \kappa \mathcal{S} \cdot \bar{\mathsf{h}}(a),$$

where $\mathcal{T}$ is the covariant energy–momentum tensor and $\mathcal{S}$ is the covariant spin tensor. Importantly, these equations do not give an evolving curvature of spacetime but rather merely give the evolution of the gauge fields within the flat spacetime. Moreover, the existence of the spin tensor does *not* endow spacetime with torsion.

26.3 Relation to general relativity

For those more familiar with general relativity, it is possible to define a metric tensor from the position-gauge field in a manner similar to tetrads. In the tetrad formalism, a set of four vectors $\{e_{(a)}{}^\mu\}$ are introduced. The Greek index μ is raised or lowered by multiplying and contracting with the spacetime's metric tensor. The parenthetical Latin index (a) is a label for each of the four tetrads, which is raised and lowered as if it were multiplied and contracted with a separate Minkowski metric tensor. GTG, roughly, reverses the roles of these indices. The metric is implicitly assumed to be Minkowski in the selection of the spacetime algebra. The information contained in the other set of indices gets subsumed by the behavior of the gauge fields.

We can make the associations

$$g_\mu = \mathsf{h}^{-1}(e_\mu)$$

$$g^\mu = \bar{\mathsf{h}}(e^\mu)$$

for a covariant vector and contravariant vector in a curved spacetime, where now the unit vectors $\{e_\mu\}$ are the chosen coordinate basis. These can define the metric using the rule

$$g_{\mu\nu} = g_\mu \cdot g_\nu.$$

Following this procedure, it is possible to show that for the most part the observable predictions of GTG agree with Einstein–Cartan–Sciama–Kibble theory for non-vanishing spin and reduce to general relativity for vanishing spin. GTG does, however, make different predictions about global solutions. For example, in the study of a point mass, the choice of a "Newtonian gauge" yields a solution similar to the Schwarzschild metric in Gullstrand–Painlevé coordinates. General relativity permits an extension known as the Kruskal–Szekeres coordinates. GTG, on the other hand, forbids any such extension.

26.4 References

[1] Lasenby, Anthony; Chris Doran; Stephen Gull (1998), "Gravity, gauge theories and geometric algebra", *Philosophical Transactions of the Royal Society A* **356**: 487–582, arXiv:gr-qc/0405033, Bibcode:1998RSPTA.356..487L, doi:10.1098/rsta.1998.0178

[2] Doran, Chris; Anthony Lasenby; Stephen Gull (1993), F. Brackx, R. Delanghe, H. Serras, ed., "Gravity as a gauge theory in the spacetime algebra", *Third International Conference on Clifford Algebras and their Applications in Mathematical Physics*

26.5 External links

- David Hestenes: Spacetime calculus for gravitation theory – an account of the mathematical formalism explicitly directed to GTG

Chapter 27

Gauge gravitation theory

In quantum field theory, **gauge gravitation theory** is the effort to extend Yang–Mills theory, which provides a universal description of the fundamental interactions, to describe gravity. It should not be confused with the related but distinct gauge theory gravity.

The first gauge model of gravity was suggested by R. Utiyama in 1956[1] just two years after birth of the gauge theory itself.[2] However, the initial attempts to construct the gauge theory of gravity by analogy with the gauge models of internal symmetries encountered a problem of treating general covariant transformations and establishing the gauge status of a pseudo-Riemannian metric (a tetrad field).

In order to overcome this drawback, representing tetrad fields as gauge fields of the translation group was attempted.[3] Infinitesimal generators of general covariant transformations were considered as those of the translation gauge group, and a tetrad (coframe) field was identified with the translation part of an affine connection on a world manifold X. Any such connection is a sum $K = \Gamma + \Theta$ of a linear world connection Γ and a soldering form $\Theta = \Theta^a_\mu dx^\mu \otimes \vartheta_a$ where $\vartheta_a = \vartheta^\lambda_a \partial_\lambda$ is a non-holonomic frame. For instance, if K is the Cartan connection, then $\Theta = \theta = dx^\mu \otimes \partial_\mu$ is the canonical soldering form on X. There are different physical interpretations of the translation part Θ of affine connections. In gauge theory of dislocations, a field Θ describes a distortion.[4] At the same time, given a linear frame ϑ_a, the decomposition $\theta = \vartheta^a \otimes \vartheta_a$ motivates many authors to treat a coframe ϑ^a as a translation gauge field.[5]

Difficulties of constructing gauge gravitation theory by analogy with the Yang-Mills one result from the gauge transformations in these theories belonging to different classes. In the case of internal symmetries, the gauge transformations are just vertical automorphisms of a principal bundle $P \to X$ leaving its base X fixed. On the other hand, gravitation theory is built on the principal bundle FX of the tangent frames to X. It belongs to the category of natural bundles $T \to X$ for which diffeomorphisms of the base X canonically give rise to automorphisms of T.[6] These automorphisms are called general covariant transformations. General covariant transformations are sufficient in order to restate Einstein's General Relativity and metric-affine gravitation theory as the gauge ones.

In terms of gauge theory on natural bundles, gauge fields are linear connections on a world manifold X, defined as principal connections on the linear frame bundle FX, and a metric (tetrad) gravitational field plays the role of a Higgs field responsible for spontaneous symmetry breaking of general covariant transformations.[7]

Spontaneous symmetry breaking is a quantum effect when the vacuum is not invariant under the transformation group. In classical gauge theory, spontaneous symmetry breaking occurs if the structure group G of a principal bundle $P \to X$ is reducible to a closed subgroup H, i.e., there exists a principal subbundle of P with the structure group H.[8] By virtue of the well-known theorem, there exists one-to-one correspondence between the reduced principal subbundles of P with the structure group H and the global sections of the quotient bundle $P/H \to X$. These sections are treated as classical Higgs fields.

The idea of the pseudo-Riemannian metric as a Higgs field appeared while constructing non-linear (induced) representations of the general linear group $GL(4, \mathbb{R})$, of which the Lorentz group is a Cartan subgroup.[9] The geometric equivalence principle postulating the existence of a reference frame in which Lorentz invariants are defined on the whole world manifold is the theoretical justification of that the structure group $GL(4, \mathbb{R})$ of the linear frame bundle FX is reduced to the Lorentz group. Then the very definition of a pseudo-Riemannian metric on a manifold X as a global section of the quotient bundle $FX/O(1,3) \to X$ leads to its physical interpretation as a Higgs field. The physical reason for world symmetry breaking is the existence of Dirac fermion matter, whose symmetry group is the universal two-sheeted covering $SL(2,\mathbb{C})$ of the restricted Lorentz group, $SO^+(1,3)$.[10]

27.1   See also

- Ashtekar variables

- Metric-affine gravitation theory

- Einstein–Cartan theory

- Spontaneous symmetry breaking

- Teleparallelism

- Reduction of the structure group

- Higgs field (classical)

- General covariant transformations

- Equivalence principle (geometric)

- Affine gauge theory

27.2   Notes

[1] R Utiyama, Invariant theoretical interpretation of interaction, Physical Review **101** (1956) 1597. doi:10.1103/PhysRev.101.1597

[2] Blagojević, Milutin; Hehl, Friedrich W. (2013). *Gauge Theories of Gravitation: A Reader with Commentaries.* World Scientific. ISBN 978-184-8167-26-1.

[3] F.Hehl, J. McCrea, E. Mielke, Y. Ne'eman, Metric-affine gauge theory of gravity: field equations, Noether identities, world spinors, and breaking of dilaton invariance, Physics Reports **258** (1995) 1. doi:10.1016/0370-1573(94)00111-F

[4] C.Malyshev, The dislocation stress functions from the double curl $T(3)$ -gauge equations: Linearity and look beyond, Annals of Physics **286** (2000) 249. doi:10.1006/aphy.2000.6088

[5] M. Blagojević, Gravitation and Gauge Symmetries (IOP Publishing, Bristol, 2002).

[6] I. Kolář, P. W. Michor, J. Slovák, Natural Operations in Differential Geometry (Springer-Verlag, Berlin, Heidelberg, 1993).

[7] D.Ivanenko, G.Sardanashvily, The gauge treatment of gravity, Physics Reports **94** (1983) 1. doi:10.1016/0370-1573(83)90046-7

[8] L. Nikolova, V. Rizov, Geometrical approach to the reduction of gauge theories with spontaneous broken symmetries, Reports on Mathematical Physics **20** (1984) 287. doi:10.1016/0034-4877(84)90039-9

[9] M. Leclerk, The Higgs sector of gravitational gauge theories, Annals of Physics **321** (2006) 708. doi:10.1016/j.aop.2005.08.009

[10] G. Sardanashvily, O.Zakharov, Gauge Gravitation Theory (World Scientific, Singapore, 1992).

27.3 References

- I. Kirsch, A Higgs mechanism for gravity, Phys. Rev. **D72** (2005) 024001; arXiv: hep-th/0503024.

- G. Sardanashvily, Classical gauge gravitation theory, Int. J. Geom. Methods Mod. Phys. **8** (2011) 1869-1895; arXiv: 1110.1176.

- Yu. Obukhov, Poincaré gauge gravity: selected topics, Int. J. Geom. Methods Mod. Phys. **3** (2006) 95-138; arXiv: gr-qc/0601090.

Chapter 28

Quantum gravity

Quantum gravity (**QG**) is a field of theoretical physics that seeks to describe the force of gravity according to the principles of quantum mechanics.

The current understanding of gravity is based on Albert Einstein's general theory of relativity, which is formulated within the framework of classical physics. On the other hand, the nongravitational forces are described within the framework of quantum mechanics, a radically different formalism for describing physical phenomena based on probability.[1] The necessity of a quantum mechanical description of gravity follows from the fact that one cannot consistently couple a classical system to a quantum one.[2]

Although a quantum theory of gravity is needed in order to reconcile general relativity with the principles of quantum mechanics, difficulties arise when one attempts to apply the usual prescriptions of quantum field theory to the force of gravity.[3] From a technical point of view, the problem is that the theory one gets in this way is not renormalizable and therefore cannot be used to make meaningful physical predictions. As a result, theorists have taken up more radical approaches to the problem of quantum gravity, the most popular approaches being string theory and loop quantum gravity.[4] A recent development is the theory of causal fermion systems which gives quantum mechanics, general relativity and quantum field theory as limiting cases.[5][6][7][8][9][10]

Strictly speaking, the aim of quantum gravity is only to describe the quantum behavior of the gravitational field and should not be confused with the objective of unifying all fundamental interactions into a single mathematical framework. Although some quantum gravity theories such as string theory try to unify gravity with the other fundamental forces, others such as loop quantum gravity make no such attempt; instead, they make an effort to quantize the gravitational field while it is kept separate from the other forces. A theory of quantum gravity that is also a grand unification of all known interactions is sometimes referred to as a theory of everything (TOE).

One of the difficulties of quantum gravity is that quantum gravitational effects are only expected to become apparent near the Planck scale, a scale far smaller in distance (equivalently, far larger in energy) than what is currently accessible at high energy particle accelerators. As a result, quantum gravity is a mainly theoretical enterprise, although there are speculations about how quantum gravity effects might be observed in existing experiments.[11]

28.1 Overview

Much of the difficulty in meshing these theories at all energy scales comes from the different assumptions that these theories make on how the universe works. Quantum field theory depends on particle fields embedded in the flat space-time of special relativity. General relativity models gravity as a curvature within space-time that changes as a gravitational mass moves. Historically, the most obvious way of combining the two (such as treating gravity as simply another particle field) ran quickly into what is known as the renormalization problem. In the old-fashioned understanding of renormalization, gravity particles would attract each other and adding together all of the interactions results in many infinite values which cannot easily be cancelled out mathematically to yield sensible, finite results. This is in contrast with quantum electrodynamics where, given that the series still do not converge, the interactions sometimes evaluate to infinite results, but those

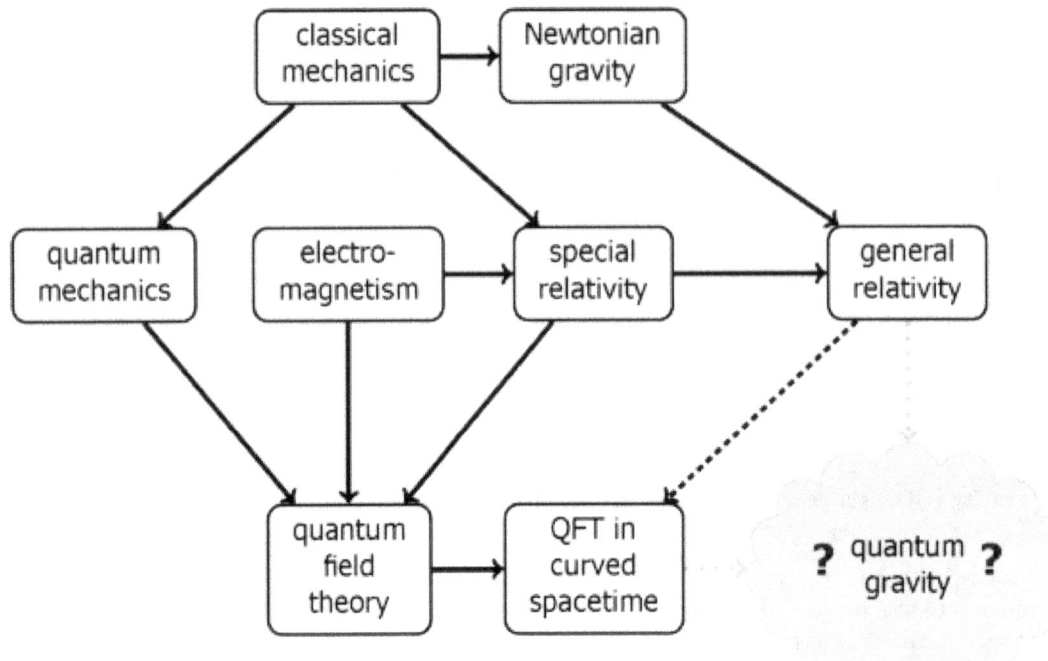

Diagram showing where quantum gravity sits in the hierarchy of physics theories

are few enough in number to be removable via renormalization.

28.1.1 Effective field theories

Quantum gravity can be treated as an effective field theory. Effective quantum field theories come with some high-energy cutoff, beyond which we do not expect that the theory provides a good description of nature. The "infinities" then become large but finite quantities depending on this finite cutoff scale, and correspond to processes that involve very high energies near the fundamental cutoff. These quantities can then be absorbed into an infinite collection of coupling constants, and at energies well below the fundamental cutoff of the theory, to any desired precision; only a finite number of these coupling constants need to be measured in order to make legitimate quantum-mechanical predictions. This same logic works just as well for the highly successful theory of low-energy pions as for quantum gravity. Indeed, the first quantum-mechanical corrections to graviton-scattering and Newton's law of gravitation have been explicitly computed[12] (although they are so astronomically small that we may never be able to measure them). In fact, gravity is in many ways a much better quantum field theory than the Standard Model, since it appears to be valid all the way up to its cutoff at the Planck scale.

While confirming that quantum mechanics and gravity are indeed consistent at reasonable energies, it is clear that near or above the fundamental cutoff of our effective quantum theory of gravity (the cutoff is generally assumed to be of the order of the Planck scale), a new model of nature will be needed. Specifically, the problem of combining quantum mechanics and gravity becomes an issue only at very high energies, and may well require a totally new kind of model.

28.1.2 Quantum gravity theory for the highest energy scales

The general approach to deriving a quantum gravity theory that is valid at even the highest energy scales is to assume that such a theory will be simple and elegant and, accordingly, to study symmetries and other clues offered by current theories that might suggest ways to combine them into a comprehensive, unified theory. One problem with this approach is that it is unknown whether quantum gravity will actually conform to a simple and elegant theory, as it should resolve the dual conundrums of special relativity with regard to the uniformity of acceleration and gravity, and general relativity with regard to spacetime curvature.

Such a theory is required in order to understand problems involving the combination of very high energy and very small dimensions of space, such as the behavior of black holes, and the origin of the universe.

28.2   Quantum mechanics and general relativity

28.2.1   The graviton

Main article: Graviton

At present, one of the deepest problems in theoretical physics is harmonizing the theory of general relativity, which describes gravitation, and applications to large-scale structures (stars, planets, galaxies), with quantum mechanics, which describes the other three fundamental forces acting on the atomic scale. This problem must be put in the proper context, however. In particular, contrary to the popular claim that quantum mechanics and general relativity are fundamentally incompatible, one can demonstrate that the structure of general relativity essentially follows inevitably from the quantum mechanics of interacting theoretical spin-2 massless particles (called gravitons).[13][14][15][16][17]

While there is no concrete proof of the existence of gravitons, quantized theories of matter may necessitate their existence. Supporting this theory is the observation that all fundamental forces except gravity have one or more known messenger particles, leading researchers to believe that at least one most likely does exist; they have dubbed this hypothetical particle the *graviton*. The predicted find would result in the classification of the graviton as a "force particle" similar to the photon of the electromagnetic field. Many of the accepted notions of a unified theory of physics since the 1970s assume, and to some degree depend upon, the existence of the graviton. These include string theory, superstring theory, M-theory, and loop quantum gravity. Detection of gravitons is thus vital to the validation of various lines of research to unify quantum mechanics and relativity theory.

28.2.2   The dilaton

Main article: Dilaton

The dilaton made its first appearance in Kaluza–Klein theory, a five-dimensional theory that combined gravitation and electromagnetism. Generally, it appears in string theory. More recently, it has appeared in the lower-dimensional many-bodied gravity problem[18] based on the field theoretic approach of Roman Jackiw. The impetus arose from the fact that complete analytical solutions for the metric of a covariant N-body system have proven elusive in general relativity. To simplify the problem, the number of dimensions was lowered to *(1+1)*, i.e. one spatial dimension and one temporal dimension. This model problem, known as $R=T$ theory[19] (as opposed to the general $G=T$ theory) was amenable to exact solutions in terms of a generalization of the Lambert W function. It was also found that the field equation governing the dilaton (derived from differential geometry) was the Schrödinger equation and consequently amenable to quantization.[20]

Thus, one had a theory which combined gravity, quantization and even the electromagnetic interaction, promising ingredients of a fundamental physical theory. It is worth noting that the outcome revealed a previously unknown and already existing *natural link* between general relativity and quantum mechanics. However, this theory needs to be generalized in *(2+1)* or *(3+1)* dimensions although, in principle, the field equations are amenable to such generalization as shown with the inclusion of a one-graviton process[21] and yielding the correct Newtonian limit in d dimensions if a dilaton is included. However, it is not yet clear what the fully generalized field equation governing the dilaton in (3+1) dimensions should be. This is further complicated by the fact that gravitons can propagate in *(3+1)* dimensions and consequently that would imply gravitons and dilatons exist in the real world. Moreover, detection of the dilaton is expected to be even more elusive than the graviton. However, since this approach allows for the combination of gravitational, electromagnetic and quantum effects, their coupling could potentially lead to a means of vindicating the theory, through cosmology and perhaps even *experimentally*.

28.2.3 Nonrenormalizability of gravity

Further information: Renormalization

General relativity, like electromagnetism, is a classical field theory. One might expect that, as with electromagnetism, there should be a corresponding quantum field theory.

However, gravity is perturbatively nonrenormalizable.[22][23] For a quantum field theory to be well-defined according to this understanding of the subject, it must be asymptotically free or asymptotically safe. The theory must be characterized by a choice of *finitely many* parameters, which could, in principle, be set by experiment. For example, in quantum electrodynamics, these parameters are the charge and mass of the electron, as measured at a particular energy scale.

On the other hand, in quantizing gravity, there are *infinitely many independent parameters* (counterterm coefficients) needed to define the theory. For a given choice of those parameters, one could make sense of the theory, but since we can never do infinitely many experiments to fix the values of every parameter, we do not have a meaningful physical theory:

- At low energies, the logic of the renormalization group tells us that, despite the unknown choices of these infinitely many parameters, quantum gravity will reduce to the usual Einstein theory of general relativity.

- On the other hand, if we could probe very high energies where quantum effects take over, then *every one* of the infinitely many unknown parameters would begin to matter, and we could make no predictions at all.

As explained below, there is a way around this problem by treating QG as an effective field theory.

Any meaningful theory of quantum gravity that makes sense and is predictive at all energy scales must have some deep principle that reduces the infinitely many unknown parameters to a finite number that can then be measured.

- One possibility is that normal perturbation theory is not a reliable guide to the renormalizability of the theory, and that there really *is* a UV fixed point for gravity. Since this is a question of non-perturbative quantum field theory, it is difficult to find a reliable answer, but some people still pursue this option.

- Another possibility is that there are new symmetry principles that constrain the parameters and reduce them to a finite set. This is the route taken by string theory, where all of the excitations of the string essentially manifest themselves as new symmetries.

28.2.4 QG as an effective field theory

Main article: Effective field theory

In an effective field theory, all but the first few of the infinite set of parameters in a non-renormalizable theory are suppressed by huge energy scales and hence can be neglected when computing low-energy effects. Thus, at least in the low-energy regime, the model is indeed a predictive quantum field theory.[12] (A very similar situation occurs for the very similar effective field theory of low-energy pions.) Furthermore, many theorists agree that even the Standard Model should really be regarded as an effective field theory as well, with "nonrenormalizable" interactions suppressed by large energy scales and whose effects have consequently not been observed experimentally.

Recent work[12] has shown that by treating general relativity as an effective field theory, one can actually make legitimate predictions for quantum gravity, at least for low-energy phenomena. An example is the well-known calculation of the tiny first-order quantum-mechanical correction to the classical Newtonian gravitational potential between two masses.

28.2.5 Spacetime background dependence

Main article: Background independence

A fundamental lesson of general relativity is that there is no fixed spacetime background, as found in Newtonian mechanics and special relativity; the spacetime geometry is dynamic. While easy to grasp in principle, this is the hardest idea to understand about general relativity, and its consequences are profound and not fully explored, even at the classical level. To a certain extent, general relativity can be seen to be a relational theory,[24] in which the only physically relevant information is the relationship between different events in space-time.

On the other hand, quantum mechanics has depended since its inception on a fixed background (non-dynamic) structure. In the case of quantum mechanics, it is time that is given and not dynamic, just as in Newtonian classical mechanics. In relativistic quantum field theory, just as in classical field theory, Minkowski spacetime is the fixed background of the theory.

String theory

String theory can be seen as a generalization of quantum field theory where instead of point particles, string-like objects propagate in a fixed spacetime background, although the interactions among closed strings give rise to space-time in a dynamical way. Although string theory had its origins in the study of quark confinement and not of quantum gravity, it was soon discovered that the string spectrum contains the graviton, and that "condensation" of certain vibration modes of strings is equivalent to a modification of the original background. In this sense, string perturbation theory exhibits exactly the features one would expect of a perturbation theory that may exhibit a strong dependence on asymptotics (as seen, for example, in the AdS/CFT correspondence) which is a weak form of background dependence.

Background independent theories

Loop quantum gravity is the fruit of an effort to formulate a background-independent quantum theory.

Topological quantum field theory provided an example of background-independent quantum theory, but with no local degrees of freedom, and only finitely many degrees of freedom globally. This is inadequate to describe gravity in 3+1 dimensions, which has local degrees of freedom according to general relativity. In 2+1 dimensions, however, gravity is a topological field theory, and it has been successfully quantized in several different ways, including spin networks.

28.2.6   Semi-classical quantum gravity

Quantum field theory on curved (non-Minkowskian) backgrounds, while not a full quantum theory of gravity, has shown many promising early results. In an analogous way to the development of quantum electrodynamics in the early part of the 20th century (when physicists considered quantum mechanics in classical electromagnetic fields), the consideration of quantum field theory on a curved background has led to predictions such as black hole radiation.

Phenomena such as the Unruh effect, in which particles exist in certain accelerating frames but not in stationary ones, do not pose any difficulty when considered on a curved background (the Unruh effect occurs even in flat Minkowskian backgrounds). The vacuum state is the state with the least energy (and may or may not contain particles). See Quantum field theory in curved spacetime for a more complete discussion.

28.2.7   Points of tension

There are other points of tension between quantum mechanics and general relativity.

- First, classical general relativity breaks down at singularities, and quantum mechanics becomes inconsistent with general relativity in the neighborhood of singularities (however, no one is certain that classical general relativity applies near singularities in the first place).

- Second, it is not clear how to determine the gravitational field of a particle, since under the Heisenberg uncertainty principle of quantum mechanics its location and velocity cannot be known with certainty. The resolution of these points may come from a better understanding of general relativity.[25]

- Third, there is the problem of time in quantum gravity. Time has a different meaning in quantum mechanics and general relativity and hence there are subtle issues to resolve when trying to formulate a theory which combines the two.[26]

28.3 Candidate theories

There are a number of proposed quantum gravity theories.[27] Currently, there is still no complete and consistent quantum theory of gravity, and the candidate models still need to overcome major formal and conceptual problems. They also face the common problem that, as yet, there is no way to put quantum gravity predictions to experimental tests, although there is hope for this to change as future data from cosmological observations and particle physics experiments becomes available.[28][29]

28.3.1 String theory

Main article: String theory
 One suggested starting point is ordinary quantum field theories which, after all, are successful in describing the other three basic fundamental forces in the context of the standard model of elementary particle physics. However, while this leads to an acceptable effective (quantum) field theory of gravity at low energies,[30] gravity turns out to be much more problematic at higher energies. Where, for ordinary field theories such as quantum electrodynamics, a technique known as renormalization is an integral part of deriving predictions which take into account higher-energy contributions,[31] gravity turns out to be nonrenormalizable: at high energies, applying the recipes of ordinary quantum field theory yields models that are devoid of all predictive power.[32]

One attempt to overcome these limitations is to replace ordinary quantum field theory, which is based on the classical concept of a point particle, with a quantum theory of one-dimensional extended objects: string theory.[33] At the energies reached in current experiments, these strings are indistinguishable from point-like particles, but, crucially, different modes of oscillation of one and the same type of fundamental string appear as particles with different (electric and other) charges. In this way, string theory promises to be a unified description of all particles and interactions.[34] The theory is successful in that one mode will always correspond to a graviton, the messenger particle of gravity; however, the price to pay are unusual features such as six extra dimensions of space in addition to the usual three for space and one for time.[35]

In what is called the second superstring revolution, it was conjectured that both string theory and a unification of general relativity and supersymmetry known as supergravity[36] form part of a hypothesized eleven-dimensional model known as M-theory, which would constitute a uniquely defined and consistent theory of quantum gravity.[37][38] As presently understood, however, string theory admits a very large number (10^{500} by some estimates) of consistent vacua, comprising the so-called "string landscape". Sorting through this large family of solutions remains a major challenge.

28.3.2 Loop quantum gravity

Main article: Loop quantum gravity
 Loop quantum gravity is based first of all on the idea to take seriously the insight of general relativity that spacetime is a dynamical field and therefore is a quantum object. The second idea is that the quantum discreteness that determines the particle-like behavior of other field theories (for instance, the photons of the electromagnetic field) also affects the structure of space.

The main result of loop quantum gravity is the derivation of a granular structure of space at the Planck length. This is derived as follows. In the case of electromagnetism, the quantum operator representing the energy of each frequency of the field has discrete spectrum. Therefore the energy of each frequency is quantized, and the quanta are the photons. In the case of gravity, the operators representing the area and the volume of each surface or space region have discrete spectrum. Therefore area and volume of any portion of space are quantized, and the quanta are elementary quanta of space. It follows that spacetime has an elementary quantum granular structure at the Planck scale, which cuts-off the ultraviolet infinities of quantum field theory.

The quantum state of spacetime is described in the theory by means of a mathematical structure called spin networks. Spin networks were initially introduced by Roger Penrose in abstract form, and later shown by Carlo Rovelli and Lee Smolin to derive naturally from a non perturbative quantization of general relativity. Spin networks do not represent quantum states of a field in spacetime: they represent directly quantum states of spacetime.

The theory is based on the reformulation of general relativity known as Ashtekar variables, which represent geometric gravity using mathematical analogues of electric and magnetic fields.[39][40] In the quantum theory space is represented by a network structure called a spin network, evolving over time in discrete steps.[41][42][43][44]

The dynamics of the theory is today constructed in several versions. One version starts with the canonical quantization of general relativity. The analogue of the Schrödinger equation is a Wheeler–DeWitt equation, which can be defined in the theory.[45] In the covariant, or spinfoam formulation of the theory, the quantum dynamics is obtained via a sum over discrete versions of spacetime, called spinfoams. These represent histories of spin networks.

28.3.3   Scale Relativity

Main article: Scale relativity
 Most quantum gravity theories assume quantum laws as a starting point. However, in the framework of scale relativity, this is not needed.[46] The theory is an extension of special and general relativity, including the relativity of scale transformations. It thus takes a geometrical approach to the problem, where quantum phenomena became a manifestation of the fractality of spacetime. This is similar to the geometrical interpretation of gravitation in general relativity, where gravitation become a manifestation of spacetime curvature instead of a force. Although much remains to be developed, validated predictions have already been obtained in physics, astrophysics and cosmology.

28.3.4   Other approaches

There are a number of other approaches to quantum gravity. The approaches differ depending on which features of general relativity and quantum theory are accepted unchanged, and which features are modified.[47][48] Examples include:

- Acoustic metric and other analog models of gravity
- Asymptotic safety in quantum gravity
- Euclidean quantum gravity
- Causal dynamical triangulation[49]
- Causal fermion systems,[5][6][7][8][9][10] giving quantum mechanics, general relativity and quantum field theory as limiting cases.
- Causal sets[50]
- Covariant Feynman path integral approach
- Group field theory[51]
- Wheeler-DeWitt equation
- Geometrodynamics
- Hořava–Lifshitz gravity
- MacDowell–Mansouri action
- Noncommutative geometry.
- Path-integral based models of quantum cosmology[52]
- Regge calculus

- String-nets giving rise to gapless helicity ±2 excitations with no other gapless excitations[53]

- Superfluid vacuum theory a.k.a. theory of BEC vacuum

- Supergravity

- Twistor theory[54]

- Canonical quantum gravity

- E8 Theory

28.4 Weinberg–Witten theorem

In quantum field theory, the Weinberg–Witten theorem places some constraints on theories of composite gravity/emergent gravity. However, recent developments attempt to show that if locality is only approximate and the holographic principle is correct, the Weinberg–Witten theorem would not be valid.

28.5 Experimental tests

As was emphasized above, quantum gravitational effects are extremely weak and therefore difficult to test. For this reason, the possibility of experimentally testing quantum gravity had not received much attention prior to the late 1990s. However, in the past decade, physicists have realized that evidence for quantum gravitational effects can guide the development of the theory. Since theoretical development has been slow, the field of phenomenological quantum gravity, which studies the possibility of experimental tests, has obtained increased attention.[55][56]

The most widely pursued possibilities for quantum gravity phenomenology include violations of Lorentz invariance, imprints of quantum gravitational effects in the cosmic microwave background (in particular its polarization), and decoherence induced by fluctuations in the space-time foam.

The BICEP2 experiment detected what was initially thought to be primordial B-mode polarization caused by gravitational waves in the early universe. If truly primordial, these waves were born as quantum fluctuations in gravity itself. Cosmologist Ken Olum (Tufts University) stated: "I think this is the only observational evidence that we have that actually shows that gravity is quantized....It's probably the only evidence of this that we will ever have."[57]

28.6 See also

28.7 References

[1] Griffiths, David J. (2004). *Introduction to Quantum Mechanics*. Pearson Prentice Hall. OCLC 803860989.

[2] Wald, Robert M. (1984). *General Relativity*. University of Chicago Press. p. 382. OCLC 471881415.

[3] Zee, Anthony (2010). *Quantum Field Theory in a Nutshell* (2nd ed.). Princeton University Press. p. 172. OCLC 659549695.

[4] Penrose, Roger (2007). *The road to reality : a complete guide to the laws of the universe*. Vintage. p. 1017. OCLC 716437154.

[5] F. Finster, J. Kleiner, Causal Fermion Systems as a Candidate for a Unified Physical Theory, http://arxiv.org/abs/1502.03587

[6] F. Finster, The Principle of the Fermionic Projector, hep-th/0001048, hep-th/0202059, hep- th/0210121, AMS/IP Studies in Advanced Mathematics, vol. **35**, American Mathematical Society, Providence, RI, 2006.

[7] F. Finster, A formulation of quantum field theory realizing a sea of interacting Dirac particles, arXiv:0911.2102 [hep-th], Lett. Math. Phys. **97** (2011), no. 2, 165–183.

[8] F. Finster, An action principle for an interacting fermion system and its analysis in the continuum limit, arXiv:0908.1542 [math-ph] (2009).

[9] F. Finster, The continuum limit of a fermion system involving neutrinos: Weak and gravitational interactions, arXiv:1211.3351 [math-ph] (2012).

[10] F. Finster, Perturbative quantum field theory in the framework of the fermionic projector, arXiv:1310.4121 [math-ph], J. Math. Phys. **55** (2014), no. 4, 042301.

[11] Quantum effects in the early universe might have an observable effect on the structure of the present universe, for example, or gravity might play a role in the unification of the other forces. Cf. the text by Wald cited above.

[12] Donoghue (1995). "Introduction to the Effective Field Theory Description of Gravity". arXiv:gr-qc/9512024. (verify against ISBN 9789810229085)

[13] Kraichnan, R. H. (1955). "Special-Relativistic Derivation of Generally Covariant Gravitation Theory". *Physical Review* **98** (4): 1118–1122. Bibcode:1955PhRv...98.1118K. doi:10.1103/PhysRev.98.1118.

[14] Gupta, S. N. (1954). "Gravitation and Electromagnetism". *Physical Review* **96** (6): 1683–1685. Bibcode:1954PhRv...96.1683G. doi:10.1103/PhysRev.96.1683.

[15] Gupta, S. N. (1957). "Einstein's and Other Theories of Gravitation". *Reviews of Modern Physics* **29** (3): 334–336. Bibcode:1957RvMP...29..334G. doi:10.1103/RevModPhys.29.334.

[16] Gupta, S. N. (1962). "Quantum Theory of Gravitation". *Recent Developments in General Relativity*. Pergamon Press. pp. 251–258.

[17] Deser, S. (1970). "Self-Interaction and Gauge Invariance". *General Relativity and Gravitation* **1**: 9–18. arXiv:gr-qc/0411023. Bibcode:1970GReGr...1....9D. doi:10.1007/BF00759198.

[18] Ohta, Tadayuki; Mann, Robert (1996). "Canonical reduction of two-dimensional gravity for particle dynamics". *Classical and Quantum Gravity* **13** (9): 2585–2602. arXiv:gr-qc/9605004. Bibcode:1996CQGra..13.2585O. doi:10.1088/0264-9381/13/9/022.

[19] Sikkema, A E; Mann, R B (1991). "Gravitation and cosmology in (1+1) dimensions". *Classical and Quantum Gravity* **8**: 219–235. Bibcode:1991CQGra...8..219S. doi:10.1088/0264-9381/8/1/022.

[20] Farrugia; Mann; Scott (2007). "N-body Gravity and the Schroedinger Equation". *Classical and Quantum Gravity* **24** (18): 4647–4659. arXiv:gr-qc/0611144. Bibcode:2007CQGra..24.4647F. doi:10.1088/0264-9381/24/18/006.

[21] Mann, R B; Ohta, T (1997). "Exact solution for the metric and the motion of two bodies in (1+1)-dimensional gravity". *Physical Review D* **55** (8): 4723–4747. arXiv:gr-qc/9611008. Bibcode:1997PhRvD..55.4723M. doi:10.1103/PhysRevD.55.4723.

[22] Feynman, R. P.; Morinigo, F. B.; Wagner, W. G.; Hatfield, B. (1995). *Feynman lectures on gravitation*. Addison-Wesley. ISBN 0-201-62734-5.

[23] Hamber, H. W. (2009). *Quantum Gravitation - The Feynman Path Integral Approach*. Springer Publishing. ISBN 978-3-540-85292-6.

[24] Smolin, Lee (2001). *Three Roads to Quantum Gravity*. Basic Books. pp. 20–25. ISBN 0-465-07835-4. Pages 220–226 are annotated references and guide for further reading.

[25] Hunter Monroe (2005). "Singularity-Free Collapse through Local Inflation". arXiv:astro-ph/0506506.

[26] Edward Anderson (2010). "The Problem of Time in Quantum Gravity". arXiv:1009.2157 [gr-qc]. (also published as chapter 4 of ISBN 9781611229578)

[27] A timeline and overview can be found in Rovelli, Carlo (2000). "Notes for a brief history of quantum gravity". arXiv:gr-qc/0006061. (verify against ISBN 9789812777386)

[28] Ashtekar, Abhay (2007). "Loop Quantum Gravity: Four Recent Advances and a Dozen Frequently Asked Questions". *11th Marcel Grossmann Meeting on Recent Developments in Theoretical and Experimental General Relativity*. p. 126. arXiv:0705.2222. Bibcode:2008mgm..conf..126A. doi:10.1142/9789812834300_0008.

[29] Schwarz, John H. (2007). "String Theory: Progress and Problems". *Progress of Theoretical Physics Supplement* **170**: 214–226. arXiv:hep-th/0702219. Bibcode:2007PThPS.170..214S. doi:10.1143/PTPS.170.214.

[30] Donoghue, John F. (editor) (1995). "Introduction to the Effective Field Theory Description of Gravity". In Cornet, Fernando. *Effective Theories: Proceedings of the Advanced School, Almunecar, Spain, 26 June–1 July 1995.* Singapore: World Scientific. arXiv:gr-qc/9512024. ISBN 981-02-2908-9.

[31] Weinberg, Steven (1996). "Chapters 17–18". *The Quantum Theory of Fields II: Modern Applications.* Cambridge University Press. ISBN 0-521-55002-5.

[32] Goroff, Marc H.; Sagnotti, Augusto; Sagnotti, Augusto (1985). "Quantum gravity at two loops". *Physics Letters B* **160**: 81–86. Bibcode:1985PhLB..160...81G. doi:10.1016/0370-2693(85)91470-4.

[33] An accessible introduction at the undergraduate level can be found in Zwiebach, Barton (2004). *A First Course in String Theory.* Cambridge University Press. ISBN 0-521-83143-1., and more complete overviews in Polchinski, Joseph (1998). *String Theory Vol. 1: An Introduction to the Bosonic String.* Cambridge University Press. ISBN 0-521-63303-6. and Polchinski, Joseph (1998b). *String Theory Vol. II: Superstring Theory and Beyond.* Cambridge University Press. ISBN 0-521-63304-4.

[34] Ibanez, L. E. (2000). "The second string (phenomenology) revolution". *Classical & Quantum Gravity* **17** (5): 1117–1128. arXiv:hep-ph/9911499. Bibcode:2000CQGra..17.1117I. doi:10.1088/0264-9381/17/5/321.

[35] For the graviton as part of the string spectrum, e.g. Green, Schwarz & Witten 1987, sec. 2.3 and 5.3; for the extra dimensions, ibid sec. 4.2.

[36] Weinberg, Steven (2000). "Chapter 31". *The Quantum Theory of Fields II: Modern Applications.* Cambridge University Press. ISBN 0-521-55002-5.

[37] Townsend, Paul K. (1996). *Four Lectures on M-Theory.* ICTP Series in Theoretical Physics. p. 385. arXiv:hep-th/9612121. Bibcode:1997hepcbconf..385T.

[38] Duff, Michael (1996). "M-Theory (the Theory Formerly Known as Strings)". *International Journal of Modern Physics A* **11** (32): 5623–5642. arXiv:hep-th/9608117. Bibcode:1996IJMPA..11.5623D. doi:10.1142/S0217751X96002583.

[39] Ashtekar, Abhay (1986). "New variables for classical and quantum gravity". *Physical Review Letters* **57** (18): 2244–2247. Bibcode:1986PhRvL..57.2244A. doi:10.1103/PhysRevLett.57.2244. PMID 10033673.

[40] Ashtekar, Abhay (1987). "New Hamiltonian formulation of general relativity". *Physical Review D* **36** (6): 1587–1602. Bibcode:1987PhRvD..36.1587A. doi:10.1103/PhysRevD.36.1587.

[41] Thiemann, Thomas (2006). "Loop Quantum Gravity: An Inside View". *Approaches to Fundamental Physics.* Lecture Notes in Physics **721**: 185. arXiv:hep-th/0608210. Bibcode:2007LNP...721..185T. doi:10.1007/978-3-540-71117-9_10. ISBN 978-3-540-71115-5.

[42] Rovelli, Carlo (1998). "Loop Quantum Gravity". *Living Reviews in Relativity* **1**. Retrieved 2008-03-13.

[43] Ashtekar, Abhay; Lewandowski, Jerzy (2004). "Background Independent Quantum Gravity: A Status Report". *Classical & Quantum Gravity* **21** (15): R53–R152. arXiv:gr-qc/0404018. Bibcode:2004CQGra..21R..53A. doi:10.1088/0264-9381/21/15/R01.

[44] Thiemann, Thomas (2003). "Lectures on Loop Quantum Gravity". *Lecture Notes in Physics.* Lecture Notes in Physics **631**: 41–135. arXiv:gr-qc/0210094. Bibcode:2003LNP...631...41T. doi:10.1007/978-3-540-45230-0_3. ISBN 978-3-540-40810-9.

[45] Rovelli, Carlo (2004). *Quantum Gravity.* Cambridge University Press. ISBN 0521715962.

[46] Nottale, L. (2011). *Scale Relativity and Fractal Space-Time: A New Approach to Unifying Relativity and Quantum Mechanics.* World Scientific Publishing Company. ISBN 1848166508.;p. 458

[47] Isham, Christopher J. (1994). "Prima facie questions in quantum gravity". In Ehlers, Jürgen; Friedrich, Helmut. *Canonical Gravity: From Classical to Quantum.* Springer. arXiv:gr-qc/9310031. ISBN 3-540-58339-4.

[48] Sorkin, Rafael D. (1997). "Forks in the Road, on the Way to Quantum Gravity". *International Journal of Theoretical Physics* **36** (12): 2759–2781. arXiv:gr-qc/9706002. Bibcode:1997IJTP...36.2759S. doi:10.1007/BF02435709.

[49] Loll, Renate (1998). "Discrete Approaches to Quantum Gravity in Four Dimensions". *Living Reviews in Relativity* **1**: 13. arXiv:gr-qc/9805049. Bibcode:1998LRR.....1...13L. doi:10.12942/lrr-1998-13. Retrieved 2008-03-09.

[50] Sorkin, Rafael D. (2005). "Causal Sets: Discrete Gravity". In Gomberoff, Andres; Marolf, Donald. *Lectures on Quantum Gravity.* Springer. arXiv:gr-qc/0309009. ISBN 0-387-23995-2.

[51] See Daniele Oriti and references therein.

[52] Hawking, Stephen W. (1987). "Quantum cosmology". In Hawking, Stephen W.; Israel, Werner. *300 Years of Gravitation*. Cambridge University Press. pp. 631–651. ISBN 0-521-37976-8.

[53] Wen 2006

[54] See ch. 33 in Penrose 2004 and references therein.

[55] Hossenfelder, Sabine (2011). "Experimental Search for Quantum Gravity". In V. R. Frignanni. *Classical and Quantum Gravity: Theory, Analysis and Applications*. Chapter 5: Nova Publishers. ISBN 978-1-61122-957-8.

[56] Hossenfelder, Sabine (2010-10-17). V. R. Frignanni, ed. "Experimental Search for Quantum Gravity". *Classical and Quantum Gravity: Theory, Analysis and Applications* (Nova Publishers) **5** (2011). arXiv:1010.3420. Bibcode:2010arXiv1010.3420H. |chapter= ignored (help)

[57] Camille Carlisle. "First Direct Evidence of Big Bang Inflation". SkyandTelescope.com. Retrieved March 18, 2014.

28.8 Further reading

- Ahluwalia, D. V. (2002). "Interface of Gravitational and Quantum Realms". *Modern Physics Letters A* **17** (15–17): 1135. arXiv:gr-qc/0205121. Bibcode:2002MPLA...17.1135A. doi:10.1142/S021773230200765X.

- Ashtekar, Abhay (2005). "The winding road to quantum gravity" (PDF). *Current Science* **89**: 2064–2074.

- Carlip, Steven (2001). "Quantum Gravity: a Progress Report". *Reports on Progress in Physics* **64** (8): 885–942. arXiv:gr-qc/0108040. Bibcode:2001RPPh...64..885C. doi:10.1088/0034-4885/64/8/301.

- Herbert W. Hamber (2009). *Quantum Gravitation*. Springer Publishing. doi:10.1007/978-3-540-85293-3. ISBN 978-3-540-85292-6.

- Kiefer, Claus (2007). *Quantum Gravity*. Oxford University Press. ISBN 0-19-921252-X.

- Kiefer, Claus (2005). "Quantum Gravity: General Introduction and Recent Developments". *Annalen der Physik* **15**: 129–148. arXiv:gr-qc/0508120. Bibcode:2006AnP...518..129K. doi:10.1002/andp.200510175.

- Lämmerzahl, Claus, ed. (2003). *Quantum Gravity: From Theory to Experimental Search*. Lecture Notes in Physics. Springer. ISBN 3-540-40810-X.

- Rovelli, Carlo (2004). *Quantum Gravity*. Cambridge University Press. ISBN 0-521-83733-2.

- Trifonov, Vladimir (2008). "GR-friendly description of quantum systems". *International Journal of Theoretical Physics* **47** (2): 492–510. arXiv:math-ph/0702095. Bibcode:2008IJTP...47..492T. doi:10.1007/s10773-007-9474-3.

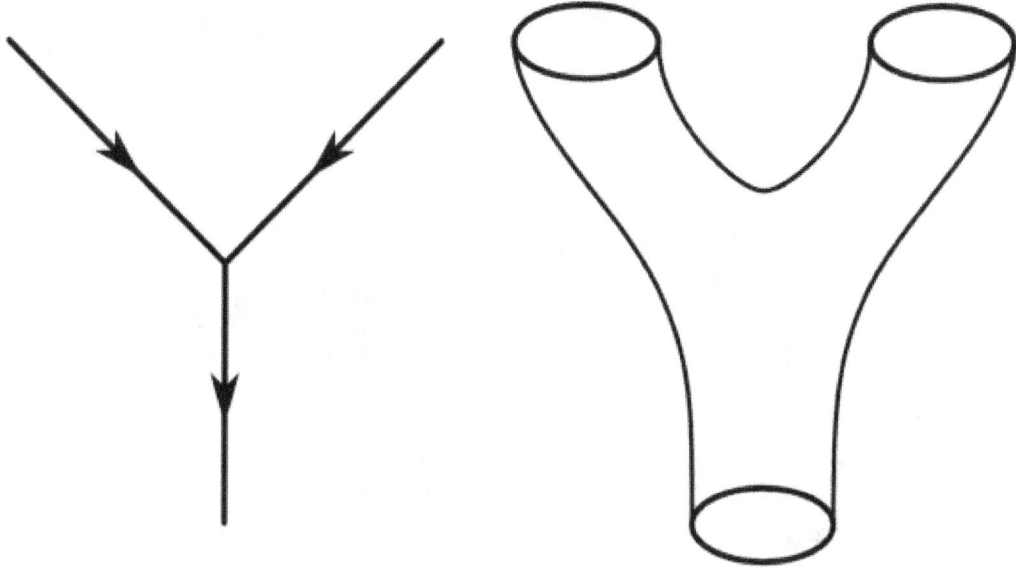

Interaction in the subatomic world: world lines of point-like particles in the Standard Model or a world sheet swept up by closed strings in string theory

Projection of a Calabi–Yau manifold, one of the ways of compactifying the extra dimensions posited by string theory

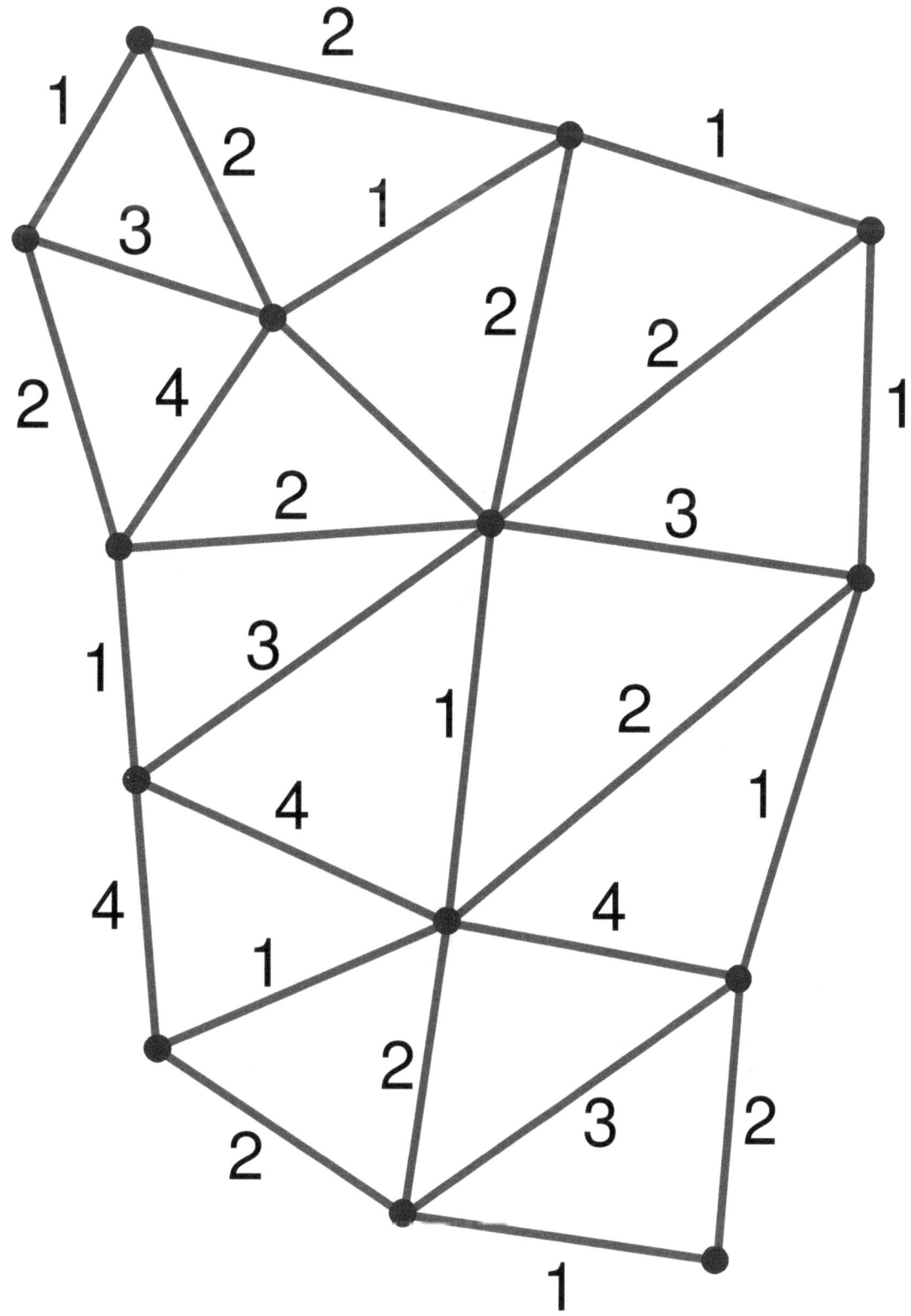

Simple spin network of the type used in loop quantum gravity

Schrödinger's flower. Morphogenesis of a flower-like structure, solution of a growth process equation that takes the form of a Schrödinger equation under fractal conditions.

Chapter 29

Circle group

For the jazz group, see Circle (jazz band).

In mathematics, the **circle group**, denoted by **T**, is the multiplicative group of all complex numbers with absolute value 1, i.e., the unit circle in the complex plane or simply the **unit complex numbers**[1]

$$\mathbb{T} = \{z \in \mathbb{C} : |z| = 1\}.$$

The circle group forms a subgroup of $\mathbf{C}^{\times}$, the multiplicative group of all nonzero complex numbers. Since $\mathbf{C}^{\times}$ is abelian, it follows that **T** is as well. The circle group is also the group **U(1)** of 1×1 unitary matrices; these act on the complex plane by rotation about the origin. The circle group can be parametrized by the angle θ of rotation by

$$\theta \mapsto z = e^{i\theta} = \cos\theta + i\sin\theta.$$

This is the exponential map for the circle group.

The circle group plays a central role in Pontryagin duality, and in the theory of Lie groups.

The notation **T** for the circle group stems from the fact that $\mathbf{T}^n$ (the direct product of **T** with itself n times) is geometrically an n-torus. The circle group is then a 1-torus.

29.1 Elementary introduction

One way to think about the circle group is that it describes how to add *angles*, where only angles between 0° and 360° are permitted. For example, the diagram illustrates how to add 150° to 270°. The answer should be 150° + 270° = 420°, but when thinking in terms of the circle group, we need to "forget" the fact that we have wrapped once around the circle. Therefore we adjust our answer by 360° which gives 420° = 60° (mod 360°).

Another description is in terms of ordinary addition, where only numbers between 0 and 1 are allowed (with 1 corresponding to a full rotation). To achieve this, we might need to throw away digits occurring before the decimal point. For example, when we work out 0.784 + 0.925 + 0.446, the answer should be 2.155, but we throw away the leading 2, so the answer (in the circle group) is just 0.155.

29.2 Topological and analytic structure

The circle group is more than just an abstract algebraic object. It has a natural topology when regarded as a subspace of the complex plane. Since multiplication and inversion are continuous functions on $\mathbf{C}^{\times}$, the circle group has the structure

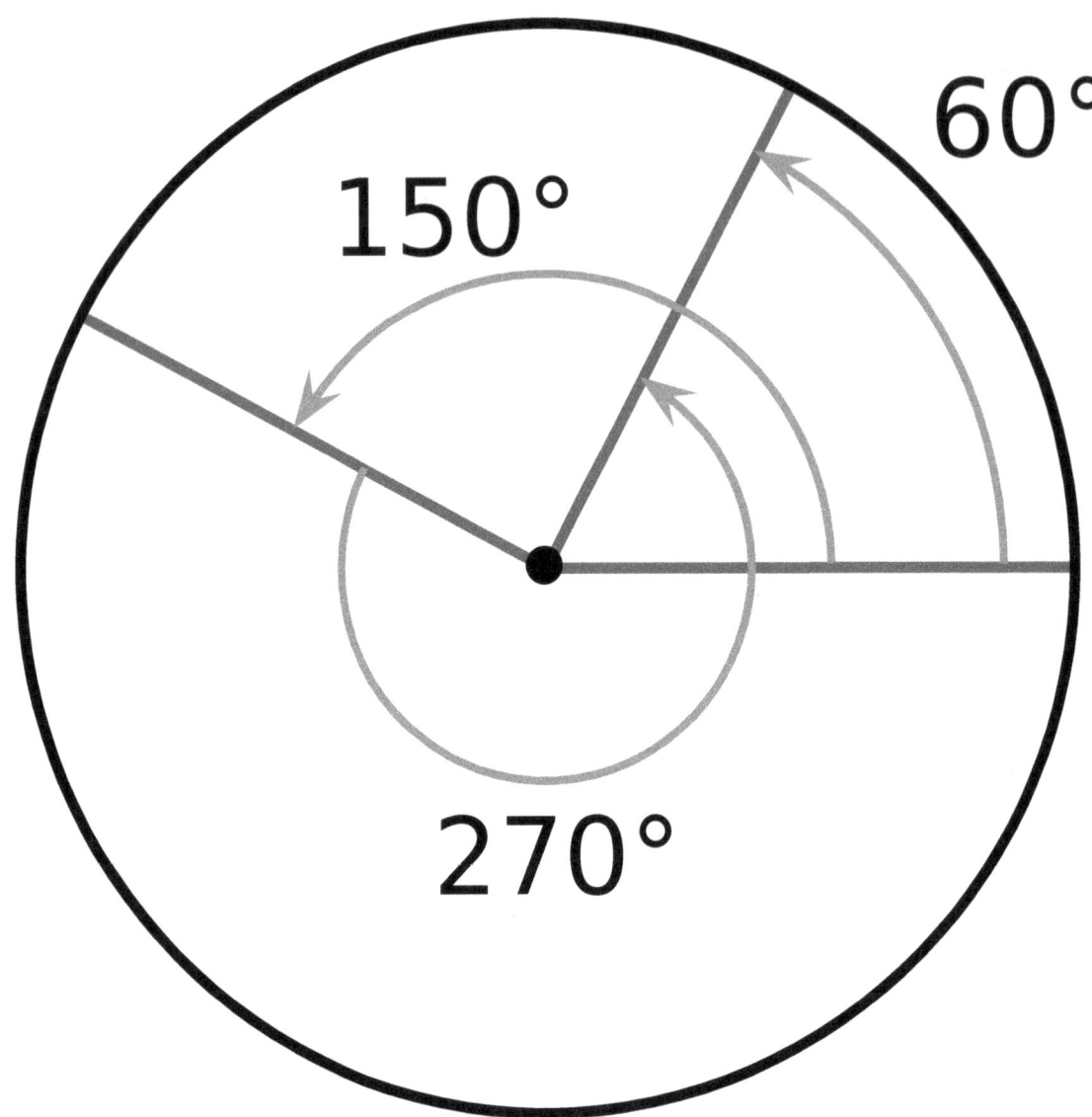

Multiplication on the circle group is equivalent to addition of angles

of a topological group. Moreover, since the unit circle is a closed subset of the complex plane, the circle group is a closed subgroup of $\mathbf{C}^{\times}$ (itself regarded as a topological group).

One can say even more. The circle is a 1-dimensional real manifold and multiplication and inversion are real-analytic maps on the circle. This gives the circle group the structure of a one-parameter group, an instance of a Lie group. In fact, up to isomorphism, it is the unique 1-dimensional compact, connected Lie group. Moreover, every n-dimensional compact, connected, abelian Lie group is isomorphic to $\mathbf{T}^n$.

29.3   Isomorphisms

The circle group shows up in a variety of forms in mathematics. We list some of the more common forms here. Specifically, we show that

$\mathbb{T} \cong \mathrm{U}(1) \cong \mathbb{R}/\mathbb{Z} \cong \mathrm{SO}(2).$

Note that the slash (/) denotes here quotient group.

The set of all 1×1 unitary matrices clearly coincides with the circle group; the unitary condition is equivalent to the condition that its element have absolute value 1. Therefore, the circle group is canonically isomorphic to U(1), the first unitary group.

The exponential function gives rise to a group homomorphism exp : $\mathbf{R} \to \mathbf{T}$ from the additive real numbers $\mathbf{R}$ to the circle group $\mathbf{T}$ via the map

$$\theta \mapsto e^{i\theta} = \cos\theta + i\sin\theta.$$

The last equality is Euler's formula or the complex exponential. The real number θ corresponds to the angle on the unit circle as measured from the positive x-axis. That this map is a homomorphism follows from the fact that the multiplication of unit complex numbers corresponds to addition of angles:

$$e^{i\theta_1}e^{i\theta_2} = e^{i(\theta_1+\theta_2)}.$$

This exponential map is clearly a surjective function from $\mathbf{R}$ to $\mathbf{T}$. It is not, however, injective. The kernel of this map is the set of all integer multiples of 2π. By the first isomorphism theorem we then have that

$$\mathbb{T} \cong \mathbb{R}/2\pi\mathbb{Z}.$$

After rescaling we can also say that $\mathbf{T}$ is isomorphic to $\mathbf{R/Z}$.

If complex numbers are realized as 2×2 real matrices (see complex number), the unit complex numbers correspond to 2×2 orthogonal matrices with unit determinant. Specifically, we have

$$e^{i\theta} \leftrightarrow \begin{bmatrix} \cos\theta & -\sin\theta \\ \sin\theta & \cos\theta \end{bmatrix}.$$

The circle group is therefore isomorphic to the special orthogonal group SO(2). This has the geometric interpretation that multiplication by a unit complex number is a proper rotation in the complex plane, and every such rotation is of this form.

29.4   Properties

Every compact Lie group G of dimension > 0 has a subgroup isomorphic to the circle group. That means that, thinking in terms of symmetry, a compact symmetry group acting *continuously* can be expected to have one-parameter circle subgroups acting; the consequences in physical systems are seen for example at rotational invariance, and spontaneous symmetry breaking.

The circle group has many subgroups, but its only proper closed subgroups consist of roots of unity: For each integer $n >$ 0, the nth roots of unity form a cyclic group of order n, which is unique up to isomorphism.

29.5   Representations

The representations of the circle group are easy to describe. It follows from Schur's lemma that the irreducible complex representations of an abelian group are all 1-dimensional. Since the circle group is compact, any representation $\rho : \mathbf{T} \to$

$GL(1, \mathbf{C}) \cong \mathbf{C}^{\times}$, must take values in $U(1) \cong \mathbf{T}$. Therefore, the irreducible representations of the circle group are just the homomorphisms from the circle group to itself.

These representations are all inequivalent. The representation $\varphi_{-}n$ is conjugate to φn,

$$\phi_{-n} = \overline{\phi_n}.$$

These representations are just the characters of the circle group. The character group of $\mathbf{T}$ is clearly an infinite cyclic group generated by φ_1:

$$\mathrm{Hom}(\mathbb{T}, \mathbb{T}) \cong \mathbb{Z}.$$

The irreducible real representations of the circle group are the trivial representation (which is 1-dimensional) and the representations

$$\rho_n(e^{i\theta}) = \begin{bmatrix} \cos n\theta & -\sin n\theta \\ \sin n\theta & \cos n\theta \end{bmatrix}, \quad n \in \mathbb{Z}^+,$$

taking values in SO(2). Here we only have positive integers n since the representation ρ_{-n} is equivalent to ρ_n.

29.6 Group structure

In this section we will forget about the topological structure of the circle group and look only at its structure as an abstract group.

The circle group $\mathbf{T}$ is a divisible group. Its torsion subgroup is given by the set of all nth roots of unity for all n, and is isomorphic to $\mathbf{Q}/\mathbf{Z}$. The structure theorem for divisible groups and the axiom of choice together tell us that $\mathbf{T}$ is isomorphic to the direct sum of $\mathbf{Q}/\mathbf{Z}$ with a number of copies of $\mathbf{Q}$. The number of copies of $\mathbf{Q}$ must be c (the cardinality of the continuum) in order for the cardinality of the direct sum to be correct. But the direct sum of c copies of $\mathbf{Q}$ is isomorphic to $\mathbf{R}$, as $\mathbf{R}$ is a vector space of dimension c over $\mathbf{Q}$. Thus

$$\mathbb{T} \cong \mathbb{R} \oplus (\mathbb{Q}/\mathbb{Z}).$$

The isomorphism

$$\mathbb{C}^{\times} \cong \mathbb{R} \oplus (\mathbb{Q}/\mathbb{Z})$$

can be proved in the same way, as $\mathbf{C}^{\times}$ is also a divisible abelian group whose torsion subgroup is the same as the torsion subgroup of $\mathbf{T}$.

29.7 See also

- Rotation number
- Torus
- One-parameter subgroup
- Unitary group

- Orthogonal group

- Group of rational points on the unit circle

- Phase factor (application in quantum-mechanics)

29.8   Notes

[1] "a **unit complex number** is a complex number of unit absolute value" (James & James 1992, p. 436)

29.9   References

- James, Robert C.; James, Glenn (1992), *Mathematics Dictionary* (Fifth ed.), Chapman & Hall

29.10   Further reading

- Hua Luogeng (1981) *Starting with the unit circle*, Springer Verlag, ISBN 0-387-90589-8 .

29.11   External links

- Homeomorphism and the Group Structure on a Circle

Chapter 30

Spinor

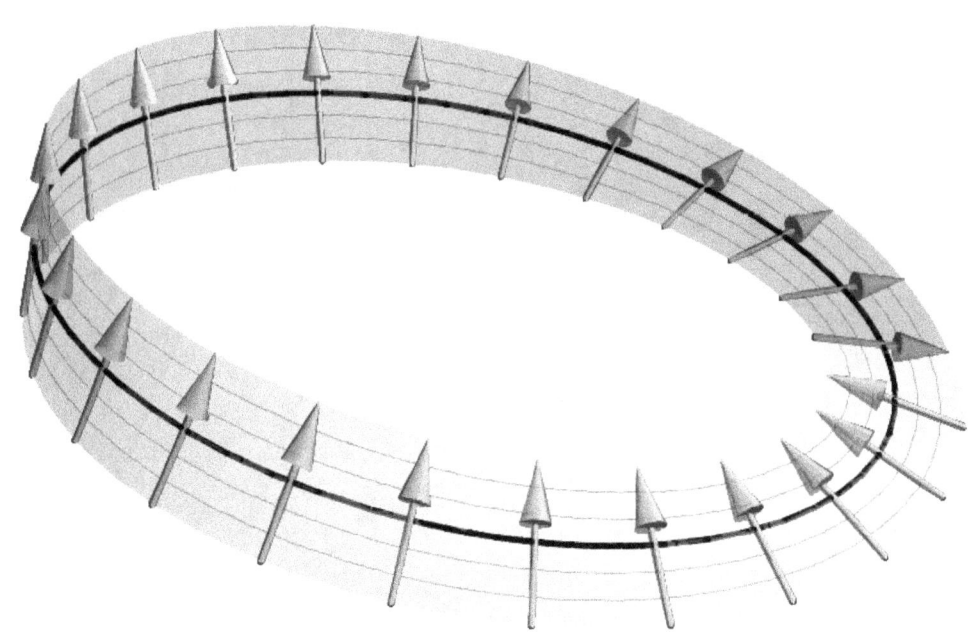

A spinor visualized as a vector pointing along the Möbius band, exhibiting a sign inversion when the circle (the "physical system") is rotated through a full turn of 360°.[nb 1]

In geometry and physics, **spinors** are elements of a (complex) vector space that can be associated with Euclidean space.[nb 2] Like geometric vectors and more general tensors, spinors transform linearly when the Euclidean space is subjected to a slight (infinitesimal) rotation.[nb 3] When a sequence of such small rotations is composed (integrated) to form an overall final rotation, however, the resulting spinor transformation depends on which sequence of small rotations was used, *unlike* for vectors and tensors. A spinor transforms to its negative when the space is rotated through a complete turn from 0° to 360° (see picture), and it is this property that characterizes spinors. It is also possible to associate a substantially similar notion of spinor to Minkowski space in which case the Lorentz transformations of special relativity play the role of rotations. Spinors were introduced in geometry by Élie Cartan in 1913.[1][2] In the 1920s physicists discovered that spinors are essential to describe the intrinsic angular momentum, or "spin", of the electron and other subatomic particles.[nb 4]

Spinors are characterized by the specific way in which they behave under rotations. They change in different ways depending not just on the overall final rotation, but the details of how that rotation was achieved (by a continuous path in the rotation group). There are two topologically distinguishable classes (homotopy classes) of paths through rotations that result in the same overall rotation, as famously illustrated by the belt trick puzzle (below). These two inequivalent classes yield spinor transformations of opposite sign. The spin group is the group of all rotations keeping track of the class.[nb 5] It doubly-covers the rotation group, since each rotation can be obtained in two inequivalent ways as the endpoint of a path. The space of spinors by definition is equipped with a (complex) linear representation of the spin group, meaning that elements of the spin group act as linear transformations on the space of spinors, in a way that genuinely depends on the homotopy class.[nb 6]

Although spinors can be defined purely as elements of a representation space of the spin group (or its Lie algebra of infinitesimal rotations), they are typically defined as elements of a vector space that carries a linear representation of the Clifford algebra. The Clifford algebra is an associative algebra that can be constructed from Euclidean space and its inner product in a basis independent way. Both the spin group and its Lie algebra are embedded inside the Clifford algebra in a natural way, and in applications the Clifford algebra is often the easiest to work with.[nb 7] After choosing an orthonormal basis of Euclidean space, a representation of the Clifford algebra is generated by gamma matrices, matrices that satisfy a set of canonical anti-commutation relations. The spinors are the column vectors on which these matrices act. In three Euclidean dimensions, for instance, the Pauli spin matrices are a set of gamma matrices,[nb 8] and the two-component complex column vectors on which these matrices act are spinors. However, the particular matrix representation of the Clifford algebra, and hence what precisely constitutes a "column vector" (or spinor), involves the choice of basis and gamma matrices in an essential way. As a representation of the spin group, this realization of spinors as (complex[nb 9]) column vectors will either be irreducible if the dimension is odd, or it will decompose into a pair of so-called "half-spin" or Weyl representations if the dimension is even.[nb 10]

30.1   Introduction

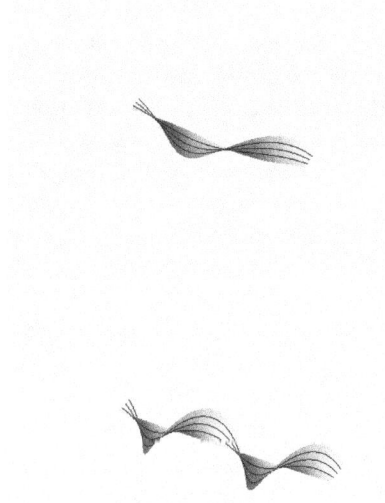

A gradual rotation can be visualized as a ribbon in space (the TNB frame of the ribbon defines a rotation continuously for each value of the arc length parameter). Two gradual rotations with different classes, one through 2π and one through

4π, are illustrated here in the belt trick puzzle. A solution of the puzzle is a (continuous) manipulation of the belt, fixing the endpoints, that untwists it. This is impossible with the 2π rotation, but possible with the 4π rotation. A solution, shown in the second animation, actually gives an explicit homotopy in the rotation group between the 4π rotation and the trivial (identity) rotation.

What characterizes spinors and distinguishes them from geometric vectors and other tensors is subtle. Consider applying a rotation to the coordinates of a system. No object in the system itself has moved, only the coordinates have, so there will always be a compensating change in those coordinate values when applied to any object of the system. Geometrical vectors, for example, have components that will undergo *the same* rotation as the coordinates. More broadly, any tensor associated with the system (for instance, the stress of some medium) also has coordinate descriptions that adjust to compensate for changes to the coordinate system itself. Spinors do not appear at this level of the description of a physical system, when one is concerned only with the properties of a single isolated rotation of the coordinates. Rather, spinors appear when we imagine that instead of a single rotation, the coordinate system is gradually (continuously) rotated between some initial and final configuration. For any of the familiar and intuitive ("tensorial") quantities associated with the system, the transformation law does not depend on the precise details of how the coordinates arrived at their final configuration. Spinors, on the other hand, are constructed in such a way that makes them *sensitive* to how the gradual rotation of the coordinates arrived there: they exhibit path-dependence. It turns out that, for any final configuration of the coordinates, there are actually two ("topologically") inequivalent *gradual* (continuous) rotations of the coordinate system that result in this same configuration. This ambiguity is called the homotopy class of the gradual rotation. The belt trick puzzle (shown) famously demonstrates two different rotations, one through an angle of 2π and the other through an angle of 4π, having the same final configurations but different classes. Spinors actually exhibit a sign-reversal that genuinely depends on this homotopy class. This distinguishes them from vectors and other tensors, none of which can feel the class.

Spinors can be exhibited as concrete objects using a choice of Cartesian coordinates. In three Euclidean dimensions, for instance, spinors can be constructed by making a choice of Pauli spin matrices corresponding to (angular momenta about) the three coordinate axes. These are 2×2 matrices with complex entries, and the two-component complex column vectors on which these matrices act by matrix multiplication are the spinors. In this case, the spin group is isomorphic to the group of 2×2 unitary matrices with determinant one, which naturally sits inside the matrix algebra. This group acts by conjugation on the real vector space spanned by the Pauli matrices themselves,[nb 11] realizing it as a group of rotations among them,[nb 12] but it also acts on the column vectors (that is, the spinors).

More generally, a Clifford algebra can be constructed from any vector space V equipped with a (nondegenerate) quadratic form, such as Euclidean space with its standard dot product or Minkowski space with its standard Lorentz metric. Given a suitably normalized basis of V, the Clifford algebra is generated by gamma matrices, matrices that satisfy a set of canonical anti-commutation relations, and the space of spinors is the space of column vectors with $2^{\lfloor \dim V/2 \rfloor}$ components on which those matrices act. Although the Clifford algebra can be defined abstractly in a coordinate-independent way, its particular realization as a specific algebra of matrices depends on which orthogonal axes the gamma matrices represent. So what precisely constitutes a "column vector" (or spinor) also depends on such arbitrary choices.[nb 13] The orthogonal Lie algebra (i.e., the infinitesimal "rotations") and the spin group associated to the quadratic form are both (canonically) contained in the Clifford algebra, so every Clifford algebra representation also defines a representation of the Lie algebra and the spin group.[nb 14] Depending on the dimension and metric signature, this realization of spinors as column vectors may be irreducible or it may decompose into a pair of so-called "half-spin" or Weyl representations.[nb 15]

30.2 Overview

There are essentially two frameworks for viewing the notion of a spinor.

One is representation theoretic. In this point of view, one knows beforehand that there are some representations of the Lie algebra of the orthogonal group that cannot be formed by the usual tensor constructions. These missing representations are then labeled the **spin representations**, and their constituents *spinors*. In this view, a spinor must belong to a representation of the double cover of the rotation group SO(n, **R**), or more generally of double cover of the generalized special orthogonal group SO$^+$(p, q, **R**) on spaces with metric signature (p, q). These double covers are Lie groups, called the spin groups Spin(n) or Spin(p, q). All the properties of spinors, and their applications and derived objects, are manifested first in the spin group. Representations of the double covers of these groups yield projective representations of the groups themselves, which do not meet the full definition of a representation.

The other point of view is geometrical. One can explicitly construct the spinors, and then examine how they behave under the action of the relevant Lie groups. This latter approach has the advantage of providing a concrete and elementary description of what a spinor is. However, such a description becomes unwieldy when complicated properties of spinors, such as Fierz identities, are needed.

30.2.1 Clifford algebras

For more details on this topic, see Clifford algebra.

The language of Clifford algebras[3] (sometimes called geometric algebras) provides a complete picture of the spin representations of all the spin groups, and the various relationships between those representations, via the classification of Clifford algebras. It largely removes the need for *ad hoc* constructions.

In detail, let V be a finite-dimensional complex vector space with nondegenerate bilinear form g. The Clifford algebra $C\ell(V, g)$ is the algebra generated by V along with the anticommutation relation $xy + yx = 2g(x, y)$. It is an abstract version of the algebra generated by the gamma or Pauli matrices. If $V = \mathbf{C}^n$, with the standard form $g(x, y) = x^t y = x_1 y_1 + \dots + x_n y n$ we denote the Clifford algebra by $C\ell n(\mathbf{C})$. Since by the choice of an orthonormal basis every complex vectorspace with non-degenerate form is isomorphic to this standard example, this notation is abused more generally if $\dim_{\mathbf{C}}(V) = n$. If $n = 2k$ is even, $C\ell n(\mathbf{C})$ is isomorphic as an algebra (in a non-unique way) to the algebra $\mathrm{Mat}(2^k, \mathbf{C})$ of $2^k \times 2^k$ complex matrices (by the Artin-Wedderburn theorem and the easy to prove fact that the Clifford algebra is central simple). If $n = 2k + 1$ is odd, $C\ell_{2k+1}(\mathbf{C})$ is isomorphic to the algebra $\mathrm{Mat}(2^k, \mathbf{C}) \oplus \mathrm{Mat}(2^k, \mathbf{C})$ of two copies of the $2^k \times 2^k$ complex matrices. Therefore, in either case $C\ell(V, g)$ has a unique (up to isomorphism) irreducible representation (also called simple Clifford module), commonly denoted by Δ, of dimension $2^{[n/2]}$. Since the Lie algebra $\mathbf{so}(V, g)$ is embedded as a Lie subalgebra in $C\ell(V, g)$ equipped with the Clifford algebra commutator as Lie bracket, the space Δ is also a Lie algebra representation of $\mathbf{so}(V, g)$ called a spin representation. If n is odd, this Lie algebra representation is irreducible. If n is even, it splits further into two irreducible representations $\Delta = \Delta_+ \oplus \Delta_-$ called the Weyl or *half-spin representations*.

Irreducible representations over the reals in the case when V is a real vector space are much more intricate, and the reader is referred to the Clifford algebra article for more details.

30.2.2 Spin groups

Spinors form a vector space, usually over the complex numbers, equipped with a linear group representation of the spin group that does not factor through a representation of the group of rotations (see diagram). The spin group is the group of rotations keeping track of the homotopy class. Spinors are needed to encode basic information about the topology of the group of rotations because that group is not simply connected, but the simply connected spin group is its double cover. So for every rotation there are two elements of the spin group that represent it. Geometric vectors and other tensors cannot feel the difference between these two elements, but they produce *opposite* signs when they affect any spinor under the representation. Thinking of the elements of the spin group as homotopy classes of one-parameter families of rotations, each rotation is represented by two distinct homotopy classes of paths to the identity. If a one-parameter family of rotations is visualized as a ribbon in space, with the arc length parameter of that ribbon being the parameter (its tangent, normal, binormal frame actually gives the rotation), then these two distinct homotopy classes are visualized in the two states of the belt trick puzzle (above). The space of spinors is an auxiliary vector space that can be constructed explicitly in coordinates, but ultimately only exists up to isomorphism in that there is no "natural" construction of them that does not rely on arbitrary choices such as coordinate systems. A notion of spinors can be associated, as such an auxiliary mathematical object, with any vector space equipped with a quadratic form such as Euclidean space with its standard dot product, or Minkowski space with its Lorentz metric. In the latter case, the "rotations" include the Lorentz boosts, but otherwise the theory is substantially similar.

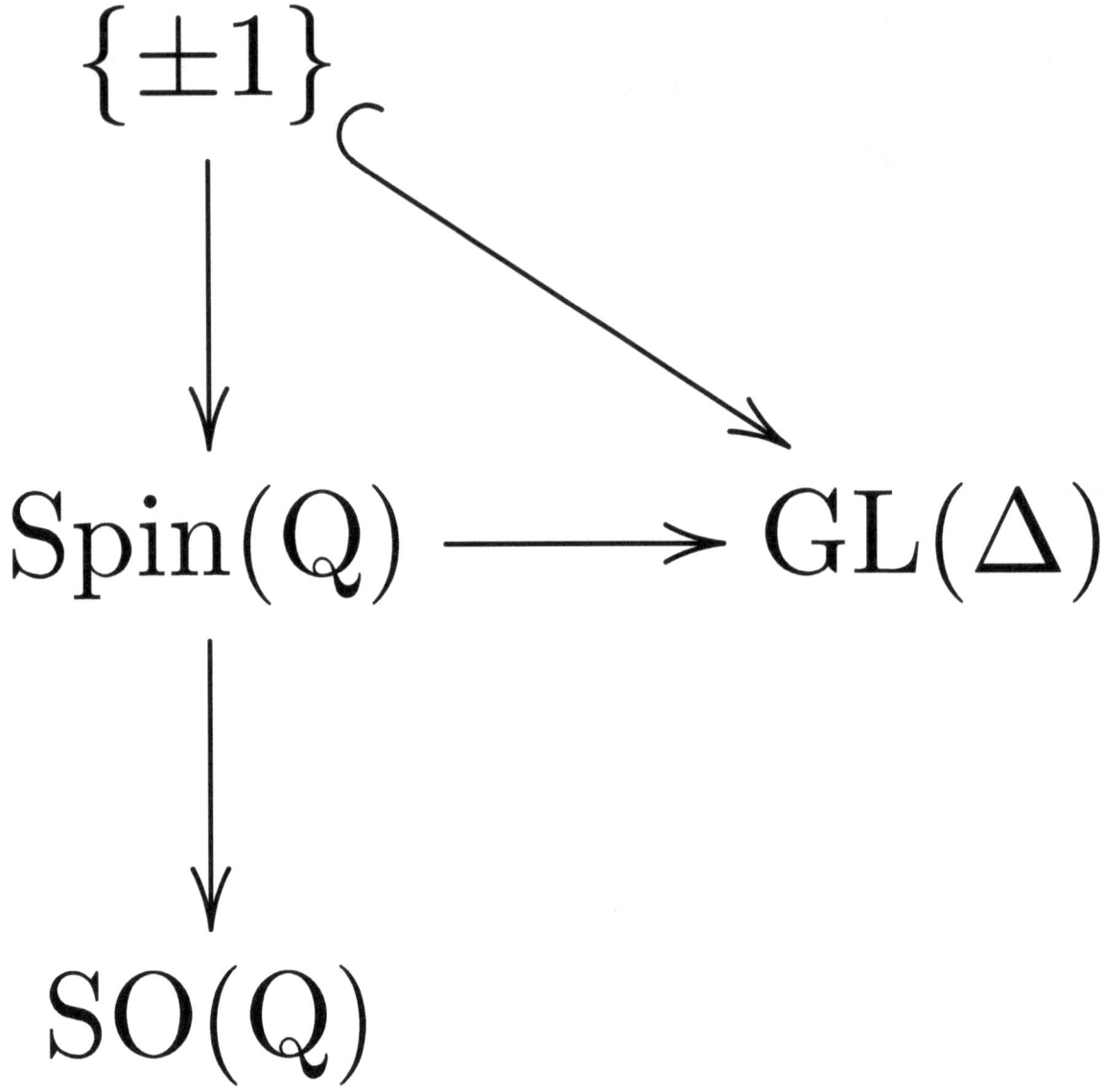

The spin representation Δ is a vector space equipped with a representation of the spin group that does not factor through a representation of the (special) orthogonal group.

30.2.3 Terminology in physics

The most typical type of spinor, the Dirac spinor,[4] is an element of the fundamental representation of $C\ell p{+}q(\mathbf{C})$, the complexification of the Clifford algebra $C\ell p, q(\mathbf{R})$, into which the spin group $\mathrm{Spin}(p, q)$ may be embedded. On a $2k$- or $2k{+}1$-dimensional space a Dirac spinor may be represented as a vector of 2^k complex numbers. (See Special unitary group.) In even dimensions, this representation is reducible when taken as a representation of $\mathrm{Spin}(p, q)$ and may be decomposed into two: the left-handed and right-handed **Weyl spinor**[5] representations. In addition, sometimes the non-complexified version of $C\ell p,q(\mathbf{R})$ has a smaller real representation, the Majorana spinor representation.[6] If this happens in an even dimension, the Majorana spinor representation will sometimes decompose into two Majorana–Weyl spinor representations. Dirac and Weyl spinors are complex representations while Majorana spinors are real representations.

The Dirac, Lorentz, Weyl, and Majorana spinors are interrelated, and their relation can be elucidated on the basis of real geometric algebra.[7]

Massive particles, such as electrons, are described as Dirac spinors. The classical neutrino of the standard model of particle physics is an example of a Weyl spinor. However, because of observed neutrino oscillation, it is now believed that they are not Weyl spinors, but perhaps instead Majorana spinors.[8] It is not known whether (spin-1/2) Weyl spinors exist in nature. In 2015, an international team led by Princeton University scientists announced that they had found a quasiparticle that behaves as a Weyl fermion.[9]

30.2.4   Spinors in representation theory

Main article: Spin representation

One major mathematical application of the construction of spinors is to make possible the explicit construction of linear representations of the Lie algebras of the special orthogonal groups, and consequently spinor representations of the groups themselves. At a more profound level, spinors have been found to be at the heart of approaches to the Atiyah–Singer index theorem, and to provide constructions in particular for discrete series representations of semisimple groups.

The spin representations of the special orthogonal Lie algebras are distinguished from the tensor representations given by Weyl's construction by the weights. Whereas the weights of the tensor representations are integer linear combinations of the roots of the Lie algebra, those of the spin representations are half-integer linear combinations thereof. Explicit details can be found in the spin representation article.

30.2.5   Attempts at intuitive understanding

The spinor can be described, in simple terms, as "vectors of a space the transformations of which are related in a particular way to rotations in physical space".[10] Stated differently:[2]

> *Spinors [...] provide a linear representation of the group of rotations in a space with any number n of dimensions, each spinor having 2^ν components where $n = 2\nu + 1$ or 2ν.*

Several ways of illustrating everyday analogies have been formulated in terms of the plate trick, tangloids and other examples of orientation entanglement.

Nonetheless, the concept is generally considered notoriously difficult to understand, as illustrated by Michael Atiyah's statement that is recounted by Dirac's biographer Graham Farmelo:[11]

> *No one fully understands spinors. Their algebra is formally understood but their general significance is mysterious. In some sense they describe the "square root" of geometry and, just as understanding the square root of −1 took centuries, the same might be true of spinors.*

30.3   History

The most general mathematical form of spinors was discovered by Élie Cartan in 1913.[12] The word "spinor" was coined by Paul Ehrenfest in his work on quantum physics.[13]

Spinors were first applied to mathematical physics by Wolfgang Pauli in 1927, when he introduced his spin matrices.[14] The following year, Paul Dirac discovered the fully relativistic theory of electron spin by showing the connection between spinors and the Lorentz group.[15] By the 1930s, Dirac, Piet Hein and others at the Niels Bohr Institute (then known as the Institute for Theoretical Physics of the University of Copenhagen) created toys such as Tangloids to teach and model the calculus of spinors.

Spinor spaces were represented as left ideals of a matrix algebra in 1930, by G. Juvet[16] and by Fritz Sauter.[17][18] More specifically, instead of representing spinors as complex-valued 2D column vectors as Pauli had done, they represented them as complex-valued 2×2 matrices in which only the elements of the left column are non-zero. In this manner the spinor space became a minimal left ideal in Mat(2, **C**).[19][20]

In 1947 Marcel Riesz constructed spinor spaces as elements of a minimal left ideal of Clifford algebras. In 1966/1967, David Hestenes[21][22] replaced spinor spaces by the even subalgebra $C\ell^0_{1,3}(\mathbf{R})$ of the spacetime algebra $C\ell_{1,3}(\mathbf{R})$.[18][20] As of the 1980s, the theoretical physics group at Birkbeck College around David Bohm and Basil Hiley has been developing algebraic approaches to quantum theory that build on Sauter and Riesz' identification of spinors with minimal left ideals.

30.4 Examples

Some simple examples of spinors in low dimensions arise from considering the even-graded subalgebras of the Clifford algebra $C\ell p, q(\mathbf{R})$. This is an algebra built up from an orthonormal basis of $n = p + q$ mutually orthogonal vectors under addition and multiplication, p of which have norm +1 and q of which have norm −1, with the product rule for the basis vectors

$$e_i e_j = \begin{cases} +1 & i = j,\, i \in (1 \ldots p) \\ -1 & i = j,\, i \in (p+1 \ldots n) \\ -e_j e_i & i \neq j. \end{cases}$$

30.4.1 Two dimensions

The Clifford algebra $C\ell_{2,0}(\mathbf{R})$ is built up from a basis of one unit scalar, 1, two orthogonal unit vectors, σ_1 and σ_2, and one unit pseudoscalar $i = \sigma_1\sigma_2$. From the definitions above, it is evident that $(\sigma_1)^2 = (\sigma_2)^2 = 1$, and $(\sigma_1\sigma_2)(\sigma_1\sigma_2) = -\sigma_1\sigma_1\sigma_2\sigma_2 = -1$.

The even subalgebra $C\ell^0_{2,0}(\mathbf{R})$, spanned by *even-graded* basis elements of $C\ell_{2,0}(\mathbf{R})$, determines the space of spinors via its representations. It is made up of real linear combinations of 1 and $\sigma_1\sigma_2$. As a real algebra, $C\ell^0_{2,0}(\mathbf{R})$ is isomorphic to field of complex numbers $\mathbf{C}$. As a result, it admits a conjugation operation (analogous to complex conjugation), sometimes called the *reverse* of a Clifford element, defined by

$$(a + b\sigma_1\sigma_2)^* = a + b\sigma_2\sigma_1$$

which, by the Clifford relations, can be written

$$(a + b\sigma_1\sigma_2)^* = a + b\sigma_2\sigma_1 = a - b\sigma_1\sigma_2$$

The action of an even Clifford element $\gamma \in C\ell^0_{2,0}(\mathbf{R})$ on vectors, regarded as 1-graded elements of $C\ell_{2,0}(\mathbf{R})$, is determined by mapping a general vector $u = a_1\sigma_1 + a_2\sigma_2$ to the vector

$$\gamma(u) = \gamma u \gamma^*$$

where γ^* is the conjugate of γ, and the product is Clifford multiplication. In this situation, a **spinor**[23] is an ordinary complex number. The action of γ on a spinor φ is given by ordinary complex multiplication:

$$\gamma(\phi) = \gamma\phi$$

An important feature of this definition is the distinction between ordinary vectors and spinors, manifested in how the even-graded elements act on each of them in different ways. In general, a quick check of the Clifford relations reveals that even-graded elements conjugate-commute with ordinary vectors:

$$\gamma(u) = \gamma u \gamma^* = \gamma^2 u$$

On the other hand, comparing with the action on spinors $\gamma'(\varphi) = \gamma\varphi$, γ on ordinary vectors acts as the *square* of its action on spinors.

Consider, for example, the implication this has for plane rotations. Rotating a vector through an angle of θ corresponds to $\gamma^2 = \exp(\theta\,\sigma_1\sigma_2)$, so that the corresponding action on spinors is via $\gamma = \pm \exp(\theta\,\sigma_1\sigma_2/2)$. In general, because of logarithmic branching, it is impossible to choose a sign in a consistent way. Thus the representation of plane rotations on spinors is two-valued.

In applications of spinors in two dimensions, it is common to exploit the fact that the algebra of even-graded elements (that is just the ring of complex numbers) is identical to the space of spinors. So, by abuse of language, the two are often conflated. One may then talk about "the action of a spinor on a vector." In a general setting, such statements are meaningless. But in dimensions 2 and 3 (as applied, for example, to computer graphics) they make sense.

Examples

- The even-graded element

$$\gamma = \tfrac{1}{\sqrt{2}}(1 - \sigma_1\sigma_2)$$

corresponds to a vector rotation of $90°$ from σ_1 around towards σ_2, which can be checked by confirming that

$$\tfrac{1}{2}(1 - \sigma_1\sigma_2)\{a_1\sigma_1 + a_2\sigma_2\}(1 - \sigma_2\sigma_1) = a_1\sigma_2 - a_2\sigma_1$$

It corresponds to a spinor rotation of only $45°$, however:

$$\tfrac{1}{\sqrt{2}}(1 - \sigma_1\sigma_2)\{a_1 + a_2\sigma_1\sigma_2\} = \frac{a_1 + a_2}{\sqrt{2}} + \frac{-a_1 + a_2}{\sqrt{2}}\sigma_1\sigma_2$$

- Similarly the even-graded element $\gamma = -\sigma_1\sigma_2$ corresponds to a vector rotation of $180°$:

$$(-\sigma_1\sigma_2)\{a_1\sigma_1 + a_2\sigma_2\}(-\sigma_2\sigma_1) = -a_1\sigma_1 - a_2\sigma_2$$

but a spinor rotation of only $90°$:

$$(-\sigma_1\sigma_2)\{a_1 + a_2\sigma_1\sigma_2\} = a_2 - a_1\sigma_1\sigma_2$$

- Continuing on further, the even-graded element $\gamma = -1$ corresponds to a vector rotation of $360°$:

$$(-1)\{a_1\sigma_1 + a_2\sigma_2\}(-1) = a_1\sigma_1 + a_2\sigma_2$$

but a spinor rotation of $180°$.

30.4.2 Three dimensions

Main articles Spinors in three dimensions, Quaternions and spatial rotation

The Clifford algebra $C\ell_{3,0}(\mathbf{R})$ is built up from a basis of one unit scalar, 1, three orthogonal unit vectors, σ_1, σ_2 and σ_3, the three unit bivectors $\sigma_1\sigma_2$, $\sigma_2\sigma_3$, $\sigma_3\sigma_1$ and the pseudoscalar $i = \sigma_1\sigma_2\sigma_3$. It is straightforward to show that $(\sigma_1)^2 = (\sigma_2)^2 = (\sigma_3)^2 = 1$, and $(\sigma_1\sigma_2)^2 = (\sigma_2\sigma_3)^2 = (\sigma_3\sigma_1)^2 = (\sigma_1\sigma_2\sigma_3)^2 = -1$.

The sub-algebra of even-graded elements is made up of scalar dilations,

$$u' = \rho^{(1/2)} u \rho^{(1/2)} = \rho u,$$

and vector rotations

$$u' = \gamma u \gamma^*,$$

where

$$
\left.
\begin{aligned}
\gamma &= \cos(\theta/2) - \{a_1\sigma_2\sigma_3 + a_2\sigma_3\sigma_1 + a_3\sigma_1\sigma_2\}\sin(\theta/2) \\
&= \cos(\theta/2) - i\{a_1\sigma_1 + a_2\sigma_2 + a_3\sigma_3\}\sin(\theta/2) \\
&= \cos(\theta/2) - iv\sin(\theta/2)
\end{aligned}
\right\}
$$

corresponds to a vector rotation through an angle θ about an axis defined by a unit vector $v = a_1\sigma_1 + a_2\sigma_2 + a_3\sigma_3$.

As a special case, it is easy to see that, if $v = \sigma_3$, this reproduces the $\sigma_1\sigma_2$ rotation considered in the previous section; and that such rotation leaves the coefficients of vectors in the σ_3 direction invariant, since

$$(\cos(\theta/2) - i\sigma_3\sin(\theta/2))\,\sigma_3\,(\cos(\theta/2) + i\sigma_3\sin(\theta/2)) = (\cos^2(\theta/2) + \sin^2(\theta/2))\,\sigma_3 = \sigma_3.$$

The bivectors $\sigma_2\sigma_3$, $\sigma_3\sigma_1$ and $\sigma_1\sigma_2$ are in fact Hamilton's quaternions **i**, **j** and **k**, discovered in 1843:

$$
\begin{aligned}
\mathbf{i} &= -\sigma_2\sigma_3 = -i\sigma_1 \\
\mathbf{j} &= -\sigma_3\sigma_1 = -i\sigma_2 \\
\mathbf{k} &= -\sigma_1\sigma_2 = -i\sigma_3.
\end{aligned}
$$

With the identification of the even-graded elements with the algebra **H** of quaternions, as in the case of two dimensions the only representation of the algebra of even-graded elements is on itself.[24] Thus the (real[25]) spinors in three-dimensions are quaternions, and the action of an even-graded element on a spinor is given by ordinary quaternionic multiplication.

Note that the expression (1) for a vector rotation through an angle θ, the angle appearing in γ was halved. Thus the spinor rotation $\gamma(\psi) = \gamma\psi$ (ordinary quaternionic multiplication) will rotate the spinor ψ through an angle one-half the measure of the angle of the corresponding vector rotation. Once again, the problem of lifting a vector rotation to a spinor rotation is two-valued: the expression (1) with $(180° + \theta/2)$ in place of $\theta/2$ will produce the same vector rotation, but the negative of the spinor rotation.

The spinor/quaternion representation of rotations in 3D is becoming increasingly prevalent in computer geometry and other applications, because of the notable brevity of the corresponding spin matrix, and the simplicity with which they can be multiplied together to calculate the combined effect of successive rotations about different axes.

30.5 Explicit constructions

A space of spinors can be constructed explicitly with concrete and abstract constructions. The equivalence of these constructions are a consequence of the uniqueness of the spinor representation of the complex Clifford algebra. For a complete example in dimension 3, see spinors in three dimensions.

30.5.1 Component spinors

Given a vector space V and a quadratic form g an explicit matrix representation of the Clifford algebra $C\ell(V, g)$ can be defined as follows. Choose an orthonormal basis $e^1 \ldots e^n$ for V i.e. $g(e^\mu e^\nu) = \eta^{\mu\nu}$ where $\eta^{\mu\mu} = \pm 1$ and $\eta^{\mu\nu} = 0$ for $\mu \neq$

v. Let $k = \lfloor n/2 \rfloor$. Fix a set of $2^k \times 2^k$ matrices $\gamma^1 \dots \gamma^n$ such that $\gamma^\mu \gamma^\nu + \gamma^\nu \gamma^\mu = 2\eta^{\mu\nu} 1$ (i.e. fix a convention for the gamma matrices). Then the assignment $e^\mu \to \gamma^\mu$ extends uniquely to an algebra homomorphism $C\ell(V, g) \to \mathrm{Mat}(2^k, \mathbf{C})$ by sending the monomial $e^{\mu_1} \dots e^{\mu_k}$ in the Clifford algebra to the product $\gamma^{\mu_1} \dots \gamma^{\mu_k}$ of matrices and extending linearly. The space $\Delta = \mathbf{C}^{2^k}$ on which the gamma matrices act is a now a space of spinors. One needs to construct such matrices explicitly, however. In dimension 3, defining the gamma matrices to be the Pauli sigma matrices gives rise to the familiar two component spinors used in non relativistic quantum mechanics. Likewise using the 4×4 Dirac gamma matrices gives rise to the 4 component Dirac spinors used in 3+1 dimensional relativistic quantum field theory. In general, in order to define gamma matrices of the required kind, one can use the Weyl–Brauer matrices.

In this construction the representation of the Clifford algebra $C\ell(V, g)$, the Lie algebra $\mathbf{so}(V, g)$, and the Spin group $\mathrm{Spin}(V, g)$, all depend on the choice of the orthonormal basis and the choice of the gamma matrices. This can cause confusion over conventions, but invariants like traces are independent of choices. In particular, all physically observable quantities must be independent of such choices. In this construction a spinor can be represented as a vector of 2^k complex numbers and is denoted with spinor indices (usually α, β, γ). In the physics literature, abstract spinor indices are often used to denote spinors even when an abstract spinor construction is used.

30.5.2   Abstract spinors

There are at least two different, but essentially equivalent, ways to define spinors abstractly. One approach seeks to identify the minimal ideals for the left action of $C\ell(V, g)$ on itself. These are subspaces of the Clifford algebra of the form $C\ell(V, g)\omega$, admitting the evident action of $C\ell(V, g)$ by left-multiplication: $c : x\omega \to cx\omega$. There are two variations on this theme: one can either find a primitive element ω that is a nilpotent element of the Clifford algebra, or one that is an idempotent. The construction via nilpotent elements is more fundamental in the sense that an idempotent may then be produced from it.[26] In this way, the spinor representations are identified with certain subspaces of the Clifford algebra itself. The second approach is to construct a vector space using a distinguished subspace of V, and then specify the action of the Clifford algebra *externally* to that vector space.

In either approach, the fundamental notion is that of an isotropic subspace W. Each construction depends on an initial freedom in choosing this subspace. In physical terms, this corresponds to the fact that there is no measurement protocol that can specify a basis of the spin space, even if a preferred basis of V is given.

As above, we let (V, g) be an n-dimensional complex vector space equipped with a nondegenerate bilinear form. If V is a real vector space, then we replace V by its complexification $V \otimes_{\mathbf{R}} \mathbf{C}$ and let g denote the induced bilinear form on $V \otimes_{\mathbf{R}} \mathbf{C}$. Let W be a maximal isotropic subspace, i.e. a maximal subspace of V such that $g|W = 0$. If $n = 2k$ is even, then let W^* be an isotropic subspace complementary to W. If $n = 2k + 1$ is odd, let W^* be a maximal isotropic subspace with $W \cap W^* = 0$, and let U be the orthogonal complement of $W \oplus W^*$. In both the even- and odd-dimensional cases W and W^* have dimension k. In the odd-dimensional case, U is one-dimensional, spanned by a unit vector u.

30.5.3   Minimal ideals

Since W' is isotropic, multiplication of elements of W' inside $C\ell(V, g)$ is skew. Hence vectors in W' anti-commute, and $C\ell(W', g|W') = C\ell(W', 0)$ is just the exterior algebra $\Lambda^* W'$. Consequently, the k-fold product of W' with itself, W'^k, is one-dimensional. Let ω be a generator of W'^k. In terms of a basis $w'_1,\dots, w'_k$ of in W', one possibility is to set

$$\omega = w'_1 w'_2 \cdots w'_k.$$

Note that $\omega^2 = 0$ (i.e., ω is nilpotent of order 2), and moreover, $w'\omega = 0$ for all $w' \in W'$. The following facts can be proven easily:

1. If $n = 2k$, then the left ideal $\Delta = C\ell(V, g)\omega$ is a minimal left ideal. Furthermore, this splits into the two spin spaces $\Delta_+ = C\ell^{\mathrm{even}}\omega$ and $\Delta_- = C\ell^{\mathrm{odd}}\omega$ on restriction to the action of the even Clifford algebra.

2. If $n = 2k + 1$, then the action of the unit vector u on the left ideal $C\ell(V, g)\omega$ decomposes the space into a pair of isomorphic irreducible eigenspaces (both denoted by Δ), corresponding to the respective eigenvalues +1 and −1.

In detail, suppose for instance that n is even. Suppose that I is a non-zero left ideal contained in $C\ell(V, g)\omega$. We shall show that I must be equal to $C\ell(V, g)\omega$ by proving that it contains a nonzero scalar multiple of ω.

Fix a basis w_i of W and a complementary basis w_i' of W' so that

$$w_i w_j' + w_j' w_i = \delta_{ij}, \text{ and}$$
$$(w_i)^2 = 0, (w_i')^2 = 0.$$

Note that any element of I must have the form $\alpha\omega$, by virtue of our assumption that $I \subset C\ell(V, g)\,\omega$. Let $\alpha\omega \in I$ be any such element. Using the chosen basis, we may write

$$\alpha = \sum_{i_1 < i_2 < \cdots < i_p} a_{i_1\ldots i_p} w_{i_1} \cdots w_{i_p} + \sum_j B_j w_j'$$

where the $a_{i1\ldots ip}$ are scalars, and the B_j are auxiliary elements of the Clifford algebra. Observe now that the product

$$\alpha\omega = \sum_{i_1 < i_2 < \cdots < i_p} a_{i_1\ldots i_p} w_{i_1} \cdots w_{i_p}\omega.$$

Pick any nonzero monomial a in the expansion of α with maximal homogeneous degree in the elements w_i:

$$a = a_{i_1\ldots i_{max}} w_{i_1} \cdots w_{i_{max}}$$

then

$$w'_{i_{max}} \cdots w'_{i_1} \alpha\omega = a_{i_1\ldots i_{max}}\omega$$

is a nonzero scalar multiple of ω, as required.

Note that for n even, this computation also shows that

$$\Delta = C\ell(W)\omega = (\Lambda^* W)\omega$$

as a vector space. In the last equality we again used that W is isotropic. In physics terms, this shows that Δ is built up like a Fock space by creating spinors using anti-commuting creation operators in W acting on a vacuum ω.

30.5.4 Exterior algebra construction

The computations with the minimal ideal construction suggest that a spinor representation can also be defined directly using the exterior algebra $\Lambda^* W = \oplus_j \Lambda^j W$ of the isotropic subspace W. Let $\Delta = \Lambda^* W$ denote the exterior algebra of W considered as vector space only. This will be the spin representation, and its elements will be referred to as spinors.[27]

The action of the Clifford algebra on Δ is defined first by giving the action of an element of V on Δ, and then showing that this action respects the Clifford relation and so extends to a homomorphism of the full Clifford algebra into the endomorphism ring $\text{End}(\Delta)$ by the universal property of Clifford algebras. The details differ slightly according to whether the dimension of V is even or odd.

When $\dim(V)$ is even, $V = W \oplus W'$ where W' is the chosen isotropic complement. Hence any $v \in V$ decomposes uniquely as $v = w + w'$ with $w \in W$ and $w' \in W'$. The action of v on a spinor is given by

$$c(v)w_1 \wedge \cdots \wedge w_n = (\epsilon(w) + i(w'))\,(w_1 \wedge \cdots \wedge w_n)$$

where $i(w')$ is interior product with w' using the non degenerate quadratic form to identify V with V^*, and ε(w) denotes the exterior product. It may be verified that

$$c(u)c(v) + c(v)c(u) = 2\,g(u,v),$$

and so c respects the Clifford relations and extends to a homomorphism from the Clifford algebra to End(Δ).

The spin representation Δ further decomposes into a pair of irreducible complex representations of the Spin group[28] (the half-spin representations, or Weyl spinors) via

$$\Delta_+ = \Lambda^{even}W, \quad \Delta_- = \Lambda^{odd}W$$

When $\dim(V)$ is odd, $V = W \oplus U \oplus W'$, where U is spanned by a unit vector u orthogonal to W. The Clifford action c is defined as before on $W \oplus W'$, while the Clifford action of (multiples of) u is defined by

$$c(u)\alpha = \begin{cases} \alpha & \text{if } \alpha \in \Lambda^{even}W \\ -\alpha & \text{if } \alpha \in \Lambda^{odd}W \end{cases}$$

As before, one verifies that c respects the Clifford relations, and so induces a homomorphism.

30.5.5   Hermitian vector spaces and spinors

If the vector space V has extra structure that provides a decomposition of its complexification into two maximal isotropic subspaces, then the definition of spinors (by either method) becomes natural.

The main example is the case that the real vector space V is a hermitian vector space (V, h), i.e., V is equipped with a complex structure J that is an orthogonal transformation with respect to the inner product g on V. Then $V \otimes_{\mathbf{R}} \mathbf{C}$ splits in the $\pm i$ eigenspaces of J. These eigenspaces are isotropic for the complexification of g and can be identified with the complex vector space (V, J) and its complex conjugate $(V, -J)$. Therefore for a hermitian vector space (V, h) the vector space Λ.
$\mathbf{C}V$ (as well as its complex conjugate Λ.
$\mathbf{C}V$) is a spinor space for the underlying real euclidean vector space.

With the Clifford action as above but with contraction using the hermitian form, this construction gives a spinor space at every point of an almost Hermitian manifold and is the reason why every almost complex manifold (in particular every symplectic manifold) has a Spin^c structure. Likewise, every complex vector bundle on a manifold carries a Spin^c structure.[29]

30.6   Clebsch–Gordan decomposition

A number of Clebsch–Gordan decompositions are possible on the tensor product of one spin representation with another.[30] These decompositions express the tensor product in terms of the alternating representations of the orthogonal group.

For the real or complex case, the alternating representations are

- $\Gamma r = \Lambda^r V$, the representation of the orthogonal group on skew tensors of rank r.

In addition, for the real orthogonal groups, there are three characters (one-dimensional representations)

- $\sigma_+ : O(p, q) \to \{-1, +1\}$ given by $\sigma_+(R) = -1$, if R reverses the spatial orientation of V, $+1$, if R preserves the spatial orientation of V. (*The spatial character.*)

- $\sigma_- : O(p, q) \to \{-1, +1\}$ given by $\sigma_-(R) = -1$, if R reverses the temporal orientation of V, $+1$, if R preserves the temporal orientation of V. (*The temporal character.*)

- $\sigma = \sigma_+\sigma_-$. (*The orientation character.*)

The Clebsch–Gordan decomposition allows one to define, among other things:

- An action of spinors on vectors.

- A Hermitian metric on the complex representations of the real spin groups.

- A Dirac operator on each spin representation.

30.6.1 Even dimensions

If $n = 2k$ is even, then the tensor product of Δ with the contragredient representation decomposes as

$$\Delta \otimes \Delta^* \cong \bigoplus_{p=0}^{n} \Gamma_p \cong \bigoplus_{p=0}^{k-1} (\Gamma_p \oplus \sigma\Gamma_p) \oplus \Gamma_k$$

which can be seen explicitly by considering (in the Explicit construction) the action of the Clifford algebra on decomposable elements $\alpha\omega \otimes \beta\omega'$. The rightmost formulation follows from the transformation properties of the Hodge star operator. Note that on restriction to the even Clifford algebra, the paired summands $\Gamma p \oplus \sigma\Gamma p$ are isomorphic, but under the full Clifford algebra they are not.

There is a natural identification of Δ with its contragredient representation via the conjugation in the Clifford algebra:

$$(\alpha\omega)^* = \omega(\alpha^*).$$

So $\Delta \otimes \Delta$ also decomposes in the above manner. Furthermore, under the even Clifford algebra, the half-spin representations decompose

$$\Delta_+ \otimes \Delta_+^* \cong \Delta_- \otimes \Delta_-^* \cong \bigoplus_{p=0}^{k} \Gamma_{2p}$$
$$\Delta_+ \otimes \Delta_-^* \cong \Delta_- \otimes \Delta_+^* \cong \bigoplus_{p=0}^{k-1} \Gamma_{2p+1}$$

For the complex representations of the real Clifford algebras, the associated reality structure on the complex Clifford algebra descends to the space of spinors (via the explicit construction in terms of minimal ideals, for instance). In this way, we obtain the complex conjugate $\bar{\Delta}$ of the representation Δ, and the following isomorphism is seen to hold:

$$\bar{\Delta} \cong \sigma_-\Delta^*$$

In particular, note that the representation Δ of the orthochronous spin group is a unitary representation. In general, there are Clebsch–Gordan decompositions

$$\Delta \otimes \bar{\Delta} \cong \bigoplus_{p=0}^{k} (\sigma_-\Gamma_p \oplus \sigma_+\Gamma_p).$$

In metric signature (p, q), the following isomorphisms hold for the conjugate half-spin representations

- If q is even, then $\bar{\Delta}_+ \cong \sigma_- \otimes \Delta_+^*$ and $\bar{\Delta}_- \cong \sigma_- \otimes \Delta_-^*$.

- If q is odd, then $\bar{\Delta}_+ \cong \sigma_- \otimes \Delta_-^*$ and $\bar{\Delta}_- \cong \sigma_- \otimes \Delta_+^*$.

Using these isomorphisms, one can deduce analogous decompositions for the tensor products of the half-spin representations $\Delta\pm \otimes \Delta\pm$.

30.6.2   Odd dimensions

If $n = 2k + 1$ is odd, then

$$\Delta \otimes \Delta^* \cong \bigoplus_{p=0}^{k} \Gamma_{2p}.$$

In the real case, once again the isomorphism holds

$$\bar{\Delta} \cong \sigma_- \Delta^*.$$

Hence there is a Clebsch–Gordan decomposition (again using the Hodge star to dualize) given by

$$\Delta \otimes \bar{\Delta} \cong \sigma_- \Gamma_0 \oplus \sigma_+ \Gamma_1 \oplus \cdots \oplus \sigma_\pm \Gamma_k$$

30.6.3   Consequences

There are many far-reaching consequences of the Clebsch–Gordan decompositions of the spinor spaces. The most fundamental of these pertain to Dirac's theory of the electron, among whose basic requirements are

- A manner of regarding the product of two spinors $\phi\psi$ as a scalar. In physical terms, a spinor should determine a probability amplitude for the quantum state.

- A manner of regarding the product $\psi\phi$ as a vector. This is an essential feature of Dirac's theory, which ties the spinor formalism to the geometry of physical space.

- A manner of regarding a spinor as acting upon a vector, by an expression such as $\psi v \psi$. In physical terms, this represents an electric current of Maxwell's electromagnetic theory, or more generally a probability current.

30.7   Summary in low dimensions

- In 1 dimension (a trivial example), the single spinor representation is formally Majorana, a real 1-dimensional representation that does not transform.

- In 2 Euclidean dimensions, the left-handed and the right-handed Weyl spinor are 1-component complex representations, i.e. complex numbers that get multiplied by $e^{\pm i\varphi/2}$ under a rotation by angle φ.

- In 3 Euclidean dimensions, the single spinor representation is 2-dimensional and quaternionic. The existence of spinors in 3 dimensions follows from the isomorphism of the groups $SU(2) \cong \mathrm{Spin}(3)$ that allows us to define the action of $\mathrm{Spin}(3)$ on a complex 2-component column (a spinor); the generators of $SU(2)$ can be written as Pauli matrices.

- In 4 Euclidean dimensions, the corresponding isomorphism is Spin(4) $\cong$ SU(2) × SU(2). There are two inequivalent quaternionic 2-component Weyl spinors and each of them transforms under one of the SU(2) factors only.

- In 5 Euclidean dimensions, the relevant isomorphism is Spin(5) $\cong$ USp(4) $\cong$ Sp(2) that implies that the single spinor representation is 4-dimensional and quaternionic.

- In 6 Euclidean dimensions, the isomorphism Spin(6) $\cong$ SU(4) guarantees that there are two 4-dimensional complex Weyl representations that are complex conjugates of one another.

- In 7 Euclidean dimensions, the single spinor representation is 8-dimensional and real; no isomorphisms to a Lie algebra from another series (A or C) exist from this dimension on.

- In 8 Euclidean dimensions, there are two Weyl–Majorana real 8-dimensional representations that are related to the 8-dimensional real vector representation by a special property of Spin(8) called triality.

- In $d + 8$ dimensions, the number of distinct irreducible spinor representations and their reality (whether they are real, pseudoreal, or complex) mimics the structure in d dimensions, but their dimensions are 16 times larger; this allows one to understand all remaining cases. See Bott periodicity.

- In spacetimes with p spatial and q time-like directions, the dimensions viewed as dimensions over the complex numbers coincide with the case of the $(p + q)$-dimensional Euclidean space, but the reality projections mimic the structure in $|p - q|$ Euclidean dimensions. For example, in $3 + 1$ dimensions there are two non-equivalent Weyl complex (like in 2 dimensions) 2-component (like in 4 dimensions) spinors, which follows from the isomorphism SL(2, **C**) $\cong$ Spin(3,1).

30.8 See also

- Anyon

- Dirac equation in the algebra of physical space

- Einstein–Cartan theory

- Pure spinor

- Spin-½

- Spinor bundle

- Supercharge

- Twistor theory

30.9 Notes

[1] Spinors in three dimensions are points of a line bundle over a conic in the projective plane. In this picture, which is associated to spinors of a three-dimensional pseudo-Euclidean space of signature (1,2), the conic is an ordinary real conic (here the circle), the line bundle is the Möbius bundle, and the spin group is SL$_2$(**R**). In Euclidean signature, the projective plane, conic and line bundle are over the complex instead, and this picture is just a real slice.

[2] Spinors can always be defined over the complex numbers. However, in some signatures there exist real spinors. Details can be found in spin representation.

[3] A formal definition of spinors at this level is that the space of spinors is a linear representation of the Lie algebra of infinitesimal rotations of a certain kind.

[4] More precisely, it is the fermions of spin-1/2 that are described by spinors, which is true both in the relativistic and non-relativistic theory. The wavefunction of the non-relativistic electron has values in 2 component spinors transforming under three-dimensional infinitesimal rotations. The relativistic Dirac equation for the electron is an equation for 4 component spinors transforming under infinitesimal Lorentz transformations for which a substantially similar theory of spinors exists.

[5] Formally, the spin group is the group of relative homotopy classes with fixed endpoints in the rotation group.

[6] More formally, the space of spinors can be defined as an (irreducible) representation of the spin group that does not factor through a representation of the rotation group (in general, the connected component of the identity of the orthogonal group).

[7] Geometric algebra is a name for the Clifford algebra in an applied setting.

[8] the Pauli matrices correspond to angular momenta operators about the three coordinate axes. This makes them slightly a-typical gamma matrices because in addition to their anti commutation relation they also satisfy commutation relations

[9] The metric signature relevant as well if we are concerned with real spinors. See spin representation.

[10] Whether the representation decomposes depends on whether they are regarded as representations of the spin group (or its Lie algebra), in which case it decomposes in even but not odd dimensions, or the Clifford algebra when it is the other way around. Other structures than this decomposition can also exist; precise criteria are covered at spin representation and Clifford algebra.

[11] This is the set of 2×2 complex traceless hermitian matrices.

[12] Except for a kernel of $\{\pm 1\}$ corresponding to the two different elements of the spin group that go to the same rotation.

[13] Although there are several more intrinsic constructions, the spin representations are not functorial in the quadratic form, so they cannot be built up naturally within the tensor algebra.

[14] So the ambiguity in identifying the spinors themselves persists from the point of view of the group theory, and still depends on choices.

[15] The Clifford algebra can be given an even/odd grading from the parity of the degree in the gammas, and the spin group and its Lie algebra both lie in the even part. Whether here by "representation" we mean representations of the spin group or the Clifford algebra will affect the determination of their reducibility. Other structures than this splitting can also exist; precise criteria are covered at spin representation and Clifford algebra.

30.10   References

[1] Cartan 1913.

[2] Quote from Elie Cartan: *The Theory of Spinors*, Hermann, Paris, 1966, first sentence of the Introduction section of the beginning of the book (before the page numbers start): "Spinors were first used under that name, by physicists, in the field of Quantum Mechanics. In their most general form, spinors were discovered in 1913 by the author of this work, in his investigations on the linear representations of simple groups*; they provide a linear representation of the group of rotations in a space with any number n of dimensions, each spinor having 2^ν components where $n = 2\nu + 1$ or 2ν ." The star (*) refers to Cartan 1913.

[3] Named after William Kingdon Clifford,

[4] Named after Paul Dirac.

[5] Named after Hermann Weyl.

[6] Named after Ettore Majorana.

[7] Matthew R. Francis, Arthur Kosowsky: *The Construction of Spinors in Geometric Algebra*, submitted 20 March 2004, version of 18 October 2004 arXiv:math-ph/0403040

[8]   • Wilczek, Frank (2009). "Majorana returns". *Nature Phys.* (Macmillan Publishers) **5** (9): 614–618. doi:10.1038/nphys1380. ISSN 1745-2473. (subscription required (help)).

[9] Xu, Yang-Su et al. (2015). "Discovery of a Weyl Fermion semimetal and topological Fermi arcs". *Science Magazine* (AAAS). doi:10.1126/science.aaa9297. ISSN 0036-8075. (subscription required (help)).

[10] Jean Hladik: *Spinors in Physics*, translated by J. M. Cole, Springer 1999, ISBN 978-0-387-98647-0, p. 3

[11] Graham Farmelo: *The Strangest Man. The Hidden Life of Paul Dirac, Quantum Genius*, Faber & Faber, 2009, ISBN 978-0-571-22286-5, p. 430

[12] Cartan 1913

[13] Tomonaga 1998, p. 129

[14] Pauli 1927.

[15] Dirac 1928.

[16] G. Juvet: *Opérateurs de Dirac et équations de Maxwell*, Commentarii Mathematici Helvelvetici, 2 (1930), pp. 225–235, doi:10.1007/BF01214461 (abstract in French language)

[17] F. Sauter: *Lösung der Diracschen Gleichungen ohne Spezialisierung der Diracschen Operatoren*, Zeitschrift für Physik, Volume 63, Numbers 11–12, 803–814, doi:10.1007/BF01339277 (abstract in German language)

[18] Pertti Lounesto: *Crumeyrolle's bivectors and spinors*, pp. 137–166, In: Rafał Abłamowicz, Pertti Lounesto (eds.): *Clifford algebras and spinor structures: A Special Volume Dedicated to the Memory of Albert Crumeyrolle (1919–1992)*, ISBN 0-7923-3366-7, 1995, p. 151

[19] The matrices of dimension $N \times N$ in which only the elements of the left column are non-zero form a *left ideal* in the $N \times N$ matrix algebra Mat(N, **C**) – multiplying such a matrix M from the left with any $N \times N$ matrix A gives the result AM that is again an $N \times N$ matrix in which only the elements of the left column are non-zero. Moreover, it can be shown that it is a *minimal left ideal*. See also: Pertti Lounesto: *Clifford algebras and spinors*, London Mathematical Society Lecture Notes Series 286, Cambridge University Press, Second Edition 2001, DOI 978-0-521-00551-7, p. 52

[20] Pertti Lounesto: *Clifford algebras and spinors*, London Mathematical Society Lecture Notes Series 286, Cambridge University Press, Second Edition 2001, DOI 978-0-521-00551-7, p. 148 f. and p. 327 f.

[21] D. Hestenes: *Space–Time Algebra*, Gordon and Breach, New York, 1966, 1987, 1992

[22] D. Hestenes: *Real spinor fields*, J. Math. Phys. 8 (1967), pp. 798–808

[23] These are the right-handed Weyl spinors in two dimensions. For the left-handed Weyl spinors, the representation is via $\gamma(\phi) = \gamma\phi$. The Majorana spinors are the common underlying real representation for the Weyl representations.

[24] Since, for a skew field, the kernel of the representation must be trivial. So inequivalent representations can only arise via an automorphism of the skew-field. In this case, there are a pair of equivalent representations: $\gamma(\phi) = \gamma\phi$, and its quaternionic conjugate $\gamma(\phi) = \phi\gamma$.

[25] The complex spinors are obtained as the representations of the tensor product $\mathbf{H} \otimes_\mathbf{R} \mathbf{C} = \text{Mat}_2(\mathbf{C})$. These are considered in more detail in spinors in three dimensions.

[26] This construction is due to Cartan. The treatment here is based on Chevalley (1954).

[27] One source for this subsection is Fulton & Harris (1991).

[28] Via the even-graded Clifford algebra.

[29] Lawson & Michelsohn 1989, Appendix D.

[30] Brauer & Weyl 1935.

30.11   Further reading

- Brauer, Richard; Weyl, Hermann (1935), "Spinors in n dimensions", *American Journal of Mathematics* (The Johns Hopkins University Press) **57** (2): 425–449, doi:10.2307/2371218, JSTOR 2371218.

- Cartan, Élie (1913), "Les groupes projectifs qui ne laissent invariante aucune multiplicité plane" (PDF), *Bul. Soc. Math. France* **41**: 53–96.

- Cartan, Élie (1966), *The theory of spinors*, Paris, Hermann (reprinted 1981, Dover Publications), ISBN 978-0-486-64070-9

- Chevalley, Claude (1954), *The algebraic theory of spinors and Clifford algebras*, Columbia University Press (reprinted 1996, Springer), ISBN 978-3-540-57063-9.

- Dirac, Paul M. (1928), "The quantum theory of the electron", *Proceedings of the Royal Society of London* **A117**: 610–624, JSTOR 94981.

- Fulton, William; Harris, Joe (1991), *Representation theory. A first course*, Graduate Texts in Mathematics, Readings in Mathematics **129**, New York: Springer-Verlag, ISBN 0-387-97495-4, MR 1153249.

- Gilkey, Peter B. (1984), *Invariance Theory, the Heat Equation, and the Atiyah–Singer Index Theorem*, Publish or Perish, ISBN 0-914098-20-9.

- Harvey, F. Reese (1990), *Spinors and Calibrations*, Academic Press, ISBN 978-0-12-329650-4.

- Hazewinkel, Michiel, ed. (2001), "Spinor", *Encyclopedia of Mathematics*, Springer, ISBN 978-1-55608-010-4

- Hitchin, Nigel J. (1974), "Harmonic spinors", *Advances in Mathematics* **14**: 1–55, doi:10.1016/0001-8708(74)90021-8, MR 358873.

- Lawson, H. Blaine; Michelsohn, Marie-Louise (1989), *Spin Geometry*, Princeton University Press, ISBN 0-691-08542-0.

- Pauli, Wolfgang (1927), "Zur Quantenmechanik des magnetischen Elektrons", *Zeitschrift für Physik* **43** (9–10): 601–632, Bibcode:1927ZPhy...43..601P, doi:10.1007/BF01397326.

- Penrose, Roger; Rindler, W. (1988), *Spinors and Space–Time: Volume 2, Spinor and Twistor Methods in Space–Time Geometry*, Cambridge University Press, ISBN 0-521-34786-6.

- Tomonaga, Sin-Itiro (1998), "Lecture 7: The Quantity Which Is Neither Vector nor Tensor", *The story of spin*, University of Chicago Press, p. 129, ISBN 0-226-80794-0

Chapter 31

Fundamental interaction

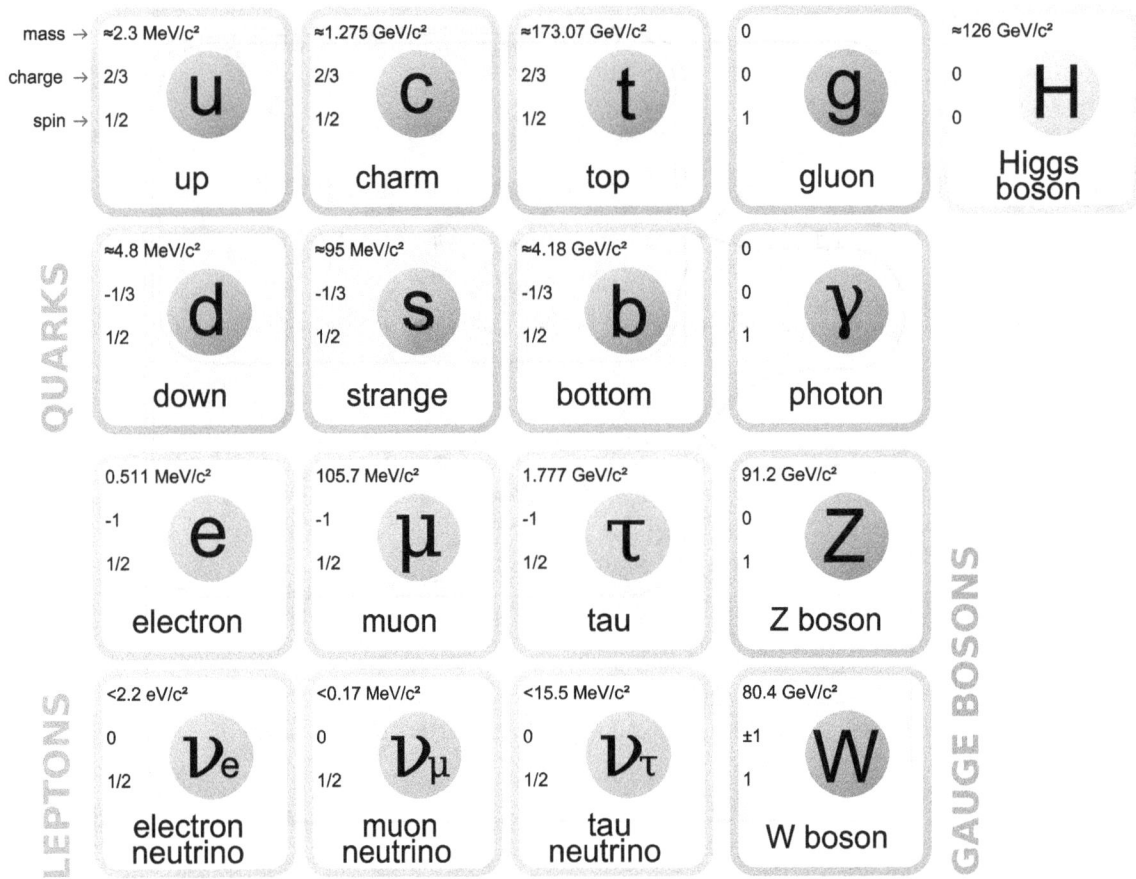

The Standard Model of elementary particles, with the fermions in the first three columns, the gauge bosons in the fourth column, and the Higgs boson in the fifth column

Fundamental interactions, also known as **fundamental forces** or **interactive forces**, are the interactions in physical systems that don't appear to be reducible to more basic interactions. There are four conventionally accepted fundamental interactions—gravitational, electromagnetic, strong nuclear, and weak nuclear. Each one is understood as the dynamics of a *field*. The gravitational force is modeled as a continuous classical field. The other three are modeled as a discrete quantum field, and exhibit a measurable unit or *elementary particle*.

Gravitation and electromagnetism act over a potentially infinite distance across the universe. They mediate macroscopic phenomena every day. The other two fields act over minuscule, subatomic distances. The strong interaction is responsible for the binding of atomic nuclei. The weak interaction also acts on the nucleus, mediating radioactive decay.

Theoretical physicists working beyond the Standard Model seek to quantize the gravitational field toward predictions that particle physicists can experimentally confirm, thus yielding acceptance to a theory of quantum gravity (QG). (Phenomena suitable to model as a fifth force—perhaps an added gravitational effect—remain widely disputed). Other theorists seek to unite the electroweak and strong fields within a Grand Unified Theory (GUT). While all four fundamental interactions are widely thought to align at an extremely minuscule scale, particle accelerators cannot produce the massive energy levels required to experimentally probe at that Planck scale (which would experimentally confirm such theories). Yet some theories, such as the string theory, seek both QG and GUT within one framework, unifying all four fundamental interactions along with mass production within a theory of everything (ToE).

31.1   General relativity

In his 1687 theory, Isaac Newton postulated space as an infinite and unalterable physical structure existing before, within, and around all objects while their states and relations unfold at a constant pace everywhere, thus absolute space and time. Inferring that all objects bearing mass approach at a constant rate, but collide by impact proportional to their masses, Newton inferred that matter exhibits an attractive force. His law of universal gravitation mathematically stated it to span the entire universe instantly (despite absolute time), or, if not actually a force, to be instant interaction among all objects (despite absolute space). As conventionally interpreted, Newton's theory of motion modeled a *central force* without a communicating medium.[2] Thus Newton's theory violated the first principle of mechanical philosophy, as stated by Descartes, *No action at a distance*. Conversely, during the 1820s, when explaining magnetism, Michael Faraday inferred a *field* filling space and transmitting that force. Faraday conjectured that ultimately, all forces unified into one.

In the early 1870s, James Clerk Maxwell unified electricity and magnetism as effects of an electromagnetic-field whose third consequence was light, traveling at constant speed in a vacuum. The electromagnetic field theory contradicted predictions of Newton's theory of motion, unless physical states of the luminiferous aether—presumed to fill all space whether within matter or in a vacuum and to manifest the electromagnetic field—aligned all phenomena and thereby held valid the Newtonian principle relativity or invariance. Disfavoring hypotheses at unobservables, Albert Einstein discarded the aether, and aligned electrodynamics with relativity by denying absolute space and time, and stating relative space and time. The two phenomena altered in the vicinity of an object measured to be in motion—length contraction and time dilation for the object experienced to be in relative motion—Einstein's principle special relativity, published in 1905.

Special relativity was accepted as a theory, too. It rendered Newton's theory of motion apparently untenable, especially since Newtonian physics postulated an object's mass to be constant. A consequence of special relativity is mass being a variant form of energy, condensed into an object. By the equivalence principle, published by Einstein in 1907, gravitation is indistinguishable from acceleration, perhaps two phenomena sharing a mechanism. That year, Hermann Minkowski modeled special relativity to a unification of space and time, 4D spacetime. So stretching the three spatial dimensions onto the single dimension of time's arrow, Einstein arrived at general theory of relativity in 1915.[3] Einstein interpreted space as a substance, *Einstein aether*, whose physical properties receive motion from an object and transmit it to other objects while modulating events' unfolding. Equivalent to energy, mass contracts space, which dilates time—events unfold more slowly—establishing local tension. The object relieves it in the likeness of a free fall at light speed along the pathway of least resistance, a straight line's equivalent on the curved surface of 4D spacetime, a pathway termed *worldline*.

Einstein abolished *action at a distance* by theorizing a gravitational field—4D spacetime—that waves while transmitting motion across the universe at light speed. All objects always travel at light speed in 4D spacetime. At zero relative speed, an object is observed to travel none through space, but age most rapidly. That is, an object at relative rest in 3D space exhibits its constant energy to an observer by exhibiting top speed along 1D time flow. Conversely, at highest relative speed, an object traverses 3D space at light speed, yet is ageless, none of its constant energy available to internal motion as flow along 1D time. Whereas Newtonian inertia is an idealized case of an object either keeping rest or holding constant velocity by hypothetical existence in a universe otherwise devoid of matter, Einsteinian inertia is indistinguishable from an object experiencing no acceleration by existing in a gravitational field possibly full of matter distributed uniformly. Conversely, even massless energy manifests gravitation—which is acceleration—on local objects by "curving" the surface of 4D spacetime. Physicists renounced belief that motion must be mediated by a *force*.

31.2 Standard Model

Main article: Standard Model
See also: Lambda-CDM model

The electromagnetic, strong, and weak interactions associate with elementary particles, whose behaviors are modeled in quantum mechanics (QM). For predictive success with QM's probabilistic outcomes, particle physics conventionally models QM events across a field set to special relativity, altogether relativistic quantum field theory (QFT).[4] Force particles, called gauge bosons—*force carriers* or *messenger particles* of underlying fields—interact with matter particles, called fermions. Everyday matter is atoms, composed of three fermion types: up-quarks and down-quarks constituting, as well as electrons orbiting, the atom's nucleus. Atoms interact, form molecules, and manifest further properties through electromagnetic interactions among their electrons absorbing and emitting photons, the electromagnetic field's force carrier, which if unimpeded traverse potentially infinite distance. Electromagnetism's QFT is quantum electrodynamics (QED).

The electromagnetic interaction was modeled with the weak interaction, whose force carriers are W and Z bosons, traversing minuscule distance, in electroweak theory (EWT). Electroweak interaction would operate at such high temperatures as soon after the presumed Big Bang, but, as the early universe cooled, split into electromagnetic and weak interactions. The strong interaction, whose force carrier is the gluon, traversing minuscule distance among quarks, is modeled in quantum chromodynamics (QCD). EWT, QCD, and the Higgs mechanism, whereby the Higgs field manifests Higgs bosons that interact with some quantum particles and thereby endow those particles with mass, comprise particle physics' Standard Model (SM). Predictions are usually made using calculational approximation methods, although such perturbation theory is inadequate to model some experimental observations (for instance bound states and solitons). Still, physicists widely accept the Standard Model as science's most experimentally confirmed theory.

Beyond the Standard Model, some theorists work to unite the electroweak and strong interactions within a Grand Unified Theory (GUT). Some attempts at GUTs hypothesize "shadow" particles, such that every known matter particle associates with an undiscovered force particle, and vice versa, altogether supersymmetry (SUSY). Other theorists seek to quantize the gravitational field by modeling behavior of its hypothetical force carrier, the graviton and achieve quantum gravity (QG). One approach to QG is loop quantum gravity (LQG). Still other theorists seek both QG and GUT within one framework, reducing all four fundamental interactions to a Theory of Everything (ToE). The most prevalent aim at a ToE is string theory, although to model matter particles, it added SUSY to force particles—and so, strictly speaking, became superstring theory. Multiple, seemingly disparate superstring theories were unified on a backbone, M theory. Theories beyond the Standard Model remain highly speculative, lacking great experimental support.

31.3 Overview of the fundamental interactions

In the conceptual model of fundamental interactions, matter consists of fermions, which carry properties called charges and spin $\pm\frac{1}{2}$ (intrinsic angular momentum $\pm\frac{\hbar}{2}$, where $\hbar$ is the reduced Planck constant). They attract or repel each other by exchanging bosons.

The interaction of any pair of fermions in perturbation theory can then be modeled thus:

Two fermions go in → *interaction* by boson exchange → Two changed fermions go out.

The exchange of bosons always carries energy and momentum between the fermions, thereby changing their speed and direction. The exchange may also transport a charge between the fermions, changing the charges of the fermions in the process (e.g., turn them from one type of fermion to another). Since bosons carry one unit of angular momentum, the fermion's spin direction will flip from $+\frac{1}{2}$ to $-\frac{1}{2}$ (or vice versa) during such an exchange (in units of the reduced Planck's constant).

Because an interaction results in fermions attracting and repelling each other, an older term for "interaction" is force.

According to the present understanding, there are four fundamental interactions or forces: gravitation, electromagnetism, the weak interaction, and the strong interaction. Their magnitude and behavior vary greatly, as described in the table below. Modern physics attempts to explain every observed physical phenomenon by these fundamental interactions.

Elementary Particles

An overview of the various families of elementary and composite particles, and the theories describing their interactions. Fermions are on the left, and Bosons are on the right.

Moreover, reducing the number of different interaction types is seen as desirable. Two cases in point are the unification of:

- Electric and magnetic force into electromagnetism;

- The electromagnetic interaction and the weak interaction into the electroweak interaction; see below.

Both magnitude ("relative strength") and "range", as given in the table, are meaningful only within a rather complex theoretical framework. It should also be noted that the table below lists properties of a conceptual scheme that is still the subject of ongoing research.

The modern (perturbative) quantum mechanical view of the fundamental forces other than gravity is that particles of matter (fermions) do not directly interact with each other, but rather carry a charge, and exchange virtual particles (gauge bosons), which are the interaction carriers or force mediators. For example, photons mediate the interaction of electric charges, and gluons mediate the interaction of color charges.

31.4   The interactions

31.4.1   Gravitation

Main article: Gravitation

Gravitation is by far the weakest of the four interactions. The weakness of gravity can easily be demonstrated by suspending a pin using a simple magnet (such as a refrigerator magnet). The magnet is able to hold the pin against the gravitational pull of the entire Earth.

Yet gravitation is very important for macroscopic objects and over macroscopic distances for the following reasons. Gravitation:

- is the only interaction that acts on all particles having mass, energy and or momentum;

- has an infinite range, like electromagnetism but unlike strong and weak interaction;

- cannot be absorbed, transformed, or shielded against;

- always attracts and never repels.

Even though electromagnetism is far stronger than gravitation, electrostatic attraction is not relevant for large celestial bodies, such as planets, stars, and galaxies, simply because such bodies contain equal numbers of protons and electrons and so have a net electric charge of zero. Nothing "cancels" gravity, since it is only attractive, unlike electric forces which can be attractive or repulsive. On the other hand, all objects having mass are subject to the gravitational force, which only attracts. Therefore, only gravitation matters on the large scale structure of the universe.

The long range of gravitation makes it responsible for such large-scale phenomena as the structure of galaxies, black holes, and it retards the expansion of the universe. Gravitation also explains astronomical phenomena on more modest scales, such as planetary orbits, as well as everyday experience: objects fall; heavy objects act as if they were glued to the ground; and animals can only jump so high.

Gravitation was the first interaction to be described mathematically. In ancient times, Aristotle hypothesized that objects of different masses fall at different rates. During the Scientific Revolution, Galileo Galilei experimentally determined that this was not the case — neglecting the friction due to air resistance, and buoyancy forces if an atmosphere is present (e.g. the case of a dropped air filled balloon vs a water filled balloon) all objects accelerate toward the Earth at the same rate. Isaac Newton's law of Universal Gravitation (1687) was a good approximation of the behaviour of gravitation. Our present-day understanding of gravitation stems from Albert Einstein's General Theory of Relativity of 1915, a more accurate (especially for cosmological masses and distances) description of gravitation in terms of the geometry of space-time.

Merging general relativity and quantum mechanics (or quantum field theory) into a more general theory of quantum gravity is an area of active research. It is hypothesized that gravitation is mediated by a massless spin-2 particle called the graviton.

Although general relativity has been experimentally confirmed (at least, in the weak field or Post-Newtonian case) on all but the smallest scales, there are rival theories of gravitation. Those taken seriously by the physics community all reduce to general relativity in some limit, and the focus of observational work is to establish limitations on what deviations from general relativity are possible.

Proposed extra dimensions could explain why the gravity force is so weak.[6]

31.4.2 Electroweak interaction

Main article: Electroweak interaction

Electromagnetism and weak interaction appear to be very different at everyday low energies. They can be modeled using two different theories. However, above unification energy, on the order of 100 GeV, they would merge into a single electroweak force.

Electroweak theory is very important for modern cosmology, particularly on how the universe evolved. This is because shortly after the Big Bang, the temperature was approximately above 10^{15} K. Electromagnetic force and weak force were merged into a combined electroweak force.

For contributions to the unification of the weak and electromagnetic interaction between elementary particles, Abdus Salam, Sheldon Glashow and Steven Weinberg were awarded the Nobel Prize in Physics in 1979.[7][8]

Electromagnetism

Main article: Electromagnetism

Electromagnetism is the force that acts between electrically charged particles. This phenomenon includes the electrostatic force acting between charged particles at rest, and the combined effect of electric and magnetic forces acting between charged particles moving relative to each other.

Electromagnetism is infinite-ranged like gravity, but vastly stronger, and therefore describes a number of macroscopic phenomena of everyday experience such as friction, rainbows, lightning, and all human-made devices using electric current, such as television, lasers, and computers. Electromagnetism fundamentally determines all macroscopic, and many atomic level, properties of the chemical elements, including all chemical bonding.

In a four kilogram (~1 gallon) jug of water there are

$$4000 \text{ g } H_2O \cdot \frac{1 \text{ mol } H_2O}{18 \text{ g } H_2O} \cdot \frac{10 \text{ mol } e^-}{1 \text{ mol } H_2O} \cdot \frac{96,000 \text{ C}}{1 \text{ mol } e^-} = 2.1 \times 10^8 C$$

of total electron charge. Thus, if we place two such jugs a meter apart, the electrons in one of the jugs repel those in the other jug with a force of

$$\frac{1}{4\pi\varepsilon_0} \frac{(2.1 \times 10^8 C)^2}{(1m)^2} = 4.1 \times 10^{26} N.$$

This is larger than what the planet Earth would weigh if weighed on another Earth. The atomic nuclei in one jug also repel those in the other with the same force. However, these repulsive forces are cancelled by the attraction of the electrons in jug A with the nuclei in jug B and the attraction of the nuclei in jug A with the electrons in jug B, resulting in no net force. Electromagnetic forces are tremendously stronger than gravity but cancel out so that for large bodies gravity dominates.

Electrical and magnetic phenomena have been observed since ancient times, but it was only in the 19th century that it was discovered that electricity and magnetism are two aspects of the same fundamental interaction. By 1864, Maxwell's equations had rigorously quantified this unified interaction. Maxwell's theory, restated using vector calculus, is the classical theory of electromagnetism, suitable for most technological purposes.

The constant speed of light in a vacuum (customarily described with the letter "c") can be derived from Maxwell's equations, which are consistent with the theory of special relativity. Einstein's 1905 theory of special relativity, however, which flows from the observation that the speed of light is constant no matter how fast the observer is moving, showed that the theoretical result implied by Maxwell's equations has profound implications far beyond electro-magnetism on the very nature of time and space.

In other work that departed from classical electro-magnetism, Einstein also explained the photoelectric effect by hypothesizing that light was transmitted in quanta, which we now call photons. Starting around 1927, Paul Dirac combined quantum mechanics with the relativistic theory of electromagnetism. Further work in the 1940s, by Richard Feynman, Freeman Dyson, Julian Schwinger, and Sin-Itiro Tomonaga, completed this theory, which is now called quantum electrodynamics, the revised theory of electromagnetism. Quantum electrodynamics and quantum mechanics provide a theoretical basis for electromagnetic behavior such as quantum tunneling, in which a certain percentage of electrically charged particles move in ways that would be impossible under classical electromagnetic theory, that is necessary for everyday electronic devices such as transistors to function.

Weak interaction

Main article: Weak interaction

The *weak interaction* or *weak nuclear force* is responsible for some nuclear phenomena such as beta decay. Electromagnetism and the weak force are now understood to be two aspects of a unified electroweak interaction — this discovery was the first step toward the unified theory known as the Standard Model. In the theory of the electroweak interaction, the carriers of the weak force are the massive gauge bosons called the W and Z bosons. The weak interaction is the only known interaction which does not conserve parity; it is left-right asymmetric. The weak interaction even violates CP symmetry but does conserve CPT.

31.4.3 Strong interaction

Main article: Strong interaction

The *strong interaction*, or *strong nuclear force*, is the most complicated interaction, mainly because of the way it varies with distance. At distances greater than 10 femtometers, the strong force is practically unobservable. Moreover, it holds only inside the atomic nucleus.

After the nucleus was discovered in 1908, it was clear that a new force was needed to overcome the electrostatic repulsion, a manifestation of electromagnetism, of the positively charged protons. Otherwise the nucleus could not exist. Moreover, the force had to be strong enough to squeeze the protons into a volume that is 10^{-15} of that of the entire atom. From the short range of this force, Hideki Yukawa predicted that it was associated with a massive particle, whose mass is approximately 100 MeV.

The 1947 discovery of the pion ushered in the modern era of particle physics. Hundreds of hadrons were discovered from the 1940s to 1960s, and an extremely complicated theory of hadrons as strongly interacting particles was developed. Most notably:

- The pions were understood to be oscillations of vacuum condensates;

- Jun John Sakurai proposed the rho and omega vector bosons to be force carrying particles for approximate symmetries of isospin and hypercharge;

- Geoffrey Chew, Edward K. Burdett and Steven Frautschi grouped the heavier hadrons into families that could be understood as vibrational and rotational excitations of strings.

While each of these approaches offered deep insights, no approach led directly to a fundamental theory.

Murray Gell-Mann along with George Zweig first proposed fractionally charged quarks in 1961. Throughout the 1960s, different authors considered theories similar to the modern fundamental theory of quantum chromodynamics (QCD) as simple models for the interactions of quarks. The first to hypothesize the gluons of QCD were Moo-Young Han and Yoichiro Nambu, who introduced the quark color charge and hypothesized that it might be associated with a force-carrying field. At that time, however, it was difficult to see how such a model could permanently confine quarks. Han and Nambu also assigned each quark color an integer electrical charge, so that the quarks were fractionally charged only on average, and they did not expect the quarks in their model to be permanently confined.

In 1971, Murray Gell-Mann and Harald Fritzsch proposed that the Han/Nambu color gauge field was the correct theory of the short-distance interactions of fractionally charged quarks. A little later, David Gross, Frank Wilczek, and David Politzer discovered that this theory had the property of asymptotic freedom, allowing them to make contact with experimental evidence. They concluded that QCD was the complete theory of the strong interactions, correct at all distance scales. The discovery of asymptotic freedom led most physicists to accept QCD, since it became clear that even the long-distance properties of the strong interactions could be consistent with experiment, if the quarks are permanently confined.

Assuming that quarks are confined, Mikhail Shifman, Arkady Vainshtein, and Valentine Zakharov were able to compute the properties of many low-lying hadrons directly from QCD, with only a few extra parameters to describe the vacuum. In 1980, Kenneth G. Wilson published computer calculations based on the first principles of QCD, establishing, to a level of confidence tantamount to certainty, that QCD will confine quarks. Since then, QCD has been the established theory of the strong interactions.

QCD is a theory of fractionally charged quarks interacting by means of 8 photon-like particles called gluons. The gluons interact with each other, not just with the quarks, and at long distances the lines of force collimate into strings. In this way, the mathematical theory of QCD not only explains how quarks interact over short distances, but also the string-like behavior, discovered by Chew and Frautschi, which they manifest over longer distances.

31.4.4   Beyond the Standard Model

Main article: Physics beyond the Standard Model
See also: Elementary particle § Beyond the Standard Model

Numerous theoretical efforts have been made to systematize the existing four fundamental interactions on the model of electro-weak unification.

Grand Unified Theories (GUTs) are proposals to show that all of the fundamental interactions, other than gravity, arise from a single interaction with symmetries that break down at low energy levels. GUTs predict relationships among constants of nature that are unrelated in the SM. GUTs also predict gauge coupling unification for the relative strengths of the electromagnetic, weak, and strong forces, a prediction verified at the Large Electron–Positron Collider in 1991 for supersymmetric theories.

Theories of everything, which integrate GUTs with a quantum gravity theory face a greater barrier, because no quantum gravity theories, which include string theory, loop quantum gravity, and twistor theory, have secured wide acceptance. Some theories look for a graviton to complete the Standard Model list of force carrying particles, while others, like loop quantum gravity, emphasize the possibility that time-space itself may have a quantum aspect to it.

Some theories beyond the Standard Model include a hypothetical fifth force, and the search for such a force is an ongoing line of experimental research in physics. In supersymmetric theories, there are particles that acquire their masses only through supersymmetry breaking effects and these particles, known as moduli can mediate new forces. Another reason to look for new forces is the recent discovery that the expansion of the universe is accelerating (also known as dark energy), giving rise to a need to explain a nonzero cosmological constant, and possibly to other modifications of general relativity. Fifth forces have also been suggested to explain phenomena such as CP violations, dark matter, and dark flow.

31.5   See also

- Standard Model

  - Strong interaction

  - Electroweak interaction

  - Weak interaction

  - Gravity

    - Quantum gravity

    - String Theory

    - Theory of Everything

- Grand Unified Theory

  - Gauge coupling unification

  - Unified Field Theory

- Quintessence, a hypothesized fifth force.

- *People*: Isaac Newton, James Clerk Maxwell, Albert Einstein, Richard Feynman, Sheldon Glashow, Abdus Salam, Steven Weinberg, Gerardus 't Hooft, David Gross, Edward Witten, Howard Georgi.

31.6   References

[1] http://www.pha.jhu.edu/~{}dfehling/particle.gif

[2] Newton's absolute space was a medium, but not one transmitting gravitation.

[3] Special relativity holds for objects at vast speed but of negligible mass, for instance elementary particles. Yet by yielding gravitation, which is a manner of acceleration, notable mass breaks inertia—that is, constant speed and direction—and thereby violates special relativity. Special relativity could approximately predict a massive object's motion during barely an instant, however, and thus is a temporally limited case of general relativity.

[4] Meinard Kuhlmann, "Physicists debate whether the world is made of particles or fields—or something else entirely", *Scientific American*, 24 Jul 2013.

[5] Approximate. See Coupling constant for more exact strengths, depending on the particles and energies involved.

[6] CERN (20 January 2012). "Extra dimensions, gravitons, and tiny black holes".

[7] Bais, Sander (2005), *The Equations. Icons of knowledge*, ISBN 0-674-01967-9 p.84

[8] "The Nobel Prize in Physics 1979". The Nobel Foundation. Retrieved 2008-12-16.

Bibliography General:

- Davies, Paul (1986), *The Forces of Nature*, Cambridge Univ. Press 2nd ed.

- Feynman, Richard (1967), *The Character of Physical Law*, MIT Press, ISBN 0-262-56003-8

- Schumm, Bruce A. (2004), *Deep Down Things*, Johns Hopkins University Press While all interactions are discussed, discussion is especially thorough on the weak.

- Weinberg, Steven (1993), *The First Three Minutes: A Modern View of the Origin of the Universe*, Basic Books, ISBN 0-465-02437-8

- Weinberg, Steven (1994), *Dreams of a Final Theory*, Basic Books, ISBN 0-679-74408-8

Texts:

- Padmanabhan, T. (1998), *After The First Three Minutes: The Story of Our Universe*, Cambridge Univ. Press, ISBN 0-521-62972-1

- Perkins, Donald H. (2000), *Introduction to High Energy Physics*, Cambridge Univ. Press, ISBN 0-521-62196-8

- Riazuddin (December 29, 2009). "Non-standard interactions" (PDF). *NCP 5th Particle Physics Sypnoisis* (Islamabad: Riazuddin, Head of High-Energy Theory Group at National Center for Physics) **1** (1): 1–25. Retrieved March 19, 2011.

Chapter 32

Weak interaction

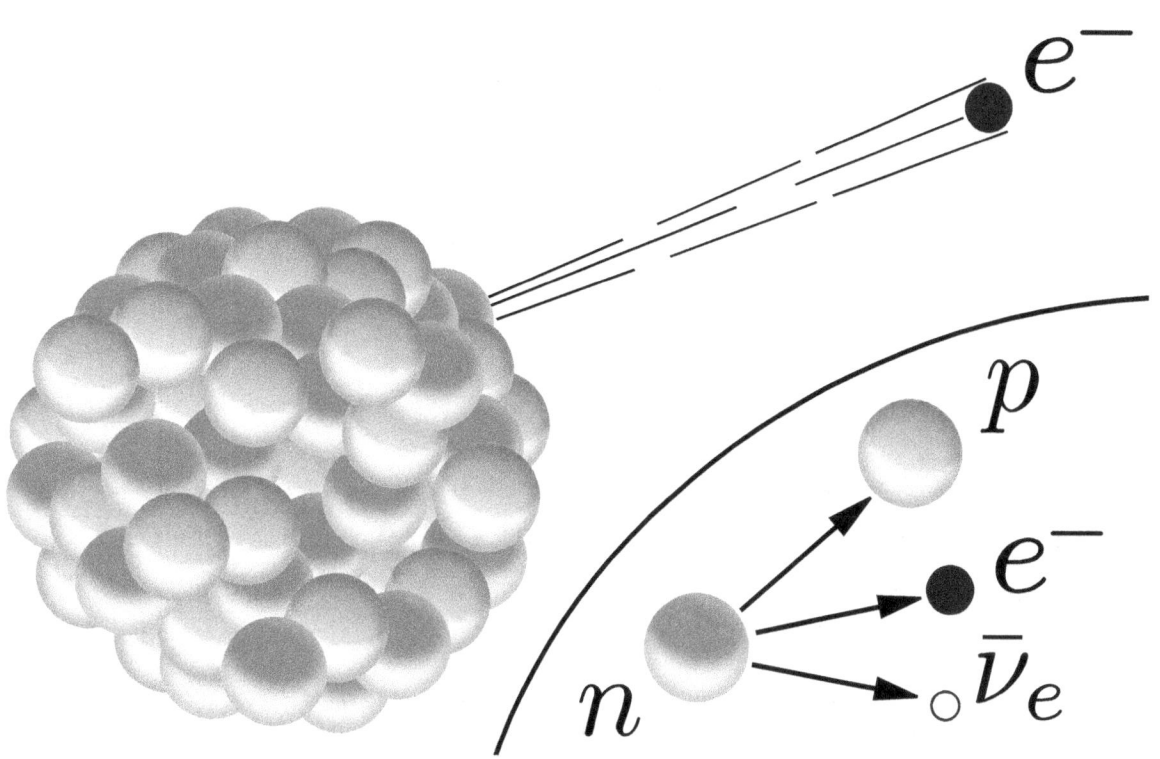

The radioactive beta decay is possible due to the weak interaction, which transforms a neutron into a proton, electron, and an electron antineutrino.

In particle physics, the **weak interaction** is the mechanism responsible for the **weak force** or **weak nuclear force**, one of the four known fundamental interactions of nature, alongside the strong interaction, electromagnetism, and gravitation. The weak interaction is responsible for the radioactive decay of subatomic particles, and it plays an essential role in nuclear fission. The theory of the weak interaction is sometimes called **quantum flavordynamics** (**QFD**), in analogy with the terms QCD and QED, but the term is rarely used because the weak force is best understood in terms of electro-weak theory (EWT).[1]

In the Standard Model of particle physics, the weak interaction is caused by the emission or absorption of W and Z bosons. All known fermions interact through the weak interaction. Fermions are particles that have half-integer spin (one of the fundamental properties of particles). A fermion can be an elementary particle, such as the electron, or it can be a composite particle, such as the proton. The masses of W$^+$, W$^-$, and Z bosons are each far greater than that of protons or neutrons, consistent with the short range of the weak force. The force is termed *weak* because its field strength over a given distance is typically several orders of magnitude less than that of the strong nuclear force and electromagnetic force.

During the quark epoch, the electroweak force split into the electromagnetic and weak forces. Most fermions will decay by a weak interaction over time. Important examples include beta decay, and the production of deuterium and then helium from hydrogen that powers the sun's thermonuclear process. Such decay also makes radiocarbon dating possible, as carbon-14 decays through the weak interaction to nitrogen-14. It can also create radioluminescence, commonly used in tritium illumination, and in the related field of betavoltaics.[2]

Quarks, which make up composite particles like neutrons and protons, come in six "flavours" – up, down, strange, charm, top and bottom – which give those composite particles their properties. The weak interaction is unique in that it allows for quarks to swap their flavour for another. For example, during beta minus decay, a down quark decays into an up quark, converting a neutron to a proton. Also the weak interaction is the only fundamental interaction that breaks parity-symmetry, and similarly, the only one to break CP-symmetry.

32.1 History

In 1933, Enrico Fermi proposed the first theory of the weak interaction, known as Fermi's interaction. He suggested that beta decay could be explained by a four-fermion interaction, involving a contact force with no range.[3][4]

However, it is better described as a non-contact force field having a finite range, albeit very short. In 1968, Sheldon Glashow, Abdus Salam and Steven Weinberg unified the electromagnetic force and the weak interaction by showing them to be two aspects of a single force, now termed the electro-weak force.

The existence of the W and Z bosons was not directly confirmed until 1983.

32.2 Properties

The weak interaction is unique in a number of respects:

1. It is the only interaction capable of changing the flavor of quarks (i.e., of changing one type of quark into another).

2. It is the only interaction that violates **P** or parity-symmetry. It is also the only one that violates **CP** symmetry.

3. It is propagated by carrier particles (known as gauge bosons) that have significant masses, an unusual feature which is explained in the Standard Model by the Higgs mechanism.

Due to their large mass (approximately 90 GeV/c^2[5]) these carrier particles, termed the W and Z bosons, are short-lived: they have a lifetime of under 1×10^{-24} seconds.[6] The weak interaction has a coupling constant (an indicator of interaction strength) of between 10^{-7} and 10^{-6}, compared to the strong interaction's coupling constant of about 1 and the electromagnetic coupling constant of about 10^{-2};[7] consequently the weak interaction is weak in terms of strength.[8] The weak interaction has a very short range (around 10^{-17}–10^{-16} m[8]).[7] At distances around 10^{-18} meters, the weak interaction has a strength of a similar magnitude to the electromagnetic force, but this starts to decrease exponentially with increasing distance. At distances of around 3×10^{-17} m, the weak interaction is 10,000 times weaker than the electromagnetic.[9]

The weak interaction affects all the fermions of the Standard Model, as well as the Higgs boson; neutrinos interact through gravity and the weak interaction only, and neutrinos were the original reason for the name *weak force*.[8] The weak interaction does not produce bound states (nor does it involve binding energy) – something that gravity does on an astronomical scale, that the electromagnetic force does at the atomic level, and that the strong nuclear force does inside nuclei.[10]

Its most noticeable effect is due to its first unique feature: flavor changing. A neutron, for example, is heavier than a proton (its sister nucleon), but it cannot decay into a proton without changing the flavor (type) of one of its two *down* quarks to

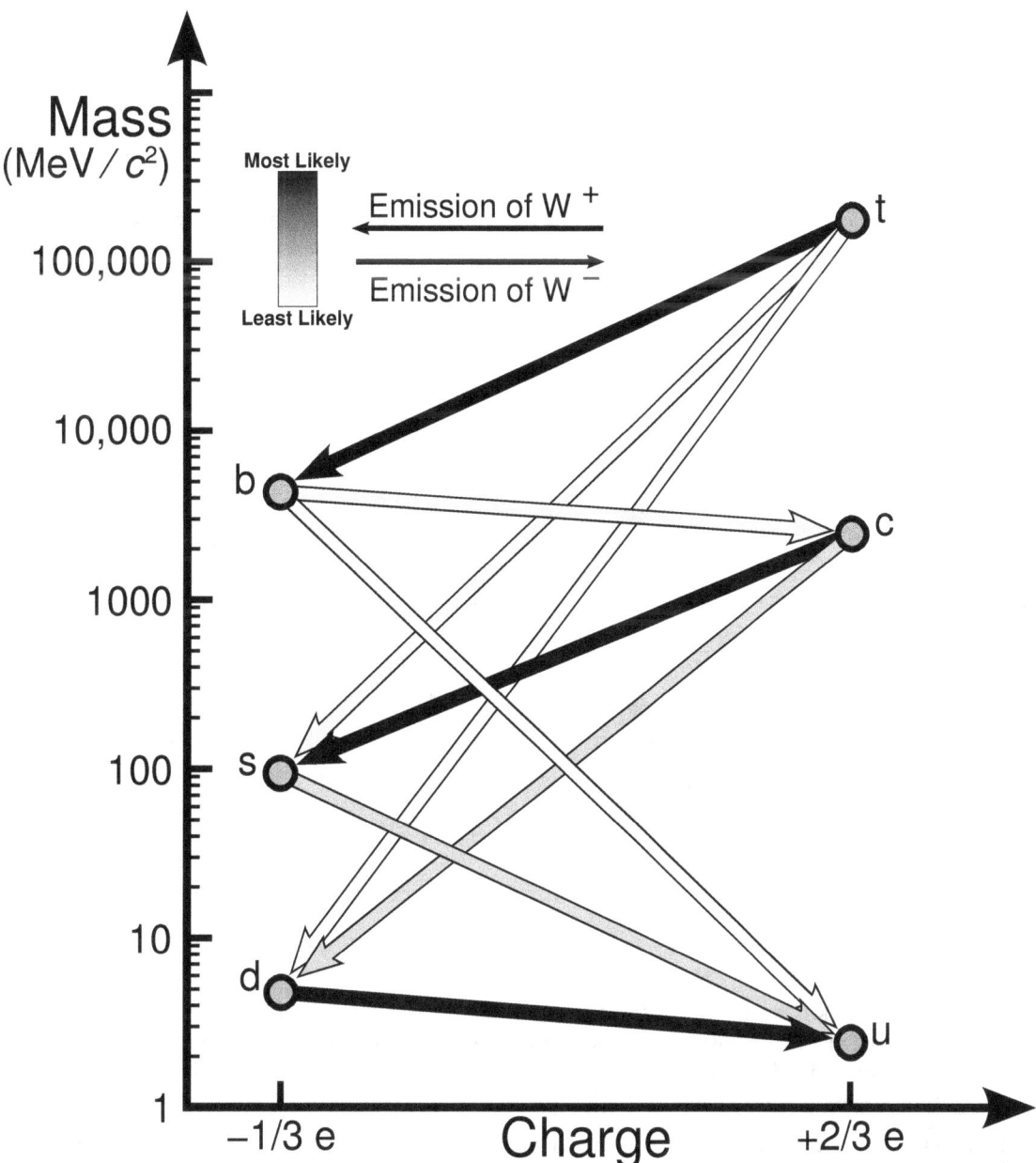

A diagram depicting the various decay routes due to the weak interaction and some indication of their likelihood. The intensity of the lines are given by the CKM parameters.

up. Neither the strong interaction nor electromagnetism permit flavour changing, so this must proceed by **weak decay**; without weak decay, quark properties such as strangeness and charm (associated with the quarks of the same name) would also be conserved across all interactions. All mesons are unstable because of weak decay.[11] In the process known as beta decay, a *down* quark in the neutron can change into an *up* quark by emitting a virtual W− boson which is then converted into an electron and an electron antineutrino.[12]

Due to the large mass of a boson, weak decay is much more unlikely than strong or electromagnetic decay, and hence occurs less rapidly. For example, a neutral pion (which decays electromagnetically) has a life of about 10^{-16} seconds, while a charged pion (which decays through the weak interaction) lives about 10^{-8} seconds, a hundred million times longer.[13] In contrast, a free neutron (which also decays through the weak interaction) lives about 15 minutes.[12]

32.2.1 Weak isospin and weak hypercharge

Main article: Weak isospin

All particles have a property called weak isospin (T_3), which serves as a quantum number and governs how that particle interacts in the weak interaction. Weak isospin therefore plays the same role in the weak interaction as electric charge does in electromagnetism, and color charge in the strong interaction. All fermions have a weak isospin value of either $+\frac{1}{2}$ or $-\frac{1}{2}$. For example, the up quark has a T_3 of $+\frac{1}{2}$ and the down quark $-\frac{1}{2}$. A quark never decays through the weak interaction into a quark of the same T_3: quarks with a T_3 of $+\frac{1}{2}$ decay into quarks with a T_3 of $-\frac{1}{2}$ and vice versa.

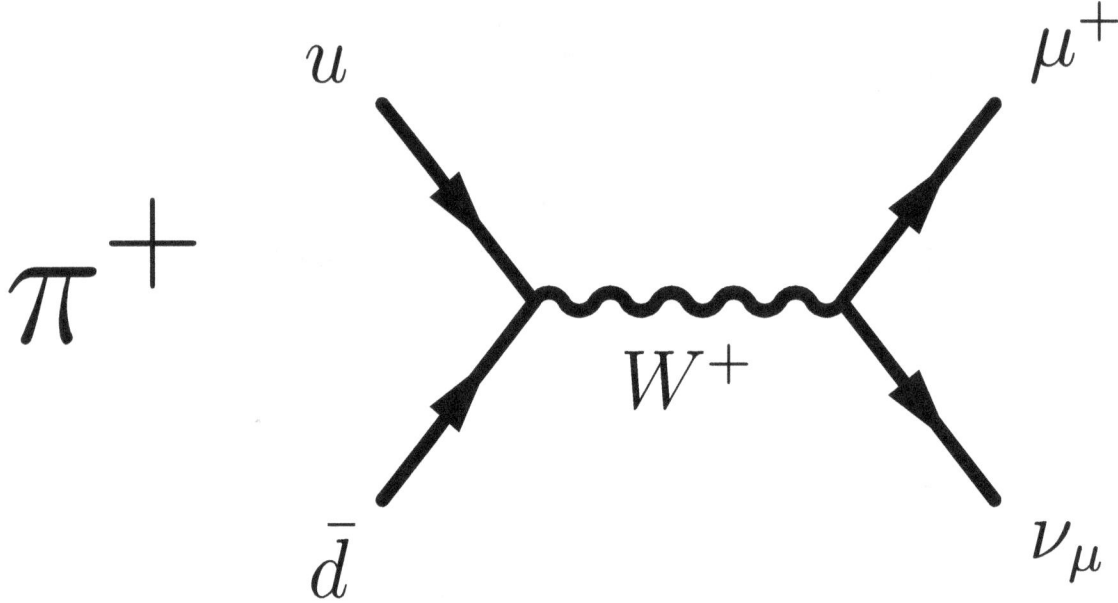

π+ decay through the weak interaction

In any given interaction, weak isospin is conserved: the sum of the weak isospin numbers of the particles entering the interaction equals the sum of the weak isospin numbers of the particles exiting that interaction. For example, a (left-handed) π+, with a weak isospin of 1 normally decays into a ν
μ (+1/2) and a μ+ (as a right-handed antiparticle, +1/2).[13]

Following the development of the electroweak theory, another property, weak hypercharge, was developed. It is dependent on a particle's electrical charge and weak isospin, and is defined as:

$$Y_W = 2(Q - T_3)$$

where YW is the weak hypercharge of a given type of particle, Q is its electrical charge (in elementary charge units) and T_3 is its weak isospin. Whereas some particles have a weak isospin of zero, all particles, except gluons, have non-zero weak hypercharge. Weak hypercharge is the generator of the U(1) component of the electroweak gauge group.

32.3 Interaction types

There are two types of weak interaction (called *vertices*). The first type is called the "charged-current interaction" because it is mediated by particles that carry an electric charge (the W+ or W− bosons), and is responsible for the beta decay phenomenon. The second type is called the "neutral-current interaction" because it is mediated by a neutral particle, the Z boson.

32.3.1   Charged-current interaction

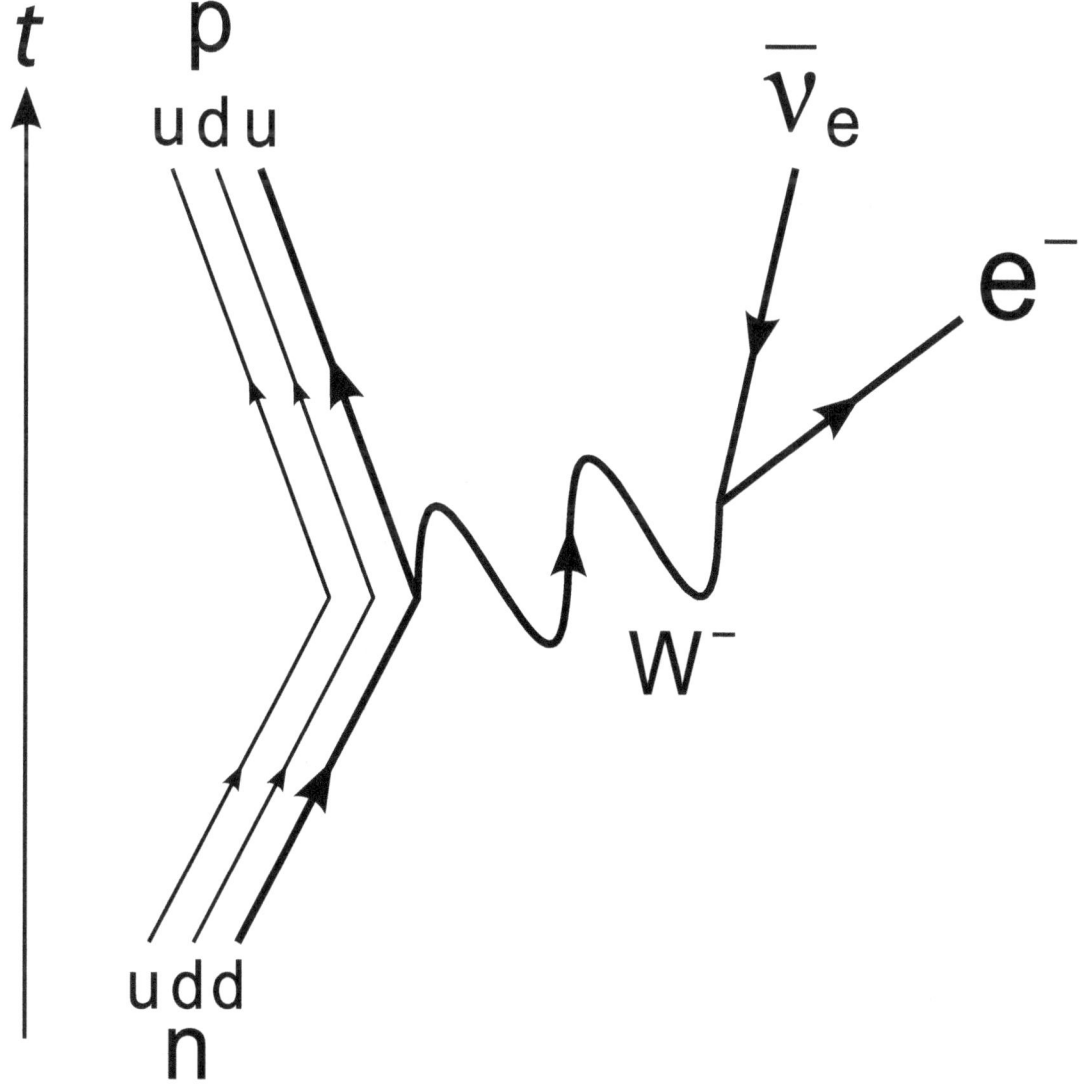

The Feynman diagram for beta-minus decay of a neutron into a proton, electron and electron anti-neutrino, via an intermediate heavy W– boson

In one type of charged current interaction, a charged lepton (such as an electron or a muon, having a charge of −1) can absorb a W+ boson (a particle with a charge of +1) and be thereby converted into a corresponding neutrino (with a charge of 0), where the type ("family") of neutrino (electron, muon or tau) is the same as the type of lepton in the interaction, for example:

$$\mu^- + W^+ \to \nu_\mu$$

Similarly, a down-type quark (d with a charge of $-\frac{1}{3}$) can be converted into an up-type quark (u, with a charge of $+\frac{2}{3}$), by emitting a W– boson or by absorbing a W+ boson. More precisely, the down-type quark becomes a quantum superposition of up-type quarks: that is to say, it has a possibility of becoming any one of the three up-type quarks, with the probabilities given in the CKM matrix tables. Conversely, an up-type quark can emit a W+ boson – or absorb a W–boson – and thereby be converted into a down-type quark, for example:

$$d \rightarrow u + W^-$$
$$d + W^+ \rightarrow u$$
$$c \rightarrow s + W^+$$
$$c + W^- \rightarrow s$$

The W boson is unstable so will rapidly decay, with a very short lifetime. For example:

$$W^- \rightarrow e^- + \bar{\nu}_e$$
$$W^+ \rightarrow e^+ + \nu_e$$

Decay of the W boson to other products can happen, with varying probabilities.[15]

In the so-called beta decay of a neutron (see picture, above), a down quark within the neutron emits a virtual W– boson and is thereby converted into an up quark, converting the neutron into a proton. Because of the energy involved in the process (i.e., the mass difference between the down quark and the up quark), the W– boson can only be converted into an electron and an electron-antineutrino.[16] At the quark level, the process can be represented as:

$$d \rightarrow u + e^- + \bar{\nu}_e$$

32.3.2 Neutral-current interaction

In neutral current interactions, a quark or a lepton (e.g., an electron or a muon) emits or absorbs a neutral Z boson. For example:

$$e^- \rightarrow e^- + Z^0$$

Like the W boson, the Z boson also decays rapidly,[15] for example:

$$Z^0 \rightarrow b + \bar{b}$$

32.4 Electroweak theory

Main article: Electroweak interaction

The Standard Model of particle physics describes the electromagnetic interaction and the weak interaction as two different aspects of a single electroweak interaction, the theory of which was developed around 1968 by Sheldon Glashow, Abdus Salam and Steven Weinberg. They were awarded the 1979 Nobel Prize in Physics for their work.[17] The Higgs mechanism provides an explanation for the presence of three massive gauge bosons (the three carriers of the weak interaction) and the massless photon of the electromagnetic interaction.[18]

According to the electroweak theory, at very high energies, the universe has four massless gauge boson fields similar to the photon and a complex scalar Higgs field doublet. However, at low energies, gauge symmetry is spontaneously broken down to the **U**(1) symmetry of electromagnetism (one of the Higgs fields acquires a vacuum expectation value). This symmetry breaking would produce three massless bosons, but they become integrated by three photon-like fields (through the Higgs mechanism) giving them mass. These three fields become the W+, W− and Z bosons of the weak interaction, while the fourth gauge field, which remains massless, is the photon of electromagnetism.[18]

This theory has made a number of predictions, including a prediction of the masses of the Z and W bosons before their discovery. On 4 July 2012, the CMS and the ATLAS experimental teams at the Large Hadron Collider independently announced that they had confirmed the formal discovery of a previously unknown boson of mass between 125–127 GeV/c^2, whose behaviour so far was "consistent with" a Higgs boson, while adding a cautious note that further data and analysis were needed before positively identifying the new boson as being a Higgs boson of some type. By 14 March 2013, the Higgs boson was tentatively confirmed to exist .[19]

32.5   Violation of symmetry

Left- and right-handed particles: p is the particle's momentum and S is its spin. Note the lack of reflective symmetry between the states.

The laws of nature were long thought to remain the same under mirror reflection, the reversal of one spatial axis. The results of an experiment viewed via a mirror were expected to be identical to the results of a mirror-reflected copy of the experimental apparatus. This so-called law of parity conservation was known to be respected by classical gravitation, electromagnetism and the strong interaction; it was assumed to be a universal law.[20] However, in the mid-1950s Chen Ning Yang and Tsung-Dao Lee suggested that the weak interaction might violate this law. Chien Shiung Wu and collaborators in 1957 discovered that the weak interaction violates parity, earning Yang and Lee the 1957 Nobel Prize in Physics.[21]

Although the weak interaction used to be described by Fermi's theory, the discovery of parity violation and renormalization theory suggested that a new approach was needed. In 1957, Robert Marshak and George Sudarshan and, somewhat later, Richard Feynman and Murray Gell-Mann proposed a **V−A** (vector minus axial vector or left-handed) Lagrangian for weak interactions. In this theory, the weak interaction acts only on left-handed particles (and right-handed antiparticles). Since the mirror reflection of a left-handed particle is right-handed, this explains the maximal violation of parity. Interestingly, the **V−A** theory was developed before the discovery of the Z boson, so it did not include the right-handed fields that enter in the neutral current interaction.

However, this theory allowed a compound symmetry **CP** to be conserved. **CP** combines parity **P** (switching left to right) with charge conjugation **C** (switching particles with antiparticles). Physicists were again surprised when in 1964, James Cronin and Val Fitch provided clear evidence in kaon decays that CP symmetry could be broken too, winning them the 1980 Nobel Prize in Physics.[22] In 1973, Makoto Kobayashi and Toshihide Maskawa showed that CP violation in the weak interaction required more than two generations of particles,[23] effectively predicting the existence of a then unknown third generation. This discovery earned them half of the 2008 Nobel Prize in Physics.[24] Unlike parity violation, CP violation occurs in only a small number of instances, but remains widely held as an answer to the difference between the amount of matter and antimatter in the universe; it thus forms one of Andrei Sakharov's three conditions for baryogenesis.[25]

32.6 See also

- Weakless Universe – the postulate that weak interactions are not anthropically necessary

- Gravity

- Nuclear force

- Electromagnetism

32.7 References

32.7.1 Citations

[1] Griffiths, David (2009). *Introduction to Elementary Particles.* pp. 59–60. ISBN 978-3-527-40601-2.

[2] "The Nobel Prize in Physics 1979: Press Release". *NobelPrize.org.* Nobel Media. Retrieved 22 March 2011.

[3] Fermi, Enrico (1934). "Versuch einer Theorie der β-Strahlen. I". *Zeitschrift für Physik A* **88** (3–4): 161–177. Bibcode:1934ZPhy...88..161F. doi:10.1007/BF01351864.

[4] Wilson, Fred L. (December 1968). "Fermi's Theory of Beta Decay". *American Journal of Physics* **36** (12): 1150–1160. Bibcode:1968AmJPh..36.1150W. doi:10.1119/1.1974382.

[5] W.-M. Yao *et al.* (Particle Data Group) (2006). "Review of Particle Physics: Quarks" (PDF). *Journal of Physics G* **33**: 1–1232. arXiv:astro-ph/0601168. Bibcode:2006JPhG...33....1Y. doi:10.1088/0954-3899/33/1/001.

[6] Peter Watkins (1986). *Story of the W and Z.* Cambridge: Cambridge University Press. p. 70. ISBN 978-0-521-31875-4.

[7] "Coupling Constants for the Fundamental Forces". *HyperPhysics.* Georgia State University. Retrieved 2 March 2011.

[8] J. Christman (2001). "The Weak Interaction" (PDF). *Physnet.* Michigan State University.

[9] "Electroweak". *The Particle Adventure.* Particle Data Group. Retrieved 3 March 2011.

[10] Walter Greiner; Berndt Müller (2009). *Gauge Theory of Weak Interactions.* Springer. p. 2. ISBN 978-3-540-87842-1.

[11] Cottingham & Greenwood (1986, 2001), p.29

[12] Cottingham & Greenwood (1986, 2001), p.28

[13] Cottingham & Greenwood (1986, 2001), p.30

[14] Baez, John C.; Huerta, John (2009). "The Algebra of Grand Unified Theories". *Bull.Am.Math.Soc.* **0904**: 483–552. arXiv:0904.1556. Bibcode:2009arXiv0904.1556B. doi:10.1090/s0273-0979-10-01294-2. Retrieved 15 October 2013.

[15] K. Nakamura *et al.* (Particle Data Group) (2010). "Gauge and Higgs Bosons" (PDF). *Journal of Physics G* **37**.

[16] K. Nakamura *et al.* (Particle Data Group) (2010). "n" (PDF). *Journal of Physics G* **37**: 7.

[17] "The Nobel Prize in Physics 1979". *NobelPrize.org.* Nobel Media. Retrieved 26 February 2011.

[18] C. Amsler *et al.* (Particle Data Group) (2008). "Review of Particle Physics – Higgs Bosons: Theory and Searches" (PDF). *Physics Letters B* **667**: 1–6. Bibcode:2008PhLB..667....1P. doi:10.1016/j.physletb.2008.07.018.

[19] "New results indicate that new particle is a Higgs boson I CERN". Home.web.cern.ch. Retrieved 20 September 2013.

[20] Charles W. Carey (2006). "Lee, Tsung-Dao". *American scientists.* Facts on File Inc. p. 225. ISBN 9781438108070.

[21] "The Nobel Prize in Physics 1957". *NobelPrize.org.* Nobel Media. Retrieved 26 February 2011.

[22] "The Nobel Prize in Physics 1980". *NobelPrize.org.* Nobel Media. Retrieved 26 February 2011.

[23] M. Kobayashi, T. Maskawa (1973). "CP-Violation in the Renormalizable Theory of Weak Interaction". *Progress of Theoretical Physics* **49** (2): 652–657. Bibcode:1973PThPh..49..652K. doi:10.1143/PTP.49.652.

[24] "The Nobel Prize in Physics 1980". *NobelPrize.org*. Nobel Media. Retrieved 17 March 2011.

[25] Paul Langacker (2001) [1989]. "Cp Violation and Cosmology". In Cecilia Jarlskog. *CP violation*. London, River Edge: World Scientific Publishing Co. p. 552. ISBN 9789971505615.

32.7.2   General readers

- R. Oerter (2006). *The Theory of Almost Everything: The Standard Model, the Unsung Triumph of Modern Physics*. Plume. ISBN 978-0-13-236678-6.

- B.A. Schumm (2004). *Deep Down Things: The Breathtaking Beauty of Particle Physics*. Johns Hopkins University Press. ISBN 0-8018-7971-X.

32.7.3   Texts

- D.A. Bromley (2000). *Gauge Theory of Weak Interactions*. Springer. ISBN 3-540-67672-4.

- G.D. Coughlan, J.E. Dodd, B.M. Gripaios (2006). *The Ideas of Particle Physics: An Introduction for Scientists* (3rd ed.). Cambridge University Press. ISBN 978-0-521-67775-2.

- W. N. Cottingham; D. A. Greenwood (2001) [1986]. *An introduction to nuclear physics* (2nd ed.). Cambridge University Press. p. 30. ISBN 978-0-521-65733-4.

- D.J. Griffiths (1987). *Introduction to Elementary Particles*. John Wiley & Sons. ISBN 0-471-60386-4.

- G.L. Kane (1987). *Modern Elementary Particle Physics*. Perseus Books. ISBN 0-201-11749-5.

- D.H. Perkins (2000). *Introduction to High Energy Physics*. Cambridge University Press. ISBN 0-521-62196-8.

Chapter 33

Electromagnetism

Electromagnetism is the study of the **electromagnetic force** which is a type of physical interaction that occurs between electrically charged particles. The electromagnetic force usually shows electromagnetic fields, such as electric fields, magnetic fields, and light. The electromagnetic force is one of the four fundamental interactions in nature. The other three fundamental interactions are the strong interaction, the weak interaction, and gravitation.[1]

Lightning is an electrostatic discharge that travels between two charged regions.

The word *electromagnetism* is a compound form of two Greek terms, ἤλεκτρον, *ēlektron*, "amber", and μαγνήτης, *magnetic*, from "magnítis líthos" (μαγνήτης λίθος), which means "magnesian stone", a type of iron ore. The science of electromagnetic phenomena is defined in terms of the electromagnetic force, sometimes called the Lorentz force, which includes both electricity and magnetism as elements of one phenomenon.

The electromagnetic force plays a major role in determining the internal properties of most objects encountered in daily life. Ordinary matter takes its form as a result of intermolecular forces between individual molecules in matter. Electrons are bound by electromagnetic wave mechanics into orbitals around atomic nuclei to form atoms, which are the building blocks of molecules. This governs the processes involved in chemistry, which arise from interactions between the electrons of neighboring atoms, which are in turn determined by the interaction between electromagnetic force and the momentum of the electrons.

There are numerous mathematical descriptions of the electromagnetic field. In classical electrodynamics, electric fields are described as electric potential and electric current in Ohm's law, magnetic fields are associated with electromagnetic induction and magnetism, and Maxwell's equations describe how electric and magnetic fields are generated and altered by each other and by charges and currents.

The theoretical implications of electromagnetism, in particular the establishment of the speed of light based on properties of the "medium" of propagation (permeability and permittivity), led to the development of special relativity by Albert Einstein in 1905.

Although electromagnetism is considered one of the four fundamental forces, at high energy the weak force and electromagnetism are unified. In the history of the universe, during the quark epoch, the electroweak force split into the electromagnetic and weak forces.

33.1   History of the theory

See also: History of electromagnetic theory

Originally electricity and magnetism were thought of as two separate forces. This view changed, however, with the publication of James Clerk Maxwell's 1873 *A Treatise on Electricity and Magnetism* in which the interactions of positive and negative charges were shown to be regulated by one force. There are four main effects resulting from these interactions, all of which have been clearly demonstrated by experiments:

1. Electric charges attract or repel one another with a force inversely proportional to the square of the distance between them: unlike charges attract, like ones repel.

2. Magnetic poles (or states of polarization at individual points) attract or repel one another in a similar way and always come in pairs: every north pole is yoked to a south pole.

3. An electric current in a wire creates a circular magnetic field around the wire, its direction (clockwise or counterclockwise) depending on that of the current.

4. A current is induced in a loop of wire when it is moved towards or away from a magnetic field, or a magnet is moved towards or away from it, the direction of current depending on that of the movement.

While preparing for an evening lecture on 21 April 1820, Hans Christian Ørsted made a surprising observation. As he was setting up his materials, he noticed a compass needle deflected from magnetic north when the electric current from the battery he was using was switched on and off. This deflection convinced him that magnetic fields radiate from all sides of a wire carrying an electric current, just as light and heat do, and that it confirmed a direct relationship between electricity and magnetism.

At the time of discovery, Ørsted did not suggest any satisfactory explanation of the phenomenon, nor did he try to represent the phenomenon in a mathematical framework. However, three months later he began more intensive investigations. Soon thereafter he published his findings, proving that an electric current produces a magnetic field as it flows through a wire. The CGS unit of magnetic induction (oersted) is named in honor of his contributions to the field of electromagnetism.

Hans Christian Ørsted.

His findings resulted in intensive research throughout the scientific community in electrodynamics. They influenced French physicist André-Marie Ampère's developments of a single mathematical form to represent the magnetic forces between current-carrying conductors. Ørsted's discovery also represented a major step toward a unified concept of energy.

This unification, which was observed by Michael Faraday, extended by James Clerk Maxwell, and partially reformulated by Oliver Heaviside and Heinrich Hertz, is one of the key accomplishments of 19th century mathematical physics. It had far-reaching consequences, one of which was the understanding of the nature of light. Unlike what was proposed in Electromagnetism, light and other electromagnetic waves are at the present seen as taking the form of quantized, self-propagating oscillatory electromagnetic field disturbances which have been called photons. Different frequencies of oscillation give rise to the different forms of electromagnetic radiation, from radio waves at the lowest frequencies, to visible light at intermediate frequencies, to gamma rays at the highest frequencies.

Ørsted was not the only person to examine the relation between electricity and magnetism. In 1802 Gian Domenico Romagnosi, an Italian legal scholar, deflected a magnetic needle by electrostatic charges. Actually, no galvanic current existed in the setup and hence no electromagnetism was present. An account of the discovery was published in 1802 in an Italian newspaper, but it was largely overlooked by the contemporary scientific community.[2]

33.2   Fundamental forces

The electromagnetic force is one of the four known fundamental forces. The other fundamental forces are:

- the weak nuclear force, which binds to all known particles in the Standard Model, and causes certain forms of radioactive decay. (In particle physics though, the electroweak interaction is the unified description of two of the four known fundamental interactions of nature: electromagnetism and the weak interaction);

- the strong nuclear force, which binds quarks to form nucleons, and binds nucleons to form nuclei

- the gravitational force.

All other forces (e.g., friction) are ultimately derived from these fundamental forces and momentum carried by the movement of particles.

The electromagnetic force is the one responsible for practically all the phenomena one encounters in daily life above the nuclear scale, with the exception of gravity. Roughly speaking, all the forces involved in interactions between atoms can be explained by the electromagnetic force acting on the electrically charged atomic nuclei and electrons inside and around the atoms, together with how these particles carry momentum by their movement. This includes the forces we experience in "pushing" or "pulling" ordinary material objects, which come from the intermolecular forces between the individual molecules in our bodies and those in the objects. It also includes all forms of chemical phenomena.

A necessary part of understanding the intra-atomic to intermolecular forces is the effective force generated by the momentum of the electrons' movement, and that electrons move between interacting atoms, carrying momentum with them. As a collection of electrons becomes more confined, their minimum momentum necessarily increases due to the Pauli exclusion principle. The behaviour of matter at the molecular scale including its density is determined by the balance between the electromagnetic force and the force generated by the exchange of momentum carried by the electrons themselves.

33.3   Classical electrodynamics

Main article: Classical electrodynamics

The scientist William Gilbert proposed, in his *De Magnete* (1600), that electricity and magnetism, while both capable of causing attraction and repulsion of objects, were distinct effects. Mariners had noticed that lightning strikes had the ability to disturb a compass needle, but the link between lightning and electricity was not confirmed until Benjamin Franklin's proposed experiments in 1752. One of the first to discover and publish a link between man-made electric current and magnetism was Romagnosi, who in 1802 noticed that connecting a wire across a voltaic pile deflected a nearby compass needle. However, the effect did not become widely known until 1820, when Ørsted performed a similar experiment.[3] Ørsted's work influenced Ampère to produce a theory of electromagnetism that set the subject on a mathematical foundation.

A theory of electromagnetism, known as classical electromagnetism, was developed by various physicists over the course of the 19th century, culminating in the work of James Clerk Maxwell, who unified the preceding developments into a single theory and discovered the electromagnetic nature of light. In classical electromagnetism, the electromagnetic field obeys a set of equations known as Maxwell's equations, and the electromagnetic force is given by the Lorentz force law.

One of the peculiarities of classical electromagnetism is that it is difficult to reconcile with classical mechanics, but it is compatible with special relativity. According to Maxwell's equations, the speed of light in a vacuum is a universal constant, dependent only on the electrical permittivity and magnetic permeability of free space. This violates Galilean invariance, a long-standing cornerstone of classical mechanics. One way to reconcile the two theories (electromagnetism and classical mechanics) is to assume the existence of a luminiferous aether through which the light propagates. However, subsequent experimental efforts failed to detect the presence of the aether. After important contributions of Hendrik Lorentz and Henri Poincaré, in 1905, Albert Einstein solved the problem with the introduction of special relativity, which replaces classical kinematics with a new theory of kinematics that is compatible with classical electromagnetism. (For more information, see History of special relativity.)

In addition, relativity theory shows that in moving frames of reference a magnetic field transforms to a field with a nonzero electric component and vice versa; thus firmly showing that they are two sides of the same coin, and thus the term "electromagnetism". (For more information, see Classical electromagnetism and special relativity and Covariant formulation of classical electromagnetism.

33.4 Quantum mechanics

33.4.1 Photoelectric effect

Main article: Photoelectric effect

In another paper published in 1905, Albert Einstein undermined the very foundations of classical electromagnetism. In his theory of the photoelectric effect (for which he won the Nobel prize in physics) and inspired by the idea of Max Planck's "quanta", he posited that light could exist in discrete particle-like quantities as well, which later came to be known as photons. Einstein's theory of the photoelectric effect extended the insights that appeared in the solution of the ultraviolet catastrophe presented by Max Planck in 1900. In his work, Planck showed that hot objects emit electromagnetic radiation in discrete packets ("quanta"), which leads to a finite total energy emitted as black body radiation. Both of these results were in direct contradiction with the classical view of light as a continuous wave. Planck's and Einstein's theories were progenitors of quantum mechanics, which, when formulated in 1925, necessitated the invention of a quantum theory of electromagnetism. This theory, completed in the 1940s-1950s, is known as quantum electrodynamics (or "QED"), and, in situations where perturbation theory is applicable, is one of the most accurate theories known to physics.

33.4.2 Quantum electrodynamics

Main article: Quantum electrodynamics

All electromagnetic phenomena are underpinned by quantum mechanics, specifically by quantum electrodynamics (which includes classical electrodynamics as a limiting case) and this accounts for almost all physical phenomena observable to the unaided human senses, including light and other electromagnetic radiation, all of chemistry, most of mechanics (excepting gravitation), and, of course, magnetism and electricity.

33.4.3 Electroweak interaction

Main article: Electroweak interaction

The **electroweak interaction** is the unified description of two of the four known fundamental interactions of nature: electromagnetism and the weak interaction. Although these two forces appear very different at everyday low energies, the theory models them as two different aspects of the same force. Above the unification energy, on the order of 100 GeV, they would merge into a single **electroweak force**. Thus if the universe is hot enough (approximately 10^{15} K, a temperature exceeded until shortly after the Big Bang) then the electromagnetic force and weak force merge into a combined electroweak force. During the electroweak epoch, the electroweak force separated from the strong force. During the quark epoch, the electroweak force split into the electromagnetic and weak force.

33.5   Quantities and units

See also: List of physical quantities and List of electromagnetism equations

Electromagnetic units are part of a system of electrical units based primarily upon the magnetic properties of electric currents, the fundamental SI unit being the ampere. The units are:

- ampere (electric current)

- coulomb (electric charge)

- farad (capacitance)

- henry (inductance)

- ohm (resistance)

- tesla (magnetic flux density)

- volt (electric potential)

- watt (power)

- weber (magnetic flux)

In the electromagnetic cgs system, electric current is a fundamental quantity defined via Ampère's law and takes the permeability as a dimensionless quantity (relative permeability) whose value in a vacuum is unity. As a consequence, the square of the speed of light appears explicitly in some of the equations interrelating quantities in this system.

Formulas for physical laws of electromagnetism (such as Maxwell's equations) need to be adjusted depending on what system of units one uses. This is because there is no one-to-one correspondence between electromagnetic units in SI and those in CGS, as is the case for mechanical units. Furthermore, within CGS, there are several plausible choices of electromagnetic units, leading to different unit "sub-systems", including Gaussian, "ESU", "EMU", and Heaviside–Lorentz. Among these choices, Gaussian units are the most common today, and in fact the phrase "CGS units" is often used to refer specifically to CGS-Gaussian units.

33.6   See also

- Abraham–Lorentz force

- Aeromagnetic surveys

- Computational electromagnetics

- Double-slit experiment

- Electromagnet

- Electromagnetic wave equation

- Electromechanics

- Magnetostatics

- Magnetoquasistatic field

- Optics

- Relativistic electromagnetism

- Wheeler–Feynman absorber theory

33.7 References

[1] Ravaioli, Fawwaz T. Ulaby, Eric Michielssen, Umberto (2010). *Fundamentals of applied electromagnetics* (6th ed.). Boston: Prentice Hall. p. 13. ISBN 978-0-13-213931-1.

[2] Martins, Roberto de Andrade. "Romagnosi and Volta's Pile: Early Difficulties in the Interpretation of Voltaic Electricity". In Fabio Bevilacqua and Lucio Fregonese (eds). *Nuova Voltiana: Studies on Volta and his Times* (PDF). vol. 3. Università degli Studi di Pavia. pp. 81–102. Retrieved 2010-12-02.

[3] Stern, Dr. David P.; Peredo, Mauricio (2001-11-25). "Magnetic Fields -- History". NASA Goddard Space Flight Center. Retrieved 2009-11-27.

[4] International Union of Pure and Applied Chemistry (1993). *Quantities, Units and Symbols in Physical Chemistry*, 2nd edition, Oxford: Blackwell Science. ISBN 0-632-03583-8. pp. 14–15. Electronic version.

33.8 Further reading

33.8.1 Web sources

- Nave, R. "Electricity and magnetism". *HyperPhysics*. Georgia State University. Retrieved 2013-11-12.

33.8.2 Lecture notes

- Littlejohn, Robert (Spring 2011). "Emission and absorption of radiation" (PDF). *Physics 221B: Quantum mechanics*. University of California Berkeley. Retrieved 2013-11-12.

- Littlejohn, Robert (Spring 2011). "The Classical Electromagnetic Field Hamiltonian" (PDF). *Physics 221B: Quantum mechanics*. University of California Berkeley. Retrieved 2013-11-12.

33.8.3 Textbooks

- G.A.G. Bennet (1974). *Electricity and Modern Physics* (2nd ed.). Edward Arnold (UK). ISBN 0-7131-2459-8.

- Dibner, Bern (2012). *Oersted and the discovery of electromagnetism*. Literary Licensing, LLC. ISBN 9781258335557.

- Durney, Carl H. and Johnson, Curtis C. (1969). *Introduction to modern electromagnetics*. McGraw-Hill. ISBN 0-07-018388-0.

- Feynman, Richard P. (1970). *The Feynman Lectures on Physics Vol II*. Addison Wesley Longman. ISBN 978-0-201-02115-8.

- Fleisch, Daniel (2008). *A Student's Guide to Maxwell's Equations.* Cambridge, UK: Cambridge University Press. ISBN 978-0-521-70147-1.

- I.S. Grant, W.R. Phillips, Manchester Physics (2008). *Electromagnetism* (2nd ed.). John Wiley & Sons. ISBN 978-0-471-92712-9.

- Griffiths, David J. (1998). *Introduction to Electrodynamics* (3rd ed.). Prentice Hall. ISBN 0-13-805326-X.

- Jackson, John D. (1998). *Classical Electrodynamics* (3rd ed.). Wiley. ISBN 0-471-30932-X.

- Moliton, André (2007). *Basic electromagnetism and materials. 430 pages* (New York City: Springer-Verlag New York, LLC). ISBN 978-0-387-30284-3.

- Purcell, Edward M. (1985). *Electricity and Magnetism Berkeley Physics Course Volume 2 (2nd ed.).* McGraw-Hill. ISBN 0-07-004908-4.

- Rao, Nannapaneni N. (1994). *Elements of engineering electromagnetics (4th ed.).* Prentice Hall. ISBN 0-13-948746-8.

- Rothwell, Edward J.; Cloud, Michael J. (2001). *Electromagnetics.* CRC Press. ISBN 0-8493-1397-X.

- Tipler, Paul (1998). *Physics for Scientists and Engineers: Vol. 2: Light, Electricity and Magnetism* (4th ed.). W. H. Freeman. ISBN 1-57259-492-6.

- Wangsness, Roald K.; Cloud, Michael J. (1986). *Electromagnetic Fields (2nd Edition).* Wiley. ISBN 0-471-81186-6.

33.8.4 General references

- A. Beiser (1987). *Concepts of Modern Physics* (4th ed.). McGraw-Hill (International). ISBN 0-07-100144-1.

- L.H. Greenberg (1978). *Physics with Modern Applications.* Holt-Saunders International W.B. Saunders and Co. ISBN 0-7216-4247-0.

- R.G. Lerner, G.L. Trigg (2005). *Encyclopaedia of Physics* (2nd ed.). VHC Publishers, Hans Warlimont, Springer. pp. 12–13. ISBN 978-0-07-025734-4.

- J.B. Marion, W.F. Hornyak (1984). *Principles of Physics.* Holt-Saunders International Saunders College. ISBN 4-8337-0195-2.

- H.J. Pain (1983). *The Physics of Vibrations and Waves* (3rd ed.). John Wiley & Sons,. ISBN 0-471-90182-2.

- C.B. Parker (1994). *McGraw Hill Encyclopaedia of Physics* (2nd ed.). McGraw Hill. ISBN 0-07-051400-3.

- R. Penrose (2007). *The Road to Reality.* Vintage books. ISBN 0-679-77631-1.

- P.A. Tipler, G. Mosca (2008). *Physics for Scientists and Engineers: With Modern Physics* (6th ed.). W.H. Freeman and Co. ISBN 9-781429-202657.

- P.M. Whelan, M.J. Hodgeson (1978). *Essential Principles of Physics* (2nd ed.). John Murray. ISBN 0-7195-3382-1.

33.9 External links

- Oppelt, Arnulf (2006-11-02). "magnetic field strength". Retrieved 2007-06-04.

- "magnetic field strength converter". Retrieved 2007-06-04.

- Electromagnetic Force - from Eric Weisstein's World of Physics

- Goudarzi, Sara (2006-08-15). "Ties That Bind Atoms Weaker Than Thought". *LiveScience.com*. Retrieved 2013-11-12.

- Quarked Electromagnetic force - A good introduction for kids

- The Deflection of a Magnetic Compass Needle by a Current in a Wire (video) on YouTube

- Electromagnetism abridged

André-Marie Ampère

Michael Faraday

James Clerk Maxwell

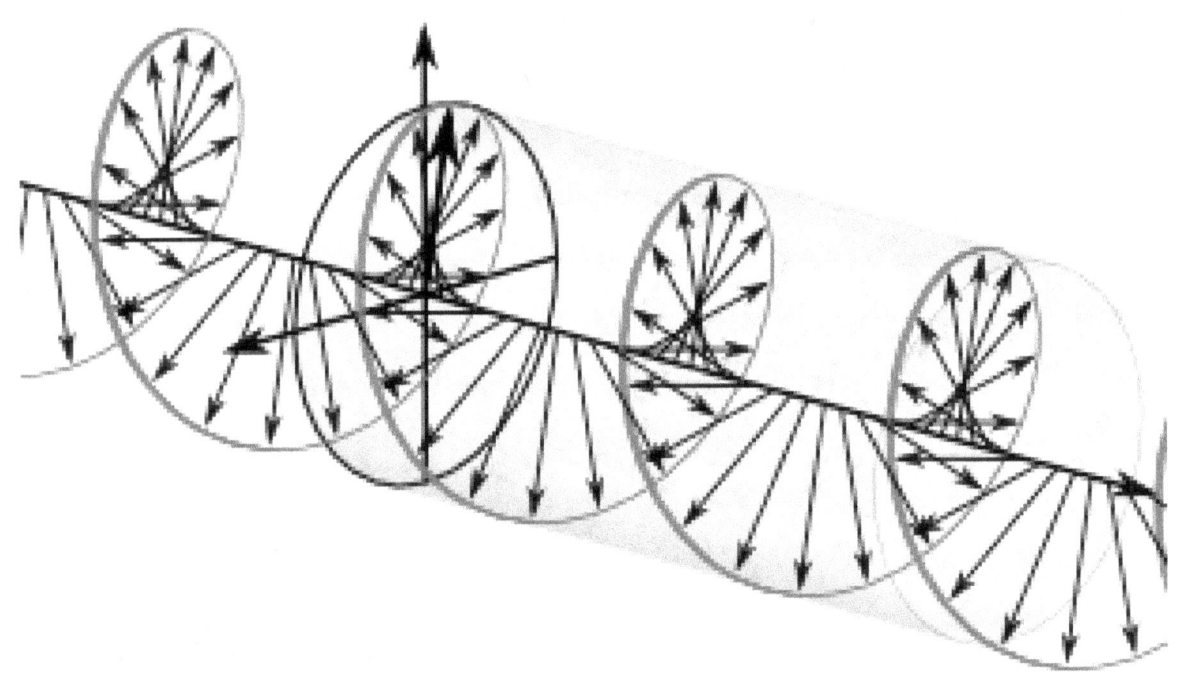

Representation of the electric field vector of a wave of circularly polarized electromagnetic radiation.

Chapter 34

Strong interaction

In particle physics, the **strong interaction** is the mechanism responsible for the strong nuclear force (also called the **strong force**, **nuclear strong force** or **colour force**), one of the four fundamental interactions of nature, the others being electromagnetism, the weak interaction and gravitation. Effective only at a distance of a femtometre, it is approximately 100 times stronger than electromagnetism, a million times stronger than the weak force interaction and 10^{38} times stronger than gravitation at that range.[1] It ensures the stability of ordinary matter, as it confines the quark elementary particles into hadron particles, such as the proton and neutron, the largest components of the mass of ordinary matter. Furthermore, most of the mass-energy of a common proton or neutron is in the form of the strong force field energy; the individual quarks provide only about 1% of the mass-energy of a proton.

The strong interaction is observable in two areas: on a larger scale (about 1 to 3 femtometers (fm)), it is the force that binds protons and neutrons (nucleons) together to form the nucleus of an atom. On the smaller scale (less than about 0.8 fm, the radius of a nucleon), it is the force (carried by gluons) that holds quarks together to form protons, neutrons, and other hadron particles. The strong force inherently has so high a strength that the energy of an object bound by the strong force (a hadron) is high enough to produce new massive particles. Thus, if hadrons are struck by high-energy particles, they give rise to new hadrons instead of emitting freely moving radiation (gluons). This property of the strong force is called colour confinement, and it prevents the free "emission" of the strong force: instead, in practice, jets of massive particles are observed.

In the context of binding protons and neutrons together to form atoms, the strong interaction is called the nuclear force (or *residual strong force*). In this case, it is the residuum of the strong interaction between the quarks that make up the protons and neutrons. As such, the residual strong interaction obeys a quite different distance-dependent behavior between nucleons, from when it is acting to bind quarks within nucleons. The binding energy that is partly released on the breakup of a nucleus is related to the residual strong force and is harnessed in nuclear power and fission-type nuclear weapons.[2][3]

The strong interaction is thought to be mediated by massless particles called gluons, that are exchanged between quarks, antiquarks, and other gluons. Gluons, in turn, are thought to interact with quarks and gluons as all carry a type of charge called colour charge. Colour charge is analogous to electromagnetic charge, but it comes in three types rather than one (+/- red, +/- green, +/- blue) that results in a different type of force, with different rules of behavior. These rules are detailed in the theory of quantum chromodynamics (QCD), which is the theory of quark-gluon interactions.

Just after the Big Bang, and during the electroweak epoch, the electroweak force separated from the strong force. Although it is expected that a Grand Unified Theory exists to describe this, no such theory has been successfully formulated, and the unification remains an unsolved problem in physics.

34.1 History

Before the 1970s, physicists were uncertain about the binding mechanism of the atomic nucleus. It was known that the nucleus was composed of protons and neutrons and that protons possessed positive electric charge, while neutrons were electrically neutral. However, these facts seemed to contradict one another. By physical understanding at that time,

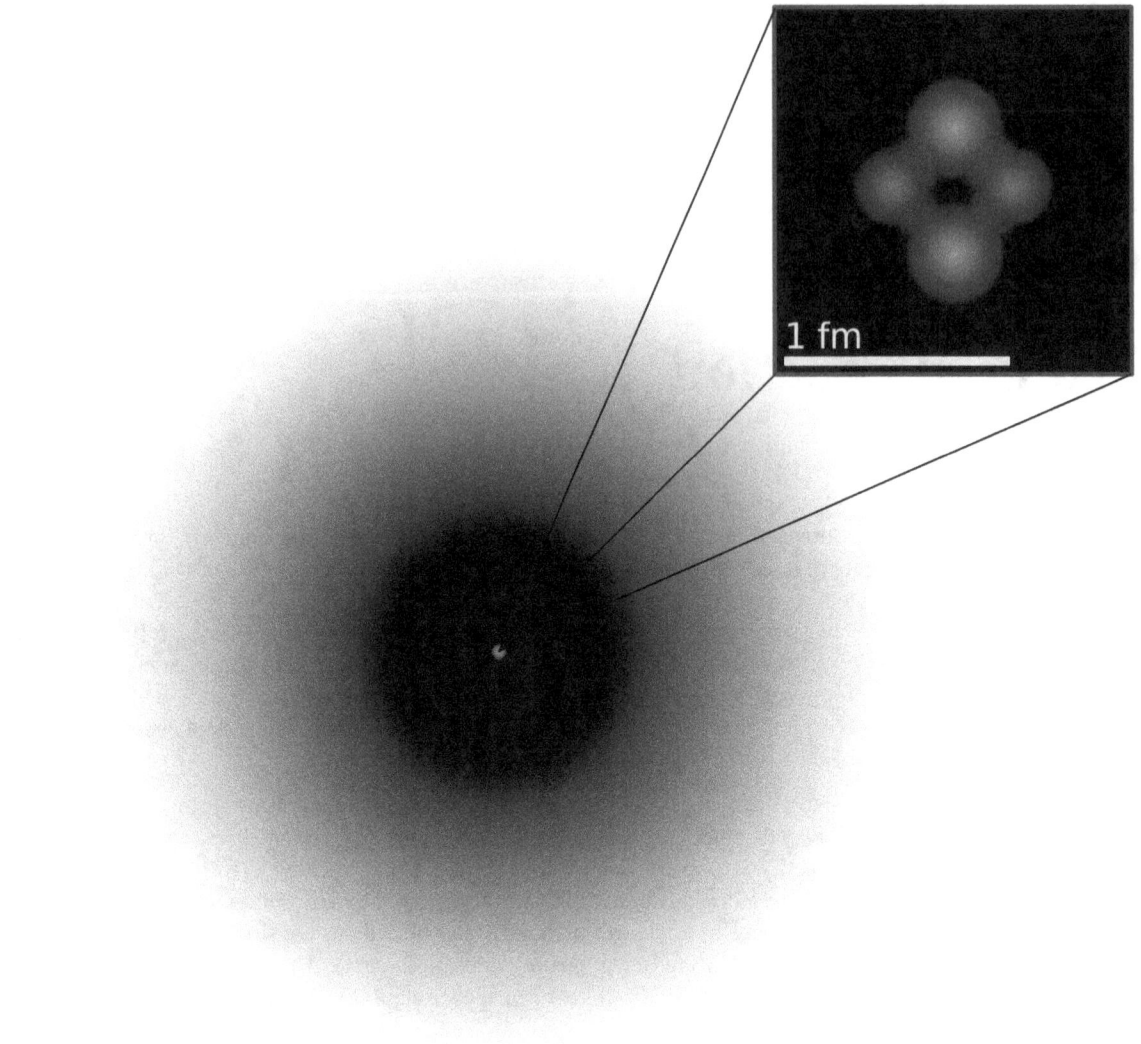

1 Å = 100,000 fm

The nucleus of a helium atom. The two protons have the same charge, but still stay together due to the residual nuclear force

positive charges would repel one another and the nucleus should therefore fly apart. However, this was never observed. New physics was needed to explain this phenomenon.

A stronger attractive force was postulated to explain how the atomic nucleus was bound together despite the protons' mutual electromagnetic repulsion. This hypothesized force was called the *strong force*, which was believed to be a fundamental force that acted on the protons and neutrons that make up the nucleus.

It was later discovered that protons and neutrons were not fundamental particles, but were made up of constituent particles called quarks. The strong attraction between nucleons was the side-effect of a more fundamental force that bound the quarks together in the protons and neutrons. The theory of quantum chromodynamics explains that quarks carry what is called a colour charge, although it has no relation to visible colour.[4] Quarks with unlike colour charge attract one another as a result of the **strong interaction**, which is mediated by particles called gluons.

34.2   Details

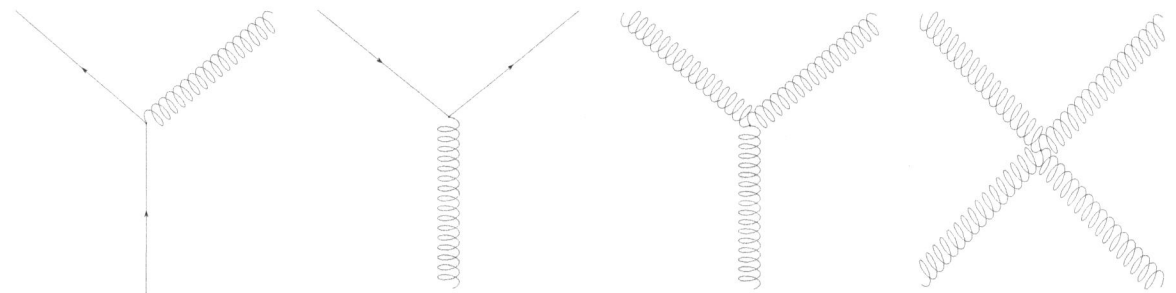

The fundamental couplings of the strong interaction, from left to right: gluon radiation, gluon splitting and gluon self-coupling.

The word *strong* is used since the strong interaction is the "strongest" of the four fundamental forces; its strength is around 10^2 times that of the electromagnetic force, some 10^6 times as great as that of the weak force, and about 10^{39} times that of gravitation, at a distance of a femtometer or less.

34.2.1   Behaviour of the strong force

The contemporary understanding of strong force is described by quantum chromodynamics (QCD), a part of the standard model of particle physics. Mathematically, QCD is a non-Abelian gauge theory based on a local (gauge) symmetry group called SU(3).

Quarks and gluons are the only fundamental particles that carry non-vanishing colour charge, and hence participate in strong interactions. The strong force itself acts directly only on elementary quark and gluon particles.

All quarks and gluons in QCD interact with each other through the strong force. The strength of interaction is parametrized by the strong coupling constant. This strength is modified by the gauge colour charge of the particle, a group theoretical property.

The strong force acts between quarks. Unlike all other forces (electromagnetic, weak, and gravitational), the strong force does not diminish in strength with increasing distance. After a limiting distance (about the size of a hadron) has been reached, it remains at a strength of about 10,000 newtons, no matter how much farther the distance between the quarks.[5] In QCD, this phenomenon is called colour confinement; it implies that only hadrons, not individual free quarks, can be observed. The explanation is that the amount of work done against a force of 10,000 newtons (about the weight of a one-metric ton mass on the surface of the Earth) is enough to create particle-antiparticle pairs within a very short distance of an interaction. In simple terms, the very energy applied to pull two quarks apart will create a pair of new quarks that will pair up with the original ones. The failure of all experiments that have searched for free quarks is considered to be evidence for this phenomenon.

The elementary quark and gluon particles affected are unobservable directly, but they instead emerge as jets of newly created hadrons, whenever energy is deposited into a quark-quark bond, as when a quark in a proton is struck by a very fast quark (in an impacting proton) during a particle accelerator experiment. However, quark–gluon plasmas have been observed.

Every quark in the universe does not attract every other quark in the above distance independent manner, since colour-confinement implies that the strong force acts without distance-diminishment only between pairs of single quarks, and that in collections of bound quarks (i.e., hadrons), the net colour-charge of the quarks cancels out, as seen from far away. Collections of quarks (hadrons) therefore appear (nearly) without colour-charge, and the strong force is therefore nearly absent between these hadrons (i.e., between baryons or mesons). However, the cancellation is not quite perfect. A small residual force remains (described below) known as the **residual strong force**. This residual force *does* diminish rapidly with distance, and is thus very short-range (effectively a few femtometers). It manifests as a force between the "colourless" hadrons, and is therefore sometimes known as the **strong nuclear force** or simply nuclear force.

34.2.2 Residual strong force

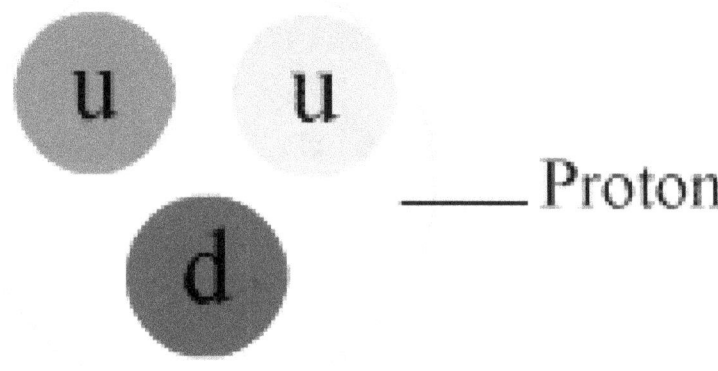

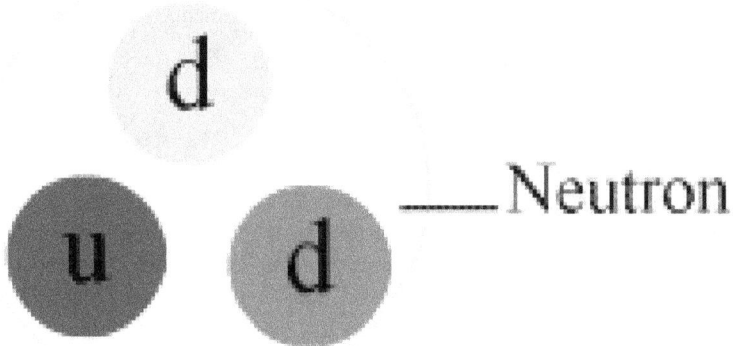

An animation of the nuclear force (or residual strong force) interaction between a proton and a neutron. The small coloured double circles are gluons, which can be seen binding the proton and neutron together. These gluons also hold the quark-antiquark combination called the pion together, and thus help transmit a residual part of the strong force even between colourless hadrons. Anticolours are shown as per this diagram. For a larger version, click here

The residual effect of the strong force is called the nuclear force. The nuclear force acts between hadrons, such as mesons or the nucleons in atomic nuclei. This "residual strong force", acting indirectly, transmits gluons that form part of the virtual pi and rho mesons, which, in turn, transmit the nuclear force between nucleons.

The residual strong force is thus a minor residuum of the strong force that binds quarks together into protons and neutrons.

This same force is much weaker *between* neutrons and protons, because it is mostly neutralized *within* them, in the same way that electromagnetic forces between neutral atoms (van der Waals forces) are much weaker than the electromagnetic forces that hold the atoms internally together.[6]

Unlike the strong force itself, the nuclear force, or residual strong force, *does* diminish in strength, and in fact diminishes rapidly with distance. The decrease is approximately as a negative exponential power of distance, though there is no simple expression known for this; see Yukawa potential. This fact, together with the less-rapid decrease of the disruptive electromagnetic force between protons with distance, causes the instability of larger atomic nuclei, such as all those with atomic numbers larger than 82 (the element lead).

34.3   See also

- Nuclear binding energy

- Colour charge

- Coupling constant

- Nuclear physics

- QCD matter

- Quantum field theory and Gauge theory

- Standard model of particle physics and Standard Model (mathematical formulation)

- Weak interaction, electromagnetism and gravity

- Intermolecular force

- Vortex

- Yukawa interaction

34.4   References

[1] Relative strength of interaction varies with distance. See for instance Matt Strassler's essay, "The strength of the known forces".

[2] on Binding energy: see Binding Energy, Mass Defect, Furry Elephant physics educational site, retr 2012 7 1

[3] on Binding energy: see Chapter 4 NUCLEAR PROCESSES, THE STRONG FORCE, M. Ragheb 1/27/2012, University of Illinois

[4] Feynman, R. P. (1985). *QED: The Strange Theory of Light and Matter*. Princeton University Press. p. 136. ISBN 0-691-08388-6. The idiot physicists, unable to come up with any wonderful Greek words anymore, call this type of polarization by the unfortunate name of 'colour,' which has nothing to do with colour in the normal sense.

[5] Fritzsch, op. cite, p. 164. The author states that the force between differently coloured quarks remains constant at any distance after they travel only a tiny distance from each other, and is equal to that need to raise one ton, which is 1000 kg x 9.8 m/s^2 = ~10,000 N.

[6] Fritzsch, H. (1983). *Quarks: The Stuff of Matter*. Basic Books. pp. 167–168. ISBN 978-0-465-06781-7.

34.5 Further reading

- Christman, J. R. (2001). "MISN-0-280: *The Strong Interaction*" (PDF). *Project PHYSNET*.

- Griffiths, David (1987). *Introduction to Elementary Particles*. John Wiley & Sons. ISBN 0-471-60386-4.

- Halzen, F.; Martin, A. D. (1984). *Quarks and Leptons: An Introductory Course in Modern Particle Physics*. John Wiley & Sons. ISBN 0-471-88741-2.

- Kane, G. L. (1987). *Modern Elementary Particle Physics*. Perseus Books. ISBN 0-201-11749-5.

- Morris, R. (2003). *The Last Sorcerers: The Path from Alchemy to the Periodic Table*. Joseph Henry Press. ISBN 0-309-50593-3.

34.6 External links

- Strong force at *Encyclopædia Britannica*

Chapter 35

Standard Model (mathematical formulation)

For a less mathematical description, see Standard Model.

This article describes the mathematics of the **Standard Model** of particle physics, a gauge quantum field theory con-

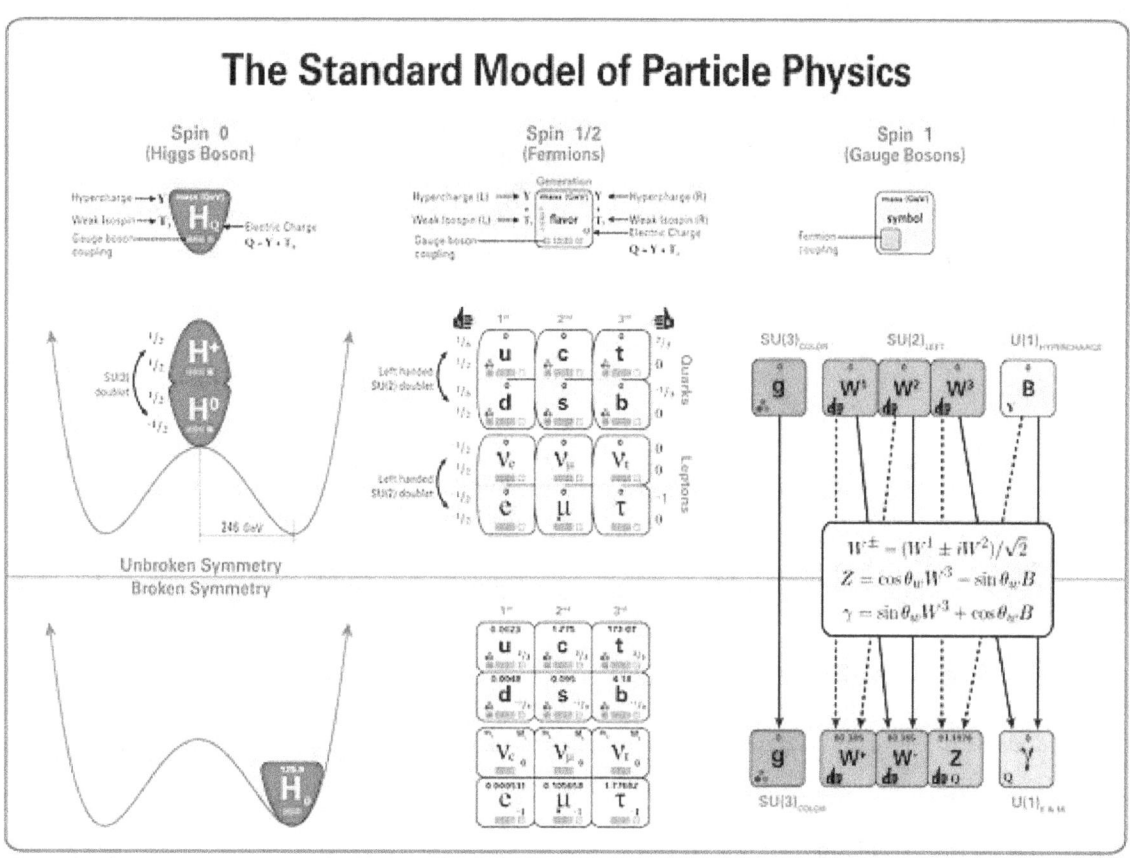

Standard Model of Particle Physics. The diagram shows the elementary particles of the Standard Model (the Higgs boson, the three generations of quarks and leptons, and the gauge bosons), including their names, masses, spins, charges, chiralities, and interactions with the strong, weak and electromagnetic forces. It also depicts the crucial role of the Higgs boson in electroweak symmetry breaking, and shows how the properties of the various particles differ in the (high-energy) symmetric phase (top) and the (low-energy) broken-symmetry phase (bottom).

taining the internal symmetries of the unitary product group SU(3) × SU(2) × U(1). The theory is commonly viewed as containing the fundamental set of particles – the leptons, quarks, gauge bosons and the Higgs particle.

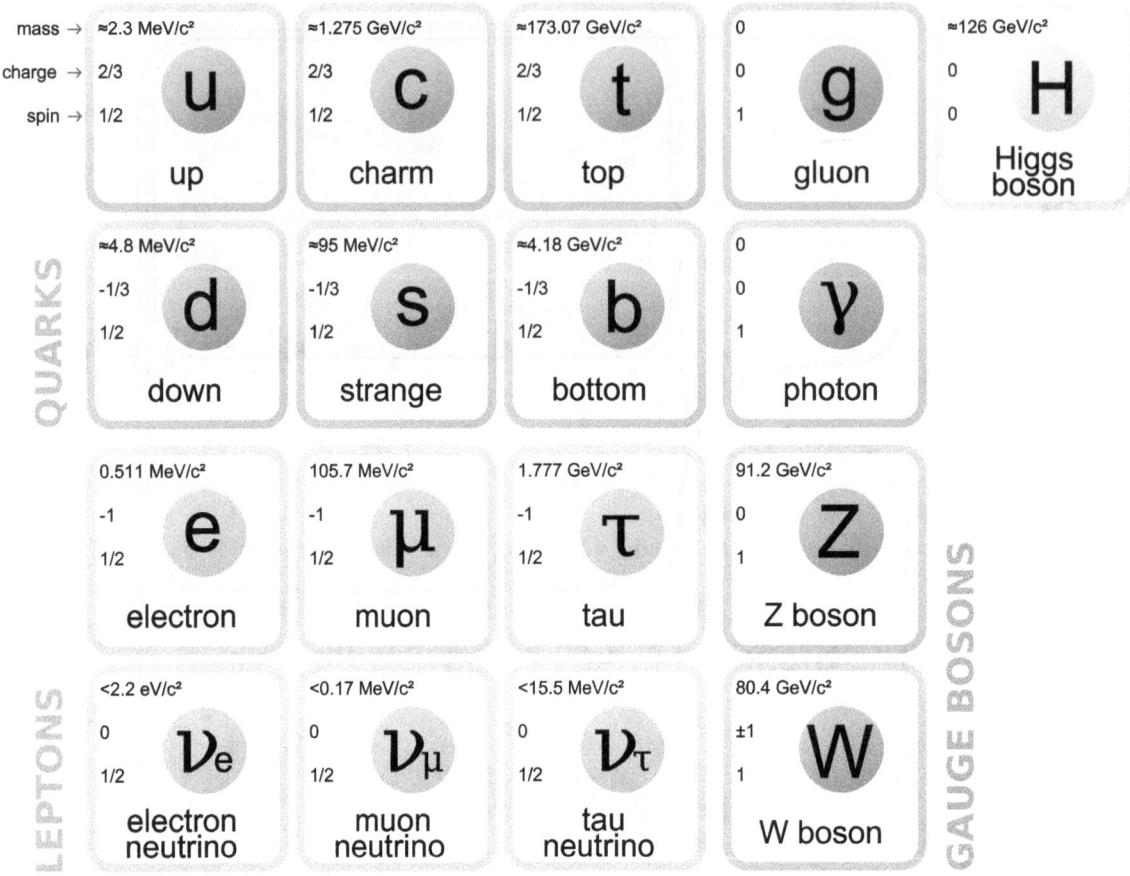

The Standard Model of Particle Physics: More Schematic Depiction

The Standard Model is renormalizable and mathematically self-consistent,[1] however despite having huge and continued successes in providing experimental predictions it does leave some unexplained phenomena. In particular, although the physics of special relativity is incorporated, general relativity is not, and the Standard Model will fail at energies or distances where the graviton is expected to emerge. Therefore in a modern field theory context, it is seen as an effective field theory.

This article requires some background in physics and mathematics, but is designed as both an introduction and a reference.

35.1 Quantum field theory

The standard model is a quantum field theory, meaning its fundamental objects are *quantum fields* which are defined at all points in spacetime. These fields are

- the fermion field, ψ, which accounts for "matter particles";

- the electroweak boson fields W_1, W_2, W_3 , and B;

- the gluon field, G_a; and

- the Higgs field, φ.

That these are *quantum* rather than *classical* fields has the mathematical consequence that they are operator-valued. In particular, values of the fields generally do not commute. As operators, they act upon the quantum state (ket vector).

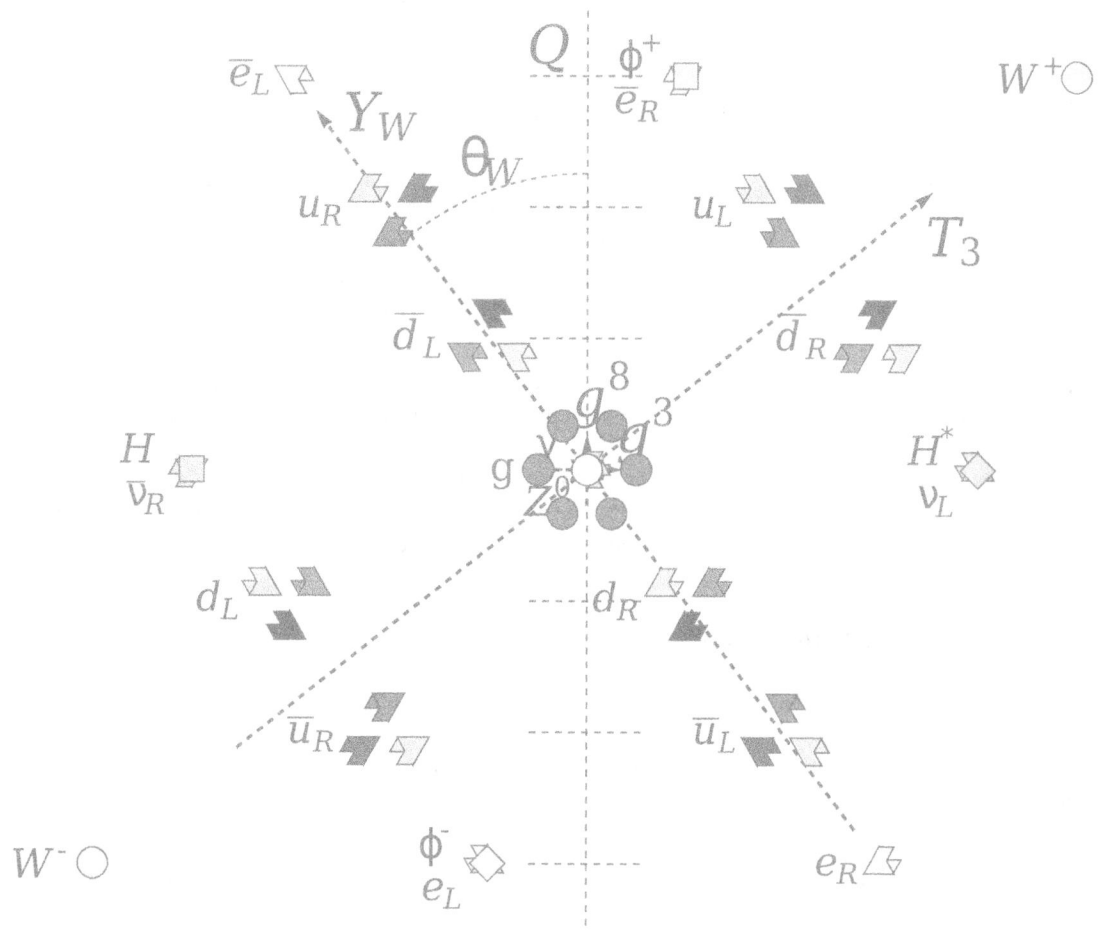

The pattern of weak isospin T_3, *weak hypercharge YW, and color charge of all known elementary particles, rotated by the weak mixing angle to show electric charge Q, roughly along the vertical. The neutral Higgs field (gray square) breaks the electroweak symmetry and interacts with other particles to give them mass.*

The dynamics of the quantum state and the fundamental fields are determined by the Lagrangian density $\mathcal{L}$ (usually for short just called the Lagrangian). This plays a role similar to that of the Schrödinger equation in non-relativistic quantum mechanics, but a Lagrangian is not an equation – rather, it is a polynomial function of the fields and their derivatives, and used with the principle of least action. While it would be possible to derive a system of differential equations governing the fields from the Langrangian, it is more common to use other techniques to compute with quantum field theories.

The standard model is furthermore a gauge theory, which means there are degrees of freedom in the mathematical formalism which do not correspond to changes in the physical state. The gauge group of the standard model is $SU(3) \times SU(2) \times U(1)$, where U(1) acts on B and φ, SU(2) acts on W and φ, and SU(3) acts on G. The fermion field ψ also transforms under these symmetries, although all of them leave some parts of it unchanged.

35.1.1 The role of the quantum fields

In classical mechanics, the state of a system can usually be captured by a small set of variables, and the dynamics of the system is thus determined by the time evolution of these variables. In classical field theory, the *field* is part of the state of the system, so in order to describe it completely one effectively introduces separate variables for every point in spacetime (even though there are many restrictions on how the values of the field "variables" may vary from point to point, for example in the form of field equations involving partial derivatives of the fields).

In quantum mechanics, the classical variables are turned into operators, but these do not capture the state of the system, which is instead encoded into a wavefunction ψ or more abstract ket vector. If ψ is an eigenstate with respect to an operator P, then $P\psi = \lambda\psi$ for the corresponding eigenvalue λ, and hence letting an operator P act on ψ is analogous to multiplying ψ by the value of the classical variable to which P corresponds. By extension, a classical formula where all variables have been replaced by the corresponding operators will behave like an operator which, when it acts upon the state of the system, multiplies it by the analogue of the quantity that the classical formula would compute. The formula as such does however not contain any information about the state of the system; it would evaluate to the same operator regardless of what state the system is in.

Quantum fields relate to quantum mechanics as classical fields do to classical mechanics, i.e., there is a separate operator for every point in spacetime, and these operators do not carry any information about the state of the system; they are merely used to exhibit some aspect of the state, at the point to which they belong. In particular, the quantum fields are *not* wavefunctions, even though the equations which govern their time evolution may be deceptively similar to those of the corresponding wavefunction in a semiclassical formulation. There is no variation in strength of the fields between different points in spacetime; the variation that happens is rather one of phase factors.

35.1.2 Vectors, scalars, and spinors

Mathematically it may look as though all of the fields are vector-valued (in addition to being operator-valued), since they all have several components, can be multiplied by matrices, etc., but physicists assign a more specific physical meaning to the word: a **vector** is something which transforms like a four-vector under Lorentz transformations, and a **scalar** is something which is invariant under Lorentz transformations. The B, W_j, and G_a fields are all vectors in this sense, so the corresponding particles are said to be vector bosons. The Higgs field φ is a scalar.

The fermion field ψ does transform under Lorentz transformations, but not like a vector should; rotations will only turn it by half the angle a proper vector should. Therefore these constitute a third kind of quantity, which is known as a spinor.

It is common to make use of abstract index notation for the vector fields, in which case the vector fields all come with a Lorentzian index μ, like so: B^μ, W_j^μ, and G_a^μ. If abstract index notation is used also for spinors then these will carry a spinorial index and the Dirac gamma will carry one Lorentzian and two spinorian indices, but it is more common to regard spinors as column matrices and the Dirac gamma $\gamma\mu$ as a matrix which additionally carries a Lorentzian index. The Feynman slash notation can be used to turn a vector field into a linear operator on spinors, like so: $\rlap{/}{B} = \gamma^\mu B_\mu$; this may involve raising and lowering indices.

35.2 Alternative presentations of the fields

As is common in quantum theory, there is more than one way to look at things. At first the basic fields given above may not seem to correspond well with the "fundamental particles" in the chart above, but there are several alternative presentations which, in particular contexts, may be more appropriate than those that are given above.

35.2.1 Fermions

Rather than having one fermion field ψ, it can be split up into separate components for each type of particle. This mirrors the historical evolution of quantum field theory, since the electron component ψ_e (describing the electron and its antiparticle the positron) is then the original ψ field of quantum electrodynamics, which was later accompanied by $\psi\mu$

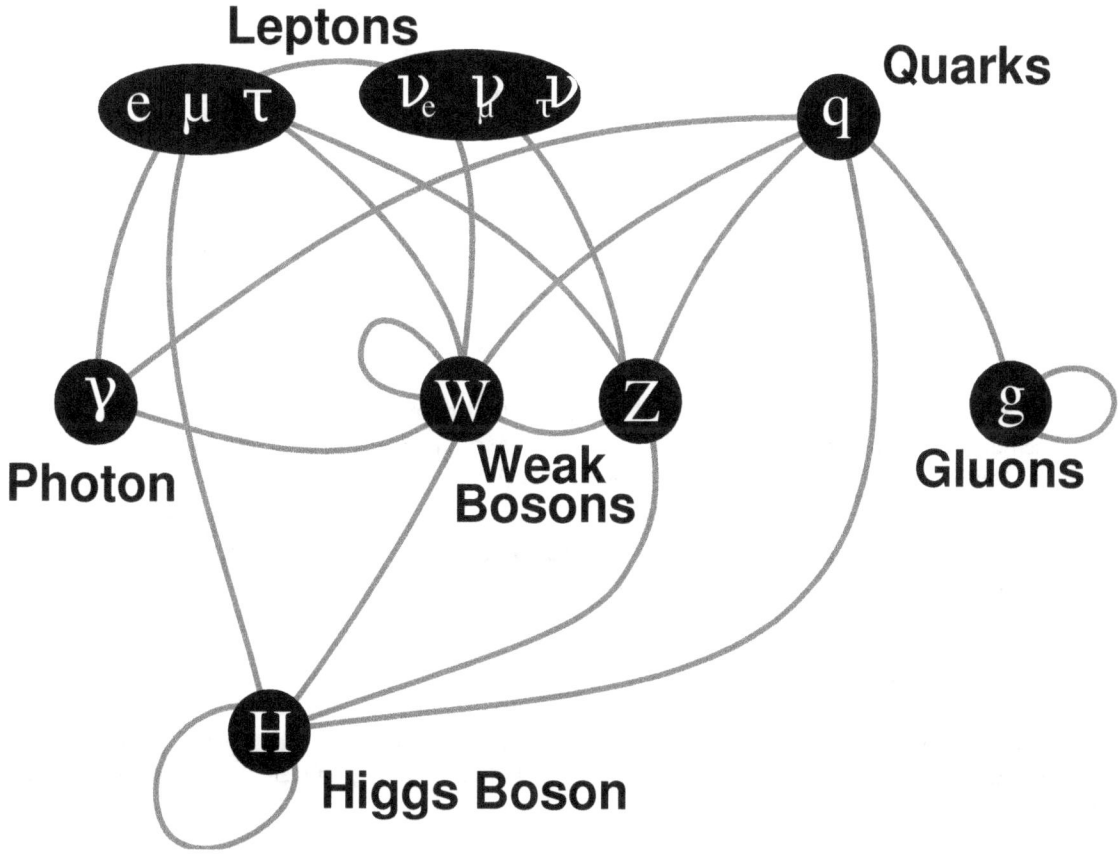

Connections denoting which particles interact with each other.

and $\psi\tau$ fields for the muon and tauon respectively (and their antiparticles). Electroweak theory added ψ_{ν_e}, ψ_{ν_μ} , and ψ_{ν_τ} for the corresponding neutrinos, and the quarks add still further components. In order to be four-spinors like the electron and other lepton components, there must be one quark component for every combination of flavour and colour, bringing the total to 24 (3 for charged leptons, 3 for neutrinos, and $2 \cdot 3 \cdot 3 = 18$ for quarks).

An important definition is the barred fermion field $\bar\psi$ is defined to be $\psi^\dagger\gamma^0$, where $\dagger$ denotes the Hermitian adjoint and γ^0 is the zeroth gamma matrix. If ψ is thought of as an $n \times 1$ matrix then $\bar\psi$ should be thought of as a $1 \times n$ matrix.

A chiral theory

An independent decomposition of ψ is that into chirality components:

"Left" chirality: $\psi^L = \frac{1}{2}(1 - \gamma_5)\psi$

"Right" chirality: $\psi^R = \frac{1}{2}(1 + \gamma_5)\psi$

where γ_5 is the fifth gamma matrix. This is very important in the Standard Model because *left and right chirality components are treated differently by the gauge interactions.*

In particular, under weak isospin SU(2) transformations the left-handed particles are weak-isospin doublets, whereas the right-handed are singlets – i.e. the weak isospin of ψR is zero. Put more simply, the weak interaction could rotate e.g. a left-handed electron into a left-handed neutrino (with emission of a W^-), but could not do so with the same right-handed particles. As an aside, the right-handed neutrino originally did not exist in the standard model – but the discovery of neutrino oscillation implies that neutrinos must have mass, and since chirality can change during the propagation of a massive particle, right-handed neutrinos must exist in reality. This does not however change the (experimentally-proven) chiral nature of the weak interaction.

Furthermore, U(1) acts differently on ψ_e^L than on ψ_e^R (because they have different weak hypercharges).

Mass and interaction eigenstates

A distinction can thus be made between, for example, the mass and interaction eigenstates of the neutrino. The former is the state which propagates in free space, whereas the latter is the *different* state that participates in interactions. Which is the "fundamental" particle? For the neutrino, it is conventional to define the "flavour" (ν e, ν μ, or ν τ) by the interaction eigenstate, whereas for the quarks we define the flavour (up, down, etc.) by the mass state. We can switch between these states using the CKM matrix for the quarks, or the PMNS matrix for the neutrinos (the charged leptons on the other hand are eigenstates of both mass and flavour).

As an aside, if a complex phase term exists within either of these matrices, it will give rise to direct CP violation, which could explain the dominance of matter over antimatter in our current universe. This has been proven for the CKM matrix, and is expected for the PMNS matrix.

Positive and negative energies

Finally, the quantum fields are sometimes decomposed into "positive" and "negative" energy parts: $\psi = \psi^+ + \psi^-$. This is not so common when a quantum field theory has been set up, but often features prominently in the process of quantizing a field theory.

35.2.2 Bosons

Due to the Higgs mechanism, the electroweak boson fields W_1, W_2, W_3 , and B "mix" to create the states which are physically observable. To retain gauge invariance, the underlying fields must be massless, but the observable states can *gain masses* in the process. These states are:

The massive neutral boson:

$$Z = \cos\theta_W W_3 - \sin\theta_W B$$

The massless neutral boson:

$$A = \sin\theta_W W_3 + \cos\theta_W B$$

The massive charged W bosons:

$$W^\pm = \frac{1}{\sqrt{2}}\left(W_1 \mp iW_2\right)$$

where θW is the Weinberg angle.

The A field is the photon, which corresponds classically to the well-known electromagnetic four-potential – i.e. the electric and magnetic fields. The Z field actually contributes in every process the photon does, but due to its large mass, the contribution is usually negligible.

35.3   Perturbative QFT and the interaction picture

Much of the qualitative descriptions of the standard model in terms of "particles" and "forces" comes from the perturbative quantum field theory view of the model. In this, the Langrangian is decomposed as $\mathcal{L} = \mathcal{L}_0 + \mathcal{L}_I$ into separate *free field* and *interaction* Langrangians. The free fields care for particles in isolation, whereas processes involving several particles arise through interactions. The idea is that the state vector should only change when particles interact, meaning a free particle is one whose quantum state is constant. This corresponds to the interaction picture in quantum mechanics.

In the more common Schrödinger picture, even the states of free particles change over time: typically the phase changes at a rate which depends on their energy. In the alternative Heisenberg picture, state vectors are kept constant, at the price of having the operators (in particular the observables) be time-dependent. The interaction picture constitutes an intermediate between the two, where some time dependence is placed in the operators (the quantum fields) and some in the state vector. In QFT, the former is called the free field part of the model, and the latter is called the interaction part. The free field model can be solved exactly, and then the solutions to the full model can be expressed as perturbations of the free field solutions, for example using the Dyson series.

It should be observed that the decomposition into free fields and interactions is in principle arbitrary. For example renormalization in QED modifies the mass of the free field electron to match that of a physical electron (with an electromagnetic field), and will in doing so add a term to the free field Lagrangian which must be cancelled by a counterterm in the interaction Lagrangian, that then shows up as a two-line vertex in the Feynman diagrams. This is also how the Higgs field is thought to give particles mass: the part of the interaction term which corresponds to the (nonzero) vacuum expectation value of the Higgs field is moved from the interaction to the free field Lagrangian, where it looks just like a mass term having nothing to do with Higgs.

35.3.1   Free fields

Under the usual free/interaction decomposition, which is suitable for low energies, the free fields obey the following equations:

- The fermion field ψ satisfies the Dirac equation; $(i\hbar\partial\!\!\!/ - m_f c)\psi_f = 0$ for each type f of fermion.

- The photon field A satisfies the wave equation $\partial_\mu \partial^\mu A^\nu = 0$.

- The Higgs field φ satisfies the Klein–Gordon equation.

- The weak interaction fields Z, $W^\pm$ also satisfy the Klein–Gordon equation.

These equations can be solved exactly. One usually does so by considering first solutions that are periodic with some period L along each spatial axis; later taking the limit: $L \to \infty$ will lift this periodicity restriction.

In the periodic case, the solution for a field F (any of the above) can be expressed as a Fourier series of the form

$$F(x) = \beta \sum_{\mathbf{p}} \sum_{r} E_{\mathbf{p}}^{-\frac{1}{2}} \left(a_r(\mathbf{p}) u_r(\mathbf{p}) e^{-\frac{ipx}{\hbar}} + b_r^\dagger(\mathbf{p}) v_r(\mathbf{p}) e^{\frac{ipx}{\hbar}} \right)$$

where:

- β is a normalization factor; for the fermion field ψ_f it is $\sqrt{m_f c^2 / V}$, where $V = L^3$ is the volume of the fundamental cell considered; for the photon field A^μ it is $\hbar c / \sqrt{2V}$.

- The sum over $\mathbf{p}$ is over all momenta consistent with the period L, i.e., over all vectors $\frac{2\pi\hbar}{L}(n_1, n_2, n_3)$ where n_1, n_2, n_3 are integers.

- The sum over r covers other degrees of freedom specific for the field, such as polarization or spin; it usually comes out as a sum from 1 to 2 or from 1 to 3.

- $E_{\mathbf{p}}$ is the relativistic energy for a momentum $\mathbf{p}$ quantum of the field, $= \sqrt{m^2c^4 + c^2\mathbf{p}^2}$ when the rest mass is m.

- $ar(\mathbf{p})$ and $b_r^\dagger(\mathbf{p})$ are annihilation and creation respectively operators for "a-particles" and "b-particles" respectively of momentum $\mathbf{p}$; "b-particles" are the antiparticles of "a-particles". Different fields have different "a-" and "b-particles". For some fields, a and b are the same.

- $ur(\mathbf{p})$ and $vr(\mathbf{p})$ are non-operators which carry the vector or spinor aspects of the field (where relevant).

- $p = (E_{\mathbf{p}}/c, \mathbf{p})$ is the four-momentum for a quanta with momentum $\mathbf{p}$. $px = p_\mu x^\mu$ denotes an inner product of four-vectors.

In the limit $L \to \infty$, the sum would turn into an integral with help from the V hidden inside β. The numeric value of β also depends on the normalization chosen for $u_r(\mathbf{p})$ and $v_r(\mathbf{p})$.

Technically, $a_r^\dagger(\mathbf{p})$ is the Hermitian adjoint of the operator $ar(\mathbf{p})$ in the inner product space of ket vectors. The identification of $a_r^\dagger(\mathbf{p})$ and $ar(\mathbf{p})$ as creation and annihilation operators comes from comparing conserved quantities for a state before and after one of these have acted upon it. $a_r^\dagger(\mathbf{p})$ can for example be seen to add one particle, because it will add 1 to the eigenvalue of the a-particle number operator, and the momentum of that particle ought to be $\mathbf{p}$ since the eigenvalue of the vector-valued momentum operator increases by that much. For these derivations, one starts out with expressions for the operators in terms of the quantum fields. That the operators with † are creation operators and the one without annihilation operators is a convention, imposed by the sign of the commutation relations postulated for them.

An important step in preparation for calculating in perturbative quantum field theory is to separate the "operator" factors a and b above from their corresponding vector or spinor factors u and v. The vertices of Feynman graphs come from the way that u and v from different factors in the interaction Lagrangian fit together, whereas the edges come from the way that the as and bs must be moved around in order to put terms in the Dyson series on normal form.

35.3.2 Interaction terms and the path integral approach

The Lagrangian can also be derived without using creation and annihilation operators (the "canonical" formalism), by using a "path integral" approach, pioneered by Feynman building on the earlier work of Dirac. See e.g. Path integral formulation on Wikipedia or A. Zee's QFT in a nutshell. This is one possible way that the Feynman diagrams, which are pictorial representations of interaction terms, can be derived relatively easily. A quick derivation is indeed presented at the article on Feynman diagrams.

35.4 Lagrangian formalism

We can now give some more detail about the aforementioned free and interaction terms appearing in the Standard Model Lagrangian density. Any such term must be both gauge and reference-frame invariant, otherwise the laws of physics would depend on an arbitrary choice or the frame of an observer. Therefore the global Poincaré symmetry, consisting of translational symmetry, rotational symmetry and the inertial reference frame invariance central to the theory of special relativity must apply. The local SU(3) × SU(2) × U(1) gauge symmetry is the internal symmetry. The three factors of the gauge symmetry together give rise to the three fundamental interactions, after some appropriate relations have been defined, as we shall see.

A complete formulation of the Standard Model Lagrangian with all the terms written together can be found e.g. here.

35.4.1 Kinetic terms

A free particle can be represented by a mass term, and a *kinetic* term which relates to the "motion" of the fields.

Standard Model Interactions
(Forces Mediated by Gauge Bosons)

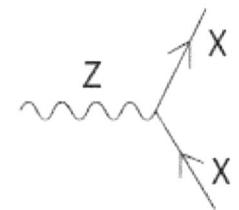

X is any fermion in
the Standard Model.

X is electrically charged.

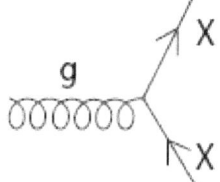

X is any quark.

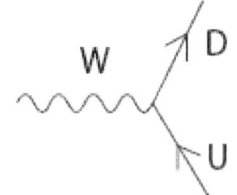

U is a up-type quark;
D is a down-type quark.

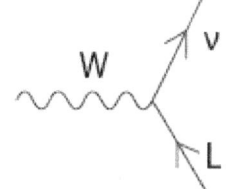

L is a lepton and v is the
corresponding neutrino.

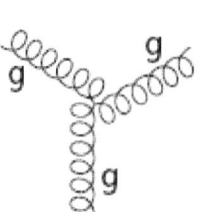

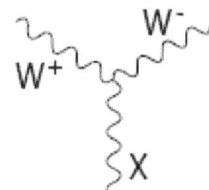

X is a photon or Z-boson.

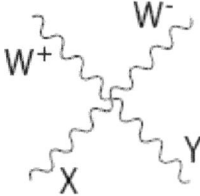

X and Y are any two
electroweak bosons such
that charge is conserved.

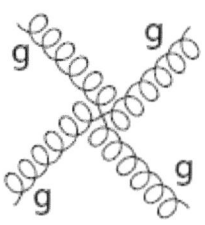

The above interactions show some basic interaction vertices – Feynman diagrams in the standard model are built from these vertices. Higgs boson interactions are however not shown, and neutrino oscillations are commonly added. The charge of the W bosons are dictated by the fermions they interact with.

Fermion fields

The kinetic term for a Dirac fermion is

$$i\bar{\psi}\gamma^{\mu}\partial_{\mu}\psi$$

where the notations are carried from earlier in the article. ψ can represent any, or all, Dirac fermions in the standard model. Generally, as below, this term is included within the couplings (creating an overall "dynamical" term).

Gauge fields

For the spin-1 fields, first define the field strength tensor

$$F^a_{\mu\nu} = \partial_\mu A^a_\nu - \partial_\nu A^a_\mu + g f^{abc} A^b_\mu A^c_\nu$$

for a given gauge field (here we use A), with gauge coupling constant g. The quantity f^{abc} is the structure constant of the particular gauge group, defined by the commutator

$$[t_a, t_b] = i f^{abc} t_c,$$

where t_i are the generators of the group. In an Abelian (commutative) group (such as the U(1) we use here), since the generators t_a all commute with each other, the structure constants vanish. Of course, this is not the case in general – the standard model includes the non-Abelian SU(2) and SU(3) groups (such groups lead to what is called a Yang–Mills gauge theory).

We need to introduce three gauge fields corresponding to each of the subgroups SU(3) × SU(2) × U(1).

- The gluon field tensor will be denoted by $G^a_{\mu\nu}$, where the index a labels elements of the **8** representation of colour SU(3). The strong coupling constant is conventionally labelled g_s (or simply g where there is no ambiguity). *The observations leading to the discovery of this part of the Standard Model are discussed in the article in quantum chromodynamics.*

- The notation $W^a_{\mu\nu}$ will be used for the gauge field tensor of SU(2) where a runs over the 3 generators of this group. The coupling can be denoted g_w or again simply g. The gauge field will be denoted by W^a_μ.

- The gauge field tensor for the U(1) of weak hypercharge will be denoted by Bμν, the coupling by g′, and the gauge field by Bμ.

The kinetic term can now be written simply as

$$\mathcal{L}_{\text{kin}} = -\frac{1}{4} B_{\mu\nu} B^{\mu\nu} - \frac{1}{2} \text{tr} W_{\mu\nu} W^{\mu\nu} - \frac{1}{2} \text{tr} G_{\mu\nu} G^{\mu\nu}$$

where the traces are over the SU(2) and SU(3) indices hidden in W and G respectively. The two-index objects are the field strengths derived from W and G the vector fields. There are also two extra hidden parameters: the theta angles for SU(2) and SU(3).

35.4.2 Coupling terms

The next step is to "couple" the gauge fields to the fermions, allowing for interactions.

Electroweak sector

Main article: Electroweak interaction

The electroweak sector interacts with the symmetry group U(1) × SU(2)L, where the subscript L indicates coupling only to left-handed fermions.

$$\mathcal{L}_{\text{EW}} = \sum_\psi \bar{\psi} \gamma^\mu \left(i\partial_\mu - g'\frac{1}{2} Y_{\text{w}} B_\mu - g\frac{1}{2} \tau \mathbf{W}_\mu \right) \psi$$

Where Bμ is the U(1) gauge field; YW is the weak hypercharge (the generator of the U(1) group); $\mathbf{W}_\mu$ is the three-component SU(2) gauge field; and the components of $\boldsymbol{\tau}$ are the Pauli matrices (infinitesimal generators of the SU(2) group) whose eigenvalues give the weak isospin. Note that we have to redefine a new U(1) symmetry of *weak hypercharge*, different from QED, in order to achieve the unification with the weak force. The electric charge Q, third component of weak isospin T_3 (also called Tz, I_3 or I_z) and weak hypercharge YW are related by

$$Q = T_3 + \tfrac{1}{2}Y_W,$$

or by the alternate convention $Q = T_3 + YW$. The first convention (used in this article) is equivalent to the earlier Gell-Mann–Nishijima formula. We can then define the conserved current for weak isospin as

$$\mathbf{j}_\mu = \frac{1}{2}\bar{\psi}_L \gamma_\mu \boldsymbol{\tau} \psi_L$$

and for weak hypercharge as

$$j_\mu^Y = 2(j_\mu^{em} - j_\mu^3)$$

where j_μ^{em} is the electric current and j_μ^3 the third weak isospin current. As explained above, *these currents mix* to create the physically observed bosons, which also leads to testable relations between the coupling constants.

To explain in a simpler way, we can see the effect of the electroweak interaction by picking out terms from the Lagrangian. We see that the SU(2) symmetry acts on each (left-handed) fermion doublet contained in ψ, for example

$$-\frac{g}{2}(\bar{\nu}_e\ \bar{e})\tau^+\gamma_\mu(W^-)^\mu\begin{pmatrix}\nu_e\\e\end{pmatrix} = -\frac{g}{2}\bar{\nu}_e\gamma_\mu(W^-)^\mu e$$

where the particles are understood to be left-handed, and where

$$\tau^\pm \equiv \frac{1}{2}(\tau^1 \pm i\tau^2) = \begin{pmatrix}0 & 1\\0 & 0\end{pmatrix}$$

This is an interaction corresponding to a "rotation in weak isospin space" or in other words, a *transformation between eL and veL via emission of a* W⁻ *boson*. The U(1) symmetry, on the other hand, is similar to electromagnetism, but acts on all "*weak hypercharged*" fermions (both left and right handed) via the neutral Z⁰, as well as the *charged* fermions via the photon.

Quantum chromodynamics sector

Main article: Quantum chromodynamics

The quantum chromodynamics (QCD) sector defines the interactions between quarks and gluons, with SU(3) symmetry, generated by T_a. Since leptons do not interact with gluons, they are not affected by this sector. The Dirac Lagrangian of the quarks coupled to the gluon fields is given by

$$\mathcal{L}_{QCD} = i\overline{U}\left(\partial_\mu - ig_s G_\mu^a T^a\right)\gamma^\mu U + i\overline{D}\left(\partial_\mu - ig_s G_\mu^a T^a\right)\gamma^\mu D.$$

where D and U are the Dirac spinors associated with up- and down-type quarks, and other notations are continued from the previous section.

35.4.3 Mass terms and the Higgs mechanism

Mass terms

The mass term arising from the Dirac Lagrangian (for any fermion ψ) is $-m\bar{\psi}\psi$ which is *not* invariant under the electroweak symmetry. This can be seen by writing ψ in terms of left and right handed components (skipping the actual calculation):

$$-m\bar{\psi}\psi = -m(\bar{\psi}_L\psi_R + \bar{\psi}_R\psi_L)$$

i.e. contribution from $\bar{\psi}_L\psi_L$ and $\bar{\psi}_R\psi_R$ terms do not appear. We see that the mass-generating interaction is achieved by constant flipping of particle chirality. The spin-half particles have no right/left helicity pair with the same SU(2) and SU(3) representation and the same weak hypercharge, so assuming these gauge charges are conserved in the vacuum, none of the spin-half particles could ever swap helicity, and must remain massless. Additionally, we know experimentally that the W and Z bosons are massive, but a boson mass term contains the combination e.g. $A^\mu A\mu$, which clearly depends on the choice of gauge. Therefore, none of the standard model fermions *or* bosons can "begin" with mass, but must acquire it by some other mechanism.

The Higgs mechanism

Main article: Higgs mechanism

The solution to both these problems comes from the Higgs mechanism, which involves scalar fields (the number of which depend on the exact form of Higgs mechanism) which (to give the briefest possible description) are "absorbed" by the massive bosons as degrees of freedom, and which couple to the fermions via Yukawa coupling to create what looks like mass terms.

In the Standard Model, the Higgs field is a complex scalar of the group SU(2)L:

$$\phi = \frac{1}{\sqrt{2}}\begin{pmatrix}\phi^+ \\ \phi^0\end{pmatrix},$$

where the superscripts + and 0 indicate the electric charge (Q) of the components. The weak isospin (YW) of both components is 1.

The Higgs part of the Lagrangian is

$$\mathcal{L}_H = \left[\left(\partial_\mu - igW_\mu^a t^a - ig'Y_\phi B_\mu\right)\phi\right]^2 + \mu^2\phi^\dagger\phi - \lambda(\phi^\dagger\phi)^2,$$

where $\lambda > 0$ and $\mu^2 > 0$, so that the mechanism of spontaneous symmetry breaking can be used. There is a parameter here, at first hidden within the shape of the potential, that is very important. In a unitarity gauge one can set $\varphi^+ = 0$ and make φ^0 real. Then $\langle\phi^0\rangle = v$ is the non-vanishing vacuum expectation value of the Higgs field. v has units of mass, and it is the only parameter in the Standard Model which is not dimensionless. It is also much smaller than the Planck scale; it is approximately equal to the Higgs mass, and sets the scale for the mass of everything else. This is the only real fine-tuning to a small nonzero value in the Standard Model, and it is called the Hierarchy problem. Quadratic terms in Wμ and Bμ arise, which give masses to the W and Z bosons:

$$M_W = \tfrac{1}{2}v|g|$$
$$M_Z = \tfrac{1}{2}v\sqrt{g^2 + g'^2}$$

The Yukawa interaction terms are

$$\mathcal{L}_{YU} = \overline{U}_L G_u U_R \phi^0 - \overline{D}_L G_u U_R \phi^- + \overline{U}_L G_d D_R \phi^+ + \overline{D}_L G_d D_R \phi^0 + hc$$

where $G_{u,d}$ are 3×3 matrices of Yukawa couplings, with the ij term giving the coupling of the generations i and j.

Neutrino masses

As previously mentioned, evidence shows neutrinos must have mass. But within the standard model, the right-handed neutrino does not exist, so even with a Yukawa coupling neutrinos remain massless. An obvious solution[2] is to simply *add a right-handed neutrino* νR resulting in a **Dirac mass** term as usual. This field however must be a sterile neutrino, since being right-handed it experimentally belongs to an isospin singlet ($T_3 = 0$) and also has charge $Q = 0$, implying $YW = 0$ (see above) i.e. it does not even participate in the weak interaction. Current experimental status is that evidence for observation of sterile neutrinos is not convincing.[3]

Another possibility to consider is that the neutrino satisfies the **Majorana equation**, which at first seems possible due to its zero electric charge. In this case the mass term is

$$-\frac{m}{2} \left(\overline{\nu}^C \nu + \overline{\nu} \nu^C \right)$$

where C denotes a charge conjugated (i.e. anti-) particle, and the terms are consistently all left (or all right) chirality (note that a left-chirality projection of an antiparticle is a right-handed field; care must be taken here due to different notations sometimes used). Here we are essentially flipping between LH neutrinos and RH anti-neutrinos (it is furthermore possible but *not* necessary that neutrinos are their own antiparticle, so these particles are the same). However for the left-chirality neutrinos, this term changes weak hypercharge by 2 units - not possible with the standard Higgs interation, requiring the Higgs field to be extended to include an extra triplet with weak hypercharge 2[4] - whereas for right-chirality neutrinos, no Higgs extensions are necessary. For both left and right chirality cases, Majorana terms violate lepton number, but possibly at a level beyond the current sensitivity of experiments to detect such violations.

It is possible to include **both** Dirac and Majorana mass terms in the same theory, which (in contrast to the Dirac-mass-only approach) can provide a "natural" explanation for the smallness of the observed neutrino masses, by linking the RH neutrinos to yet-unknown physics around the GUT scale[5] (see seesaw mechanism).

Since in any case new fields must be postulated to explain the experimental results, neutrinos are an obvious gateway to searching physics beyond the Standard Model.

35.5   Detailed Information

This section provides more detail on some aspects, and some reference material.

35.5.1   Field content in detail

The Standard Model has the following fields. These describe one *generation* of leptons and quarks, and there are three generations, so there are three copies of each field. By CPT symmetry, there is a set of right-handed fermions with the opposite quantum numbers. The column "**representation**" indicates under which representations of the gauge groups that each field transforms, in the order (SU(3), SU(2), U(1)). Symbols used are common but not universal; superscript C denotes an antiparticle; and for the U(1) group, the value of the weak hypercharge is listed. Note that there are twice as many left-handed lepton field components as left-handed antilepton field components in each generation, but an equal number of left-handed quark and antiquark fields.

35.5.2   Fermion content

This table is based in part on data gathered by the Particle Data Group.[6]

[1] These are not ordinary abelian charges, which can be added together, but are labels of group representations of Lie groups.

[2] Mass is really a coupling between a left-handed fermion and a right-handed fermion. For example, the mass of an electron is really a coupling between a left-handed electron and a right-handed electron, which is the antiparticle of a left-handed positron. Also neutrinos show large mixings in their mass coupling, so it's not accurate to talk about neutrino masses in the flavor basis or to suggest a left-handed electron antineutrino.

[3] The Standard Model assumes that neutrinos are massless. However, several contemporary experiments prove that neutrinos oscillate between their flavour states, which could not happen if all were massless. It is straightforward to extend the model to fit these data but there are many possibilities, so the mass eigenstates are still open. See neutrino mass.

[4] W.-M. Yao *et al.* (Particle Data Group) (2006). "Review of Particle Physics: Neutrino mass, mixing, and flavor change" (PDF). *Journal of Physics G* **33**: 1. arXiv:astro-ph/0601168. Bibcode:2006JPhG...33....1Y. doi:10.1088/0954-3899/33/1/001.

[5] The masses of baryons and hadrons and various cross-sections are the experimentally measured quantities. Since quarks can't be isolated because of QCD confinement, the quantity here is supposed to be the mass of the quark at the renormalization scale of the QCD scale.

35.5.3 Free parameters

Upon writing the most general Lagrangian without neutrinos, one finds that the dynamics depend on 19 parameters, whose numerical values are established by experiment. With neutrinos 7 more parameters are needed, 3 masses and 4 PMNS matrix parameters, for a total of 26 parameters.[7] The neutrino parameter values are still uncertain. The 19 certain parameters are summarized here (note: with the Higgs mass is at 125 GeV, the Higgs self-coupling strength $\lambda \sim 1/8$).

35.5.4 Additional symmetries of the Standard Model

From the theoretical point of view, the Standard Model exhibits four additional global symmetries, not postulated at the outset of its construction, collectively denoted **accidental symmetries**, which are continuous U(1) global symmetries. The transformations leaving the Lagrangian invariant are:

$$\psi_q(x) \rightarrow e^{i\alpha/3}\psi_q$$

$$E_L \rightarrow e^{i\beta}E_L \text{ and } (e_R)^c \rightarrow e^{i\beta}(e_R)^c$$

$$M_L \rightarrow e^{i\beta}M_L \text{ and } (\mu_R)^c \rightarrow e^{i\beta}(\mu_R)^c$$

$$T_L \rightarrow e^{i\beta}T_L \text{ and } (\tau_R)^c \rightarrow e^{i\beta}(\tau_R)^c$$

The first transformation rule is shorthand meaning that all quark fields for all generations must be rotated by an identical phase simultaneously. The fields ML, TL and $(\mu_R)^c$, $(\tau_R)^c$ are the 2nd (muon) and 3rd (tau) generation analogs of EL and $(e_R)^c$ fields.

By Noether's theorem, each symmetry above has an associated conservation law: the conservation of baryon number, electron number, muon number, and tau number. Each quark is assigned a baryon number of $\frac{1}{3}$, while each antiquark is assigned a baryon number of $-\frac{1}{3}$. Conservation of baryon number implies that the number of quarks minus the number of antiquarks is a constant. Within experimental limits, no violation of this conservation law has been found.

Similarly, each electron and its associated neutrino is assigned an electron number of +1, while the anti-electron and the associated anti-neutrino carry a −1 electron number. Similarly, the muons and their neutrinos are assigned a muon number of +1 and the tau leptons are assigned a tau lepton number of +1. The Standard Model predicts that each of these three numbers should be conserved separately in a manner similar to the way baryon number is conserved. These numbers are collectively known as lepton family numbers (LF).

In addition to the accidental (but exact) symmetries described above, the Standard Model exhibits several **approximate symmetries**. These are the "SU(2) custodial symmetry" and the "SU(2) or SU(3) quark flavor symmetry."

35.5.5 The U(1) symmetry

For the leptons, the gauge group can be written SU(2)$_l$ × U(1)L × U(1)R. The two U(1) factors can be combined into U(1)Y × U(1)$_l$ where l is the lepton number. Gauging of the lepton number is ruled out by experiment, leaving only the possible gauge group SU(2)L × U(1)Y. A similar argument in the quark sector also gives the same result for the electroweak theory.

35.5.6 The charged and neutral current couplings and Fermi theory

The charged currents $j^\pm = j^1 \pm ij^2$ are

$$j_\mu^+ = \overline{U}_{iL}\gamma_\mu D_{iL} + \overline{\nu}_{iL}\gamma_\mu l_{iL}.$$

These charged currents are precisely those that entered the Fermi theory of beta decay. The action contains the charge current piece

$$\mathcal{L}_{CC} = \frac{g}{\sqrt{2}}(j_\mu^+ W^{-\mu} + j_\mu^- W^{+\mu}).$$

For energy much less than the mass of the W-boson, the effective theory becomes the current–current interaction of the Fermi theory.

However, gauge invariance now requires that the component W^3 of the gauge field also be coupled to a current that lies in the triplet of SU(2). However, this mixes with the U(1), and another current in that sector is needed. These currents must be uncharged in order to conserve charge. So we require the **neutral currents**

$$j_\mu^3 = \frac{1}{2}(\overline{U}_{iL}\gamma_\mu U_{iL} - \overline{D}_{iL}\gamma_\mu D_{iL} + \overline{\nu}_{iL}\gamma_\mu \nu_{iL} - \overline{l}_{iL}\gamma_\mu l_{iL})$$

$$j_\mu^{em} = \frac{2}{3}\overline{U}_i\gamma_\mu U_i - \frac{1}{3}\overline{D}_i\gamma_\mu D_i - \overline{l}_i\gamma_\mu l_i.$$

The neutral current piece in the Lagrangian is then

$$\mathcal{L}_{NC} = ej_\mu^{em} A^\mu + \frac{g}{\cos\theta_W}(J_\mu^3 - \sin^2\theta_W J_\mu^{em})Z^\mu.$$

35.6 See also

- Overview of Standard Model of particle physics
- Fundamental interaction
- Noncommutative standard model
- Open questions: CP violation, Neutrino masses, Quark matter
- Physics beyond the Standard Model
- Strong interactions: Flavour, Quantum chromodynamics, Quark model
- Weak interactions: Electroweak interaction, Fermi's interaction
- Weinberg angle
- Symmetry in quantum mechanics

35.7 References and external links

[1] In fact, there are mathematical issues regarding quantum field theories still under debate (see e.g. Landau pole), but the predictions extracted from the Standard Model by current methods are all self-consistent. For a further discussion see e.g. R. Mann, chapter 25.

[2] https://fas.org/sgp/othergov/doe/lanl/pubs/00326607.pdf

[3] http://t2k-experiment.org/neutrinos/oscillations-today/

[4] https://fas.org/sgp/othergov/doe/lanl/pubs/00326607.pdf

[5] http://www.mpi-hd.mpg.de/personalhomes/schwetz/tueb-2.pdf

[6] W.-M. Yao *et al.* (Particle Data Group) (2006). "Review of Particle Physics: Quarks" (PDF). *Journal of Physics G* **33**: 1. arXiv:astro-ph/0601168. Bibcode:2006JPhG...33....1Y. doi:10.1088/0954-3899/33/1/001.

[7] Mark Thomson (5 September 2013). *Modern Particle Physics*. Cambridge University Press. pp. 499–500. ISBN 978-1-107-29254-3.

- *An introduction to quantum field theory*, by M.E. Peskin and D.V. Schroeder (HarperCollins, 1995) ISBN 0-201-50397-2.

- *Gauge theory of elementary particle physics*, by T.P. Cheng and L.F. Li (Oxford University Press, 1982) ISBN 0-19-851961-3.

- Standard Model Lagrangian with explicit Higgs terms (T.D. Gutierrez, ca 1999) (PDF, PostScript, and LaTeX version)

- *The quantum theory of fields* (vol 2), by S. Weinberg (Cambridge University Press, 1996) ISBN 0-521-55002-5.

- *Quantum Field Theory in a Nutshell* (Second Edition), by A. Zee (Princeton University Press, 2010) ISBN 978-1-4008-3532-4.

- *An Introduction to Particle Physics and the Standard Model*, by R. Mann (CRC Press, 2010) ISBN 978-1420082982

Chapter 36

Quantum electrodynamics

This article is about the physical theory. For the Latin phrase, see quod erat demonstrandum. For other uses, see QED (disambiguation).

In particle physics, **quantum electrodynamics (QED)** is the relativistic quantum field theory of electrodynamics. In essence, it describes how light and matter interact and is the first theory where full agreement between quantum mechanics and special relativity is achieved. QED mathematically describes all phenomena involving electrically charged particles interacting by means of exchange of photons and represents the quantum counterpart of classical electromagnetism giving a complete account of matter and light interaction.

In technical terms, QED can be described as a perturbation theory of the electromagnetic quantum vacuum. Richard Feynman called it "the jewel of physics" for its extremely accurate predictions of quantities like the anomalous magnetic moment of the electron and the Lamb shift of the energy levels of hydrogen.[1]:Ch1

36.1 History

Main article: History of quantum mechanics
 The first formulation of a quantum theory describing radiation and matter interaction is attributed to British scientist Paul Dirac, who (during the 1920s) was first able to compute the coefficient of spontaneous emission of an atom.[2]

Dirac described the quantization of the electromagnetic field as an ensemble of harmonic oscillators with the introduction of the concept of creation and annihilation operators of particles. In the following years, with contributions from Wolfgang Pauli, Eugene Wigner, Pascual Jordan, Werner Heisenberg and an elegant formulation of quantum electrodynamics due to Enrico Fermi,[3] physicists came to believe that, in principle, it would be possible to perform any computation for any physical process involving photons and charged particles. However, further studies by Felix Bloch with Arnold Nordsieck,[4] and Victor Weisskopf,[5] in 1937 and 1939, revealed that such computations were reliable only at a first order of perturbation theory, a problem already pointed out by Robert Oppenheimer.[6] At higher orders in the series infinities emerged, making such computations meaningless and casting serious doubts on the internal consistency of the theory itself. With no solution for this problem known at the time, it appeared that a fundamental incompatibility existed between special relativity and quantum mechanics.

Difficulties with the theory increased through the end of 1940. Improvements in microwave technology made it possible to take more precise measurements of the shift of the levels of a hydrogen atom,[7] now known as the Lamb shift and magnetic moment of the electron.[8] These experiments unequivocally exposed discrepancies which the theory was unable to explain.

A first indication of a possible way out was given by Hans Bethe. In 1947, while he was traveling by train to reach Schenectady from New York,[9] after giving a talk at the conference at Shelter Island on the subject, Bethe completed the first non-relativistic computation of the shift of the lines of the hydrogen atom as measured by Lamb and Retherford.[10]

Paul Dirac

Despite the limitations of the computation, agreement was excellent. The idea was simply to attach infinities to corrections of mass and charge that were actually fixed to a finite value by experiments. In this way, the infinities get absorbed in those constants and yield a finite result in good agreement with experiments. This procedure was named renormalization.

Based on Bethe's intuition and fundamental papers on the subject by Sin-Itiro Tomonaga,[11] Julian Schwinger,[12][13]

Richard Feynman[14][15][16] and Freeman Dyson,[17][18] it was finally possible to get fully covariant formulations that were finite at any order in a perturbation series of quantum electrodynamics. Sin-Itiro Tomonaga, Julian Schwinger and Richard Feynman were jointly awarded with a Nobel prize in physics in 1965 for their work in this area.[19] Their contributions, and those of Freeman Dyson, were about covariant and gauge invariant formulations of quantum electrodynamics that allow computations of observables at any order of perturbation theory. Feynman's mathematical technique, based on his diagrams, initially seemed very different from the field-theoretic, operator-based approach of Schwinger and Tomonaga, but Freeman Dyson later showed that the two approaches were equivalent.[17] Renormalization, the need to attach a physical meaning at certain divergences appearing in the theory through integrals, has subsequently become one of the fundamental aspects of quantum field theory and has come to be seen as a criterion for a theory's general acceptability. Even though renormalization works very well in practice, Feynman was never entirely comfortable with its mathematical validity, even referring to renormalization as a "shell game" and "hocus pocus".[1]:128

QED has served as the model and template for all subsequent quantum field theories. One such subsequent theory is quantum chromodynamics, which began in the early 1960s and attained its present form in the 1975 work by H. David Politzer, Sidney Coleman, David Gross and Frank Wilczek. Building on the pioneering work of Schwinger, Gerald Guralnik, Dick Hagen, and Tom Kibble,[20][21] Peter Higgs, Jeffrey Goldstone, and others, Sheldon Glashow, Steven Weinberg and Abdus Salam independently showed how the weak nuclear force and quantum electrodynamics could be merged into a single electroweak force.

36.2 Feynman's view of quantum electrodynamics

36.2.1 Introduction

Near the end of his life, Richard P. Feynman gave a series of lectures on QED intended for the lay public. These lectures were transcribed and published as Feynman (1985), *QED: The strange theory of light and matter*,[1] a classic non-mathematical exposition of QED from the point of view articulated below.

The key components of Feynman's presentation of QED are three basic actions.[1]:85

- A photon goes from one place and time to another place and time.

- An electron goes from one place and time to another place and time.

- An electron emits or absorbs a photon at a certain place and time.

These actions are represented in a form of visual shorthand by the three basic elements of Feynman diagrams: a wavy line for the photon, a straight line for the electron and a junction of two straight lines and a wavy one for a vertex representing emission or absorption of a photon by an electron. These can all be seen in the adjacent diagram.

It is important not to over-interpret these diagrams. Nothing is implied about *how* a particle gets from one point to another. The diagrams do *not* imply that the particles are moving in straight or curved lines. They do *not* imply that the particles are moving with fixed speeds. The fact that the photon is often represented, by convention, by a wavy line and not a straight one does *not* imply that it is thought that it is more wavelike than is an electron. The images are just symbols to represent the actions above: photons and electrons do, somehow, move from point to point and electrons, somehow, emit and absorb photons. We do not know how these things happen, but the theory tells us about the probabilities of these things happening.

As well as the visual shorthand for the actions Feynman introduces another kind of shorthand for the numerical quantities called probability amplitudes. The probability is the square of the total probability amplitude. If a photon moves from one place and time—in shorthand, A—to another place and time—in shorthand, B—the associated quantity is written in Feynman's shorthand as P(A to B). The similar quantity for an electron moving from C to D is written E(C to D). The quantity which tells us about the probability amplitude for the emission or absorption of a photon he calls 'j'. This is related to, but not the same as, the measured electron charge 'e'.[1]:91

QED is based on the assumption that complex interactions of many electrons and photons can be represented by fitting together a suitable collection of the above three building blocks, and then using the probability amplitudes to calculate the probability of any such complex interaction. It turns out that the basic idea of QED can be communicated while making the assumption that the square of the total of the probability amplitudes mentioned above (P(A to B), E(A to B) and 'j') acts just like our everyday probability. (A simplification made in Feynman's book.) Later on, this will be corrected to include specifically quantum-style mathematics, following Feynman.

The basic rules of probability amplitudes that will be used are that a) if an event can happen in a variety of different ways then its probability amplitude is the **sum** of the probability amplitudes of the possible ways and b) if a process involves a number of independent sub-processes then its probability amplitude is the **product** of the component probability amplitudes.[1]:93

36.2.2 Basic constructions

Suppose we start with one electron at a certain place and time (this place and time being given the arbitrary label A) and a photon at another place and time (given the label B). A typical question from a physical standpoint is: 'What is the probability of finding an electron at C (another place and a later time) and a photon at D (yet another place and time)?'. The simplest process to achieve this end is for the electron to move from A to C (an elementary action) and for the photon to move from B to D (another elementary action). From a knowledge of the probability amplitudes of each of these sub-processes – E(A to C) and P(B to D) – then we would expect to calculate the probability amplitude of both happening together by multiplying them, using rule b) above. This gives a simple estimated overall probability amplitude, which is squared to give an estimated probability.

But there are other ways in which the end result could come about. The electron might move to a place and time E where it absorbs the photon; then move on before emitting another photon at F; then move on to C where it is detected, while the new photon moves on to D. The probability of this complex process can again be calculated by knowing the probability amplitudes of each of the individual actions: three electron actions, two photon actions and two vertexes – one emission and one absorption. We would expect to find the total probability amplitude by multiplying the probability amplitudes of each of the actions, for any chosen positions of E and F. We then, using rule a) above, have to add up all these probability amplitudes for all the alternatives for E and F. (This is not elementary in practice, and involves integration.) But there is another possibility, which is that the electron first moves to G where it emits a photon which goes on to D, while the electron moves on to H, where it absorbs the first photon, before moving on to C. Again we can calculate the probability amplitude of these possibilities (for all points G and H). We then have a better estimation for the total probability amplitude by adding the probability amplitudes of these two possibilities to our original simple estimate. Incidentally the name given to this process of a photon interacting with an electron in this way is Compton scattering.

There are an *infinite number* of other intermediate processes in which more and more photons are absorbed and/or emitted. For each of these possibilities there is a Feynman diagram describing it. This implies a complex computation for the resulting probability amplitudes, but provided it is the case that the more complicated the diagram the less it contributes to the result, it is only a matter of time and effort to find as accurate an answer as one wants to the original question. This is the basic approach of QED. To calculate the probability of *any* interactive process between electrons and photons it is a matter of first noting, with Feynman diagrams, all the possible ways in which the process can be constructed from the three basic elements. Each diagram involves some calculation involving definite rules to find the associated probability amplitude.

That basic scaffolding remains when one moves to a quantum description but some conceptual changes are needed. One is that whereas we might expect in our everyday life that there would be some constraints on the points to which a particle can move, that is *not* true in full quantum electrodynamics. There is a possibility of an electron at A, or a photon at B, moving as a basic action to *any other place and time in the universe*. That includes places that could only be reached at speeds greater than that of light and also *earlier times*. (An electron moving backwards in time can be viewed as a positron moving forward in time.)[1]:89, 98–99

36.2.3 Probability amplitudes

Quantum mechanics introduces an important change in the way probabilities are computed. Probabilities are still represented by the usual real numbers we use for probabilities in our everyday world, but probabilities are computed as the square of probability amplitudes. Probability amplitudes are complex numbers.

Feynman avoids exposing the reader to the mathematics of complex numbers by using a simple but accurate representation of them as arrows on a piece of paper or screen. (These must not be confused with the arrows of Feynman diagrams which are actually simplified representations in two dimensions of a relationship between points in three dimensions of space and one of time.) The amplitude arrows are fundamental to the description of the world given by quantum theory. No satisfactory reason has been given for *why* they are needed. But pragmatically we have to accept that they are an essential part of our description of all quantum phenomena. They are related to our everyday ideas of probability by the simple rule that the probability of an event is the **square** of the length of the corresponding amplitude arrow. So, for a given process, if two probability amplitudes, **v** and **w**, are involved, the probability of the process will be given either by

$$P = |\mathbf{v} + \mathbf{w}|^2$$

or

$$P = |\mathbf{v}\,\mathbf{w}|^2.$$

The rules as regards adding or multiplying, however, are the same as above. But where you would expect to add or multiply probabilities, instead you add or multiply probability amplitudes that now are complex numbers.

Addition and multiplication are familiar operations in the theory of complex numbers and are given in the figures. The sum is found as follows. Let the start of the second arrow be at the end of the first. The sum is then a third arrow that goes directly from the start of the first to the end of the second. The product of two arrows is an arrow whose length is the product of the two lengths. The direction of the product is found by adding the angles that each of the two have been turned through relative to a reference direction: that gives the angle that the product is turned relative to the reference direction.

That change, from probabilities to probability amplitudes, complicates the mathematics without changing the basic approach. But that change is still not quite enough because it fails to take into account the fact that both photons and electrons can be polarized, which is to say that their orientations in space and time have to be taken into account. Therefore P(A to B) actually consists of 16 complex numbers, or probability amplitude arrows.[1]:120–121 There are also some minor changes to do with the quantity "j", which may have to be rotated by a multiple of 90° for some polarizations, which is only of interest for the detailed bookkeeping.

Associated with the fact that the electron can be polarized is another small necessary detail which is connected with the fact that an electron is a fermion and obeys Fermi–Dirac statistics. The basic rule is that if we have the probability amplitude for a given complex process involving more than one electron, then when we include (as we always must) the complementary Feynman diagram in which we just exchange two electron events, the resulting amplitude is the reverse – the negative – of the first. The simplest case would be two electrons starting at A and B ending at C and D. The amplitude would be calculated as the "difference", E(A to D) × E(B to C) − E(A to C) × E(B to D), where we would expect, from our everyday idea of probabilities, that it would be a sum.[1]:112–113

36.2.4 Propagators

Finally, one has to compute P (A to B) and E (C to D) corresponding to the probability amplitudes for the photon and the electron respectively. These are essentially the solutions of the Dirac Equation which describes the behavior of the electron's probability amplitude and the Klein–Gordon equation which describes the behavior of the photon's probability amplitude. These are called Feynman propagators. The translation to a notation commonly used in the standard literature is as follows:

$$P(\text{A to B}) \rightarrow D_F(x_B - x_A), \quad E(\text{C to D}) \rightarrow S_F(x_D - x_C)$$

where a shorthand symbol such as x_A stands for the four real numbers which give the time and position in three dimensions of the point labeled A.

36.2.5 Mass renormalization

Main article: Self-energy

A problem arose historically which held up progress for twenty years: although we start with the assumption of three basic "simple" actions, the rules of the game say that if we want to calculate the probability amplitude for an electron to get from A to B we must take into account **all** the possible ways: all possible Feynman diagrams with those end points. Thus there will be a way in which the electron travels to C, emits a photon there and then absorbs it again at D before moving on to B. Or it could do this kind of thing twice, or more. In short we have a fractal-like situation in which if we look closely at a line it breaks up into a collection of "simple" lines, each of which, if looked at closely, are in turn composed of "simple" lines, and so on *ad infinitum*. This is a very difficult situation to handle. If adding that detail only altered things slightly then it would not have been too bad, but disaster struck when it was found that the simple correction mentioned above led to *infinite* probability amplitudes. In time this problem was "fixed" by the technique of renormalization. However, Feynman himself remained unhappy about it, calling it a "dippy process".[1]:128

36.2.6 Conclusions

Within the above framework physicists were then able to calculate to a high degree of accuracy some of the properties of electrons, such as the anomalous magnetic dipole moment. However, as Feynman points out, it fails totally to explain why particles such as the electron have the masses they do. "There is no theory that adequately explains these numbers. We use the numbers in all our theories, but we don't understand them – what they are, or where they come from. I believe that from a fundamental point of view, this is a very interesting and serious problem."[1]:152

36.3 Mathematics

Mathematically, QED is an abelian gauge theory with the symmetry group U(1). The gauge field, which mediates the interaction between the charged spin-1/2 fields, is the electromagnetic field. The QED Lagrangian for a spin-1/2 field interacting with the electromagnetic field is given by the real part of[22]:78

where

γ^μ are Dirac matrices;

ψ a bispinor field of spin-1/2 particles (e.g. electron–positron field);

$\bar{\psi} \equiv \psi^\dagger \gamma^0$, called "psi-bar", is sometimes referred to as the Dirac adjoint;

$D_\mu \equiv \partial_\mu + ieA_\mu + ieB_\mu$ is the gauge covariant derivative;

e is the coupling constant, equal to the electric charge of the bispinor field;

$A\mu$ is the covariant four-potential of the electromagnetic field generated by the electron itself;

$B\mu$ is the external field imposed by external source;

$F_{\mu\nu} = \partial_\mu A_\nu - \partial_\nu A_\mu$ is the electromagnetic field tensor.

36.3.1 Equations of motion

To begin, substituting the definition of D into the Lagrangian gives us

$$\mathcal{L} = i\bar{\psi}\gamma^\mu \partial_\mu \psi - e\bar{\psi}\gamma_\mu (A^\mu + B^\mu)\psi - m\bar{\psi}\psi - \frac{1}{4}F_{\mu\nu}F^{\mu\nu}.$$

Next, we can substitute this Lagrangian into the Euler–Lagrange equation of motion for a field:

to find the field equations for QED.

The two terms from this Lagrangian are then

$$\partial_\mu \left(\frac{\partial \mathcal{L}}{\partial(\partial_\mu \psi)} \right) = \partial_\mu \left(i \bar{\psi} \gamma^\mu \right),$$

$$\frac{\partial \mathcal{L}}{\partial \psi} = -e \bar{\psi} \gamma_\mu (A^\mu + B^\mu) - m \bar{\psi}.$$

Substituting these two back into the Euler–Lagrange equation (**2**) results in

$$i \partial_\mu \bar{\psi} \gamma^\mu + e \bar{\psi} \gamma_\mu (A^\mu + B^\mu) + m \bar{\psi} = 0$$

with complex conjugate

$$i \gamma^\mu \partial_\mu \psi - e \gamma_\mu (A^\mu + B^\mu) \psi - m \psi = 0.$$

Bringing the middle term to the right-hand side transforms this second equation into

The left-hand side is like the original Dirac equation and the right-hand side is the interaction with the electromagnetic field.

One further important equation can be found by substituting the above Lagrangian into another Euler–Lagrange equation, this time for the field, A^μ:

The two terms this time are

$$\partial_\nu \left(\frac{\partial \mathcal{L}}{\partial(\partial_\nu A_\mu)} \right) = \partial_\nu \left(\partial^\mu A^\nu - \partial^\nu A^\mu \right),$$

$$\frac{\partial \mathcal{L}}{\partial A_\mu} = -e \bar{\psi} \gamma^\mu \psi$$

and these two terms, when substituted back into (**3**) give us

Now, if we impose the Lorenz gauge condition, that the divergence of the four potential vanishes

$$\partial_\mu A^\mu = 0$$

then we get

$$\Box A^\mu = e \bar{\psi} \gamma^\mu \psi,$$

which is a wave equation for the four potential, the QED version of the classical Maxwell equations in the Lorenz gauge. (In the above equation, the square represents the D'Alembert operator.)

36.3.2 Interaction picture

This theory can be straightforwardly quantized by treating bosonic and fermionic sectors as free. This permits us to build a set of asymptotic states which can be used to start a computation of the probability amplitudes for different processes. In order to do so, we have to compute an evolution operator that, for a given initial state $|i\rangle$, will give a final state $\langle f|$ in such a way to have[22]:5

$$M_{fi} = \langle f|U|i\rangle.$$

This technique is also known as the S-matrix. The evolution operator is obtained in the interaction picture where time evolution is given by the interaction Hamiltonian, which is the integral over space of the second term in the Lagrangian density given above:[22]:123

$$V = e \int d^3x \bar{\psi}\gamma^\mu\psi A_\mu$$

and so, one has[22]:86

$$U = T\exp\left[-\frac{i}{\hbar}\int_{t_0}^{t} dt'V(t')\right]$$

where T is the time ordering operator. This evolution operator only has meaning as a series, and what we get here is a perturbation series with the fine structure constant as the development parameter. This series is called the Dyson series.

36.3.3 Feynman diagrams

Despite the conceptual clarity of this Feynman approach to QED, almost no early textbooks follow him in their presentation. When performing calculations it is much easier to work with the Fourier transforms of the propagators. Quantum physics considers particle's momenta rather than their positions, and it is convenient to think of particles as being created or annihilated when they interact. Feynman diagrams then *look* the same, but the lines have different interpretations. The electron line represents an electron with a given energy and momentum, with a similar interpretation of the photon line. A vertex diagram represents the annihilation of one electron and the creation of another together with the absorption or creation of a photon, each having specified energies and momenta.

Using Wick theorem on the terms of the Dyson series, all the terms of the S-matrix for quantum electrodynamics can be computed through the technique of Feynman diagrams. In this case rules for drawing are the following[22]:801–802

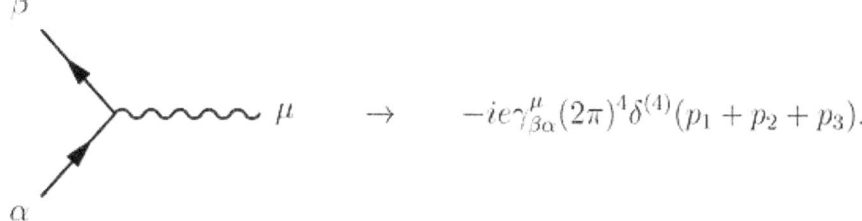

$$\alpha \longrightarrow \beta \qquad \rightarrow \qquad \left(\frac{i}{\not p - m + i\varepsilon}\right)_{\beta\alpha}$$

$$\mu \sim\!\sim\!\sim\!\sim \nu \qquad \rightarrow \qquad \frac{-i\eta_{\mu\nu}}{p^2 + i\varepsilon}$$

$$\rightarrow \qquad -ie\gamma^{\mu}_{\beta\alpha}(2\pi)^4\delta^{(4)}(p_1 + p_2 + p_3).$$

Incoming fermion: $\alpha \longrightarrow \bullet$ $\rightarrow$ $u_\alpha(\vec p, s)$

Incoming antifermion: $\alpha \longleftarrow \bullet$ $\rightarrow$ $\bar v_\alpha(\vec p, s)$

Outgoing fermion: $\bullet \longrightarrow \alpha$ $\rightarrow$ $\bar u_\alpha(\vec p, s)$

Outgoing antifermion: $\bullet \longleftarrow \alpha$ $\rightarrow$ $v_\alpha(p, s)$

Incoming photon: $\mu \sim\!\sim\!\bullet$ $\rightarrow$ $\epsilon_\mu(\vec k, \lambda)$

Outgoing photon: $\bullet\!\sim\!\sim \mu$ $\rightarrow$ $\epsilon_\mu(\vec k, \lambda)^*$

To these rules we must add a further one for closed loops that implies an integration on momenta $\int d^4p/(2\pi)^4$, since these internal ("virtual") particles are not constrained to any specific energy–momentum – even that usually required by special relativity (see this article for details). From them, computations of probability amplitudes are straightforwardly given. An example is Compton scattering, with an electron and a photon undergoing elastic scattering. Feynman diagrams are in this case[22]:158–159

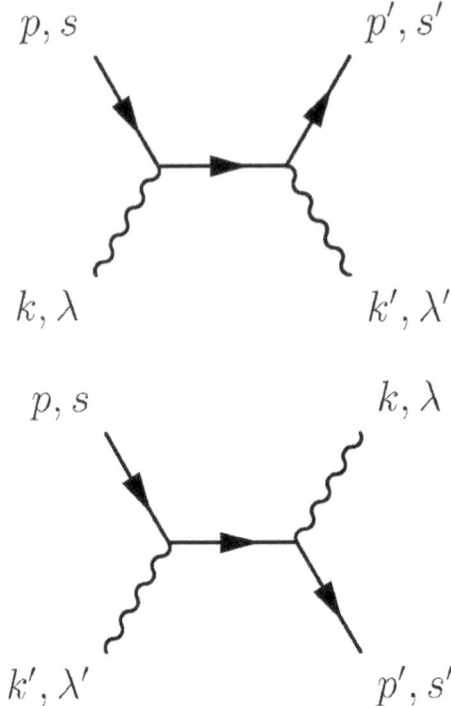

and so we are able to get the corresponding amplitude at the first order of a perturbation series for the S-matrix:

$$M_{fi} = (ie)^2 \overline{u}(\vec{p}', s') \not{\epsilon}'(\vec{k}', \lambda')^* \frac{\not{p} + \not{k} + m_e}{(p+k)^2 - m_e^2} \not{\epsilon}(\vec{k}, \lambda) u(\vec{p}, s) + (ie)^2 \overline{u}(\vec{p}', s') \not{\epsilon}(\vec{k}, \lambda) \frac{\not{p} - \not{k}' + m_e}{(p-k')^2 - m_e^2} \not{\epsilon}'(\vec{k}', \lambda')^* u(\vec{p}, s)$$

from which we are able to compute the cross section for this scattering.

36.4 Renormalizability

Higher order terms can be straightforwardly computed for the evolution operator but these terms display diagrams containing the following simpler ones[22]:ch 10

- One-loop contribution to the vacuum polarization function
- One-loop contribution to the electron self-energy function
- One-loop contribution to the vertex function

that, being closed loops, imply the presence of diverging integrals having no mathematical meaning. To overcome this difficulty, a technique called renormalization has been devised, producing finite results in very close agreement with experiments. It is important to note that a criterion for theory being meaningful after renormalization is that the number of diverging diagrams is finite. In this case the theory is said to be **renormalizable**. The reason for this is that to get observables renormalized one needs a finite number of constants to maintain the predictive value of the theory untouched. This is exactly the case of quantum electrodynamics displaying just three diverging diagrams. This procedure gives observables in very close agreement with experiment as seen e.g. for electron gyromagnetic ratio.

Renormalizability has become an essential criterion for a quantum field theory to be considered as a viable one. All the theories describing fundamental interactions, except gravitation whose quantum counterpart is presently under very active research, are renormalizable theories.

36.5 Nonconvergence of series

An argument by Freeman Dyson shows that the radius of convergence of the perturbation series in QED is zero.[23] The basic argument goes as follows: if the coupling constant were negative, this would be equivalent to the Coulomb force constant being negative. This would "reverse" the electromagnetic interaction so that *like* charges would *attract* and *unlike* charges would *repel*. This would render the vacuum unstable against decay into a cluster of electrons on one side of the universe and a cluster of positrons on the other side of the universe. Because the theory is 'sick' for any negative value of the coupling constant, the series do not converge, but are an asymptotic series.

From a modern perspective, we say that QED is not well defined as a quantum field theory to arbitrarily high energy.[24] The coupling constant runs to infinity at finite energy, signalling a Landau pole. The problem is essentially that QED is not asymptotically free. This is one of the motivations for embedding QED within a Grand Unified Theory.

36.6 See also

36.7 References

[1] Feynman, Richard (1985). *QED: The Strange Theory of Light and Matter*. Princeton University Press. ISBN 978-0-691-12575-6.

[2] P.A.M. Dirac (1927). "The Quantum Theory of the Emission and Absorption of Radiation". *Proceedings of the Royal Society of London A* **114** (767): 243–265. Bibcode:1927RSPSA.114..243D. doi:10.1098/rspa.1927.0039.

[3] E. Fermi (1932). "Quantum Theory of Radiation". *Reviews of Modern Physics* **4**: 87–132. Bibcode:1932RvMP....4...87F. doi:10.1103/RevModPhys.4.87.

[4] Bloch, F.; Nordsieck, A. (1937). "Note on the Radiation Field of the Electron". *Physical Review* **52** (2): 54–59. Bibcode:1937PhRv...52...54B. doi:10.1103/PhysRev.52.54.

[5] V. F. Weisskopf (1939). "On the Self-Energy and the Electromagnetic Field of the Electron". *Physical Review* **56**: 72–85. Bibcode:1939PhRv...56...72W. doi:10.1103/PhysRev.56.72.

[6] R. Oppenheimer (1930). "Note on the Theory of the Interaction of Field and Matter". *Physical Review* **35** (5): 461–477. Bibcode:1930PhRv...35..461O. doi:10.1103/PhysRev.35.461.

[7] Lamb, Willis; Retherford, Robert (1947). "Fine Structure of the Hydrogen Atom by a Microwave Method,". *Physical Review* **72** (3): 241–243. Bibcode:1947PhRv...72..241L. doi:10.1103/PhysRev.72.241.

[8] Foley, H.; Kusch, P. (1948). "On the Intrinsic Moment of the Electron". *Physical Review* **73** (3): 412. Bibcode:1948PhRv...73..412F. doi:10.1103/PhysRev.73.412.

[9] Schweber, Silvan (1994). "Chapter 5". *QED and the Men Who Did it: Dyson, Feynman, Schwinger, and Tomonaga*. Princeton University Press. p. 230. ISBN 978-0-691-03327-3.

[10] H. Bethe (1947). "The Electromagnetic Shift of Energy Levels". *Physical Review* **72** (4): 339–341. Bibcode:1947PhRv...72..339B. doi:10.1103/PhysRev.72.339.

[11] S. Tomonaga (1946). "On a Relativistically Invariant Formulation of the Quantum Theory of Wave Fields". *Progress of Theoretical Physics* **1** (2): 27–42. doi:10.1143/PTP.1.27.

[12] J. Schwinger (1948). "On Quantum-Electrodynamics and the Magnetic Moment of the Electron". *Physical Review* **73** (4): 416–417. Bibcode:1948PhRv...73..416S. doi:10.1103/PhysRev.73.416.

[13] J. Schwinger (1948). "Quantum Electrodynamics. I. A Covariant Formulation". *Physical Review* **74** (10): 1439–1461. Bibcode:1948PhRv...74.1439S. doi:10.1103/PhysRev.74.1439.

[14] R. P. Feynman (1949). "Space–Time Approach to Quantum Electrodynamics". *Physical Review* **76** (6): 769–789. Bibcode:1949PhRv...76..76 doi:10.1103/PhysRev.76.769.

[15] R. P. Feynman (1949). "The Theory of Positrons". *Physical Review* **76** (6): 749–759. Bibcode:1949PhRv...76..749F. doi:10.1103/PhysRev.76

[16] R. P. Feynman (1950). "Mathematical Formulation of the Quantum Theory of Electromagnetic Interaction". *Physical Review* **80** (3): 440–457. Bibcode:1950PhRv...80..440F. doi:10.1103/PhysRev.80.440.

[17] F. Dyson (1949). "The Radiation Theories of Tomonaga, Schwinger, and Feynman". *Physical Review* **75** (3): 486–502. Bibcode:1949PhRv...75..486D. doi:10.1103/PhysRev.75.486.

[18] F. Dyson (1949). "The S Matrix in Quantum Electrodynamics". *Physical Review* **75** (11): 1736–1755. Bibcode:1949PhRv...75.1736D. doi:10.1103/PhysRev.75.1736.

[19] "The Nobel Prize in Physics 1965". Nobel Foundation. Retrieved 2008-10-09.

[20] Guralnik, G. S.; Hagen, C. R.; Kibble, T. W. B. (1964). "Global Conservation Laws and Massless Particles". *Physical Review Letters* **13** (20): 585–587. Bibcode:1964PhRvL..13..585G. doi:10.1103/PhysRevLett.13.585.

[21] Guralnik, G. S. (2009). "The History of the Guralnik, Hagen and Kibble development of the Theory of Spontaneous Symmetry Breaking and Gauge Particles". *International Journal of Modern Physics A* **24** (14): 2601–2627. arXiv:0907.3466. Bibcode:2009IJMPA..24.2601G. doi:10.1142/S0217751X09045431.

[22] Peskin, Michael; Schroeder, Daniel (1995). *An introduction to quantum field theory* (Reprint ed.). Westview Press. ISBN 978-0201503975.

[23] Kinoshita, Toichiro. "Quantum Electrodynamics has Zero Radius of Convergence Summarized from Toichiro Kinoshita". Retrieved 06-10-2010. Check date values in: |accessdate= (help)

[24] Espriu and Tarrach. "Ambiguities in QED: Renormalons versus Triviality". arXiv:hep-ph/9604431.

36.8   Further reading

36.8.1   Books

- De Broglie, Louis (1925). *Recherches sur la theorie des quanta [Research on quantum theory].* France: Wiley-Interscience.

- Feynman, Richard Phillips (1998). *Quantum Electrodynamics* (New ed.). Westview Press. ISBN 978-0-201-36075-2.

- Jauch, J.M.; Rohrlich, F. (1980). *The Theory of Photons and Electrons.* Springer-Verlag. ISBN 978-0-387-07295-1.

- Greiner, Walter; Bromley, D.A.; Müller, Berndt (2000). *Gauge Theory of Weak Interactions.* Springer. ISBN 978-3-540-67672-0.

- Kane, Gordon, L. (1993). *Modern Elementary Particle Physics.* Westview Press. ISBN 978-0-201-62460-1.

- Miller, Arthur I. (1995). *Early Quantum Electrodynamics: A Sourcebook.* Cambridge University Press. ISBN 978-0-521-56891-3.

- Milonni, Peter W., (1994) *The quantum vacuum - an introduction to quantum electrodynamics.* Academic Press. ISBN 0-12-498080-5

- Schweber, Silvan S. (1994). *QED and the Men Who Made It.* Princeton University Press. ISBN 978-0-691-03327-3.

- Schwinger, Julian (1958). *Selected Papers on Quantum Electrodynamics*. Dover Publications. ISBN 978-0-486-60444-2.

- Tannoudji-Cohen, Claude; Dupont-Roc, Jacques; Grynberg, Gilbert (1997). *Photons and Atoms: Introduction to Quantum Electrodynamics*. Wiley-Interscience. ISBN 978-0-471-18433-1.

36.8.2   Journals

- Dudley, J.M.; Kwan, A.M. (1996). "Richard Feynman's popular lectures on quantum electrodynamics: The 1979 Robb Lectures at Auckland University". *American Journal of Physics* **64** (6): 694–698. Bibcode:1996AmJPh..64..694D. doi:10.1119/1.18234.

36.9   External links

- Feynman's Nobel Prize lecture describing the evolution of QED and his role in it

- Feynman's New Zealand lectures on QED for non-physicists

- http://qed.wikina.org/ - Animations demonstrating QED

Hans Bethe

Feynman (center) and Oppenheimer (right) at Los Alamos.

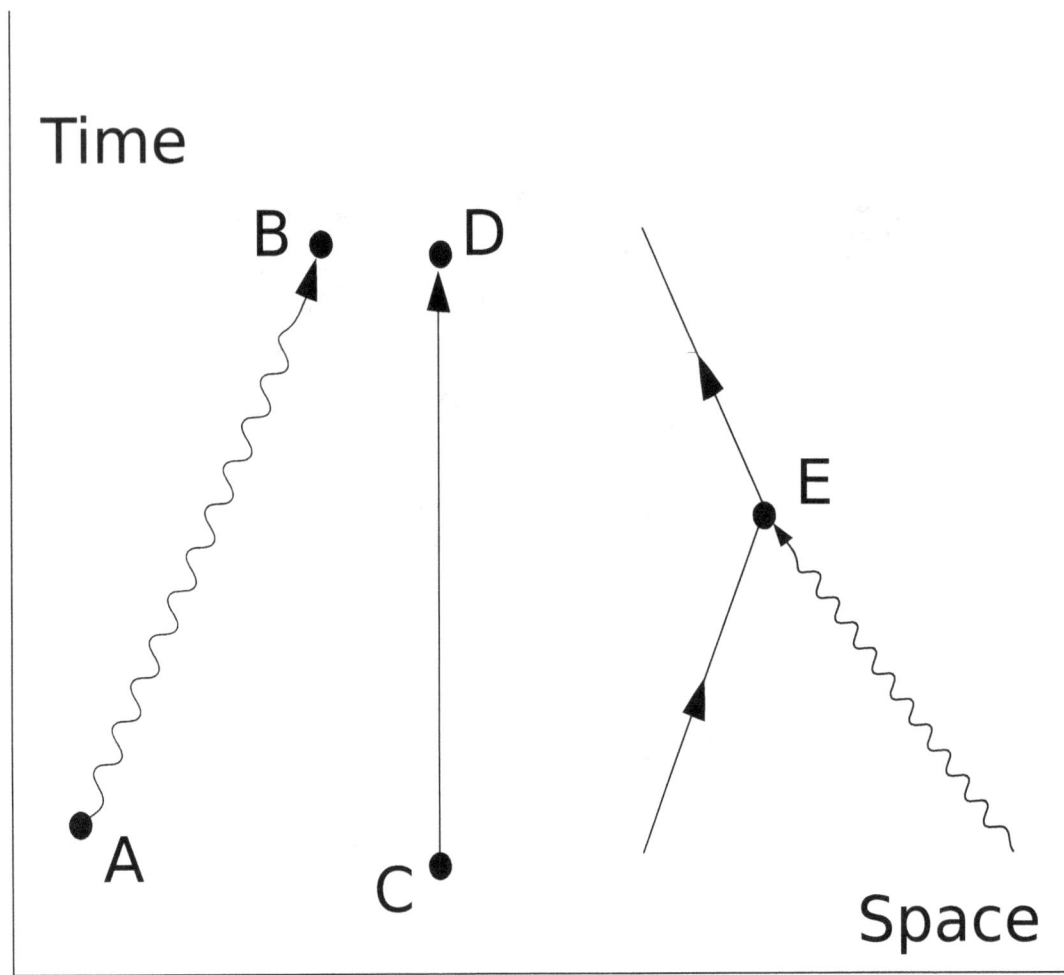

Feynman diagram elements

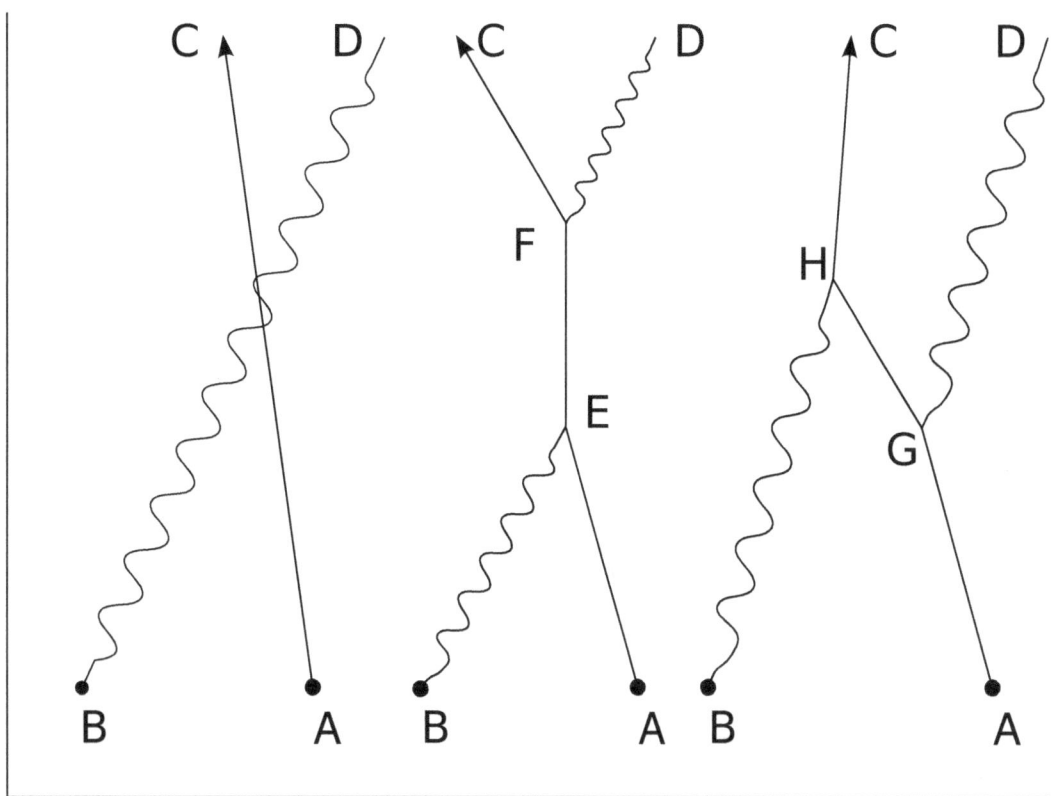

Compton scattering

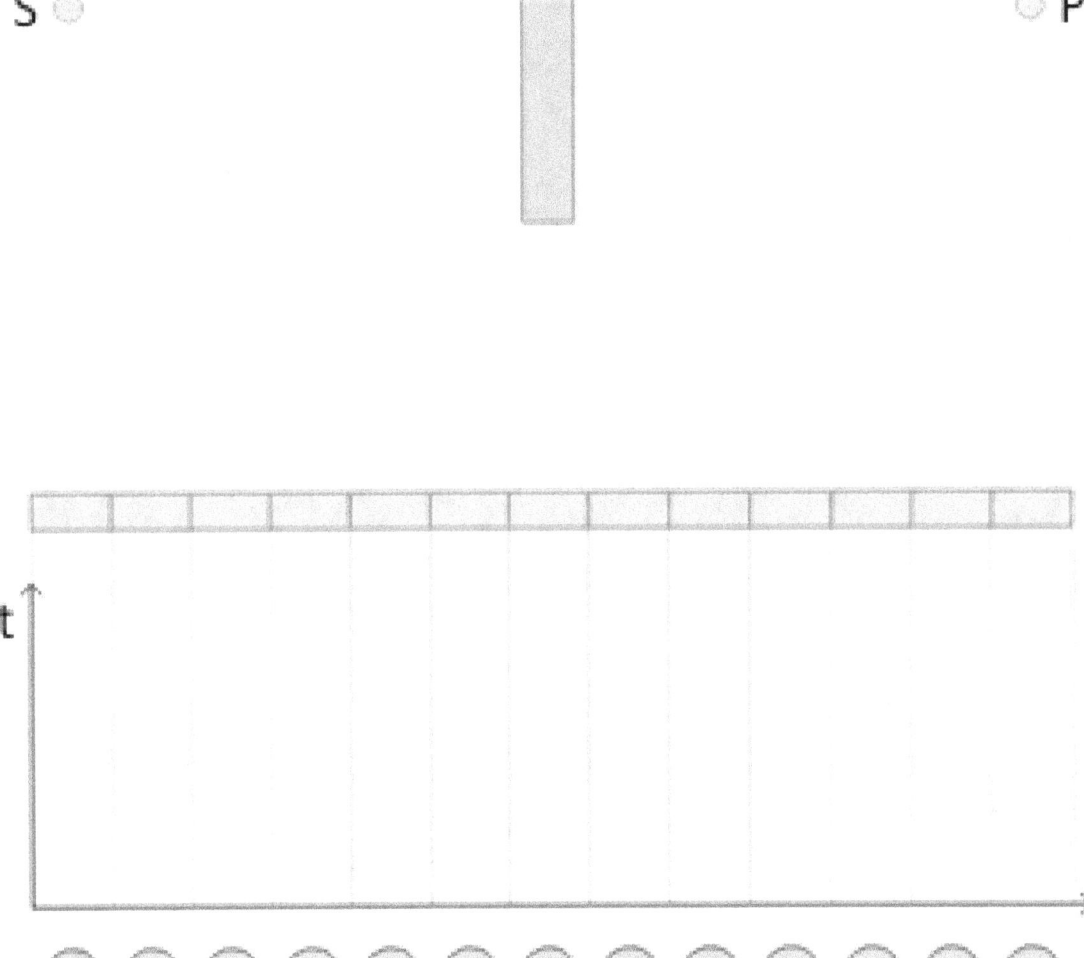

*Feynman replaces complex numbers with spinning arrows, which start at emission and end at detection of a particle. The sum of all resulting arrows represents the total probability of the event. In this diagram, light emitted by the source **S** bounces off a few segments of the mirror (in blue) before reaching the detector at **P**. The sum of all paths must be taken into account. The graph below depicts the total time spent to traverse each of the paths above.*

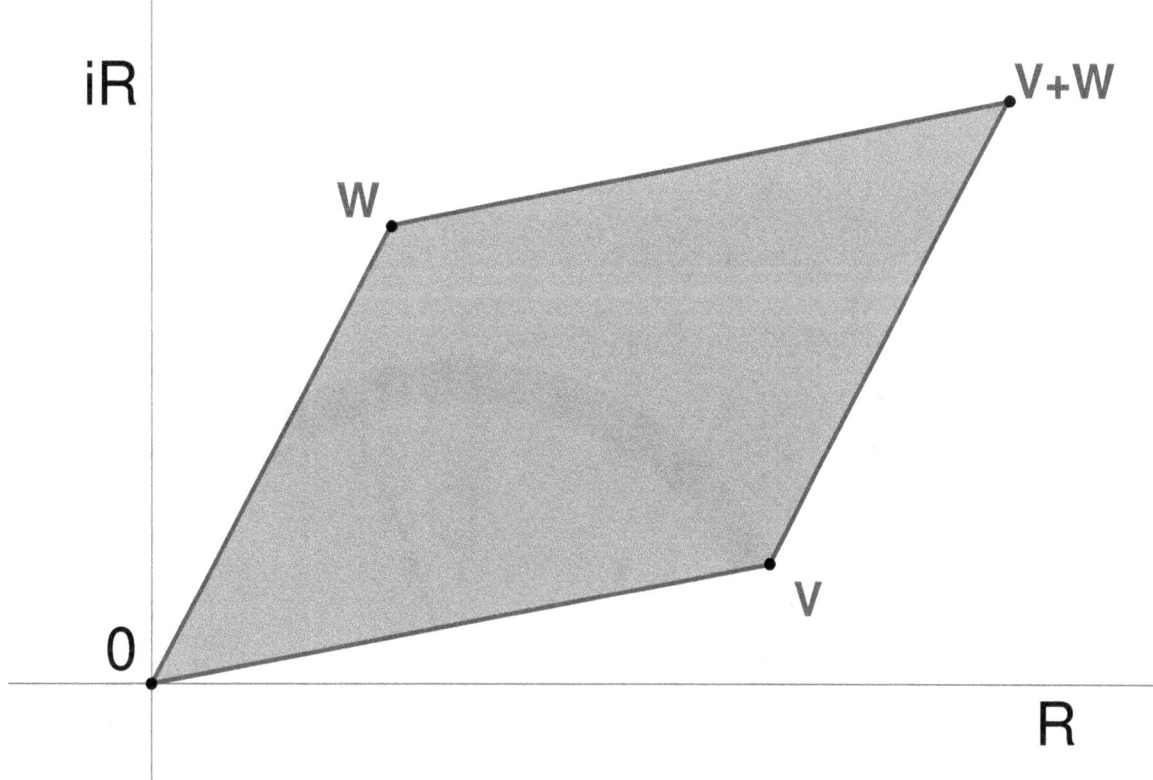

Addition of probability amplitudes as complex numbers

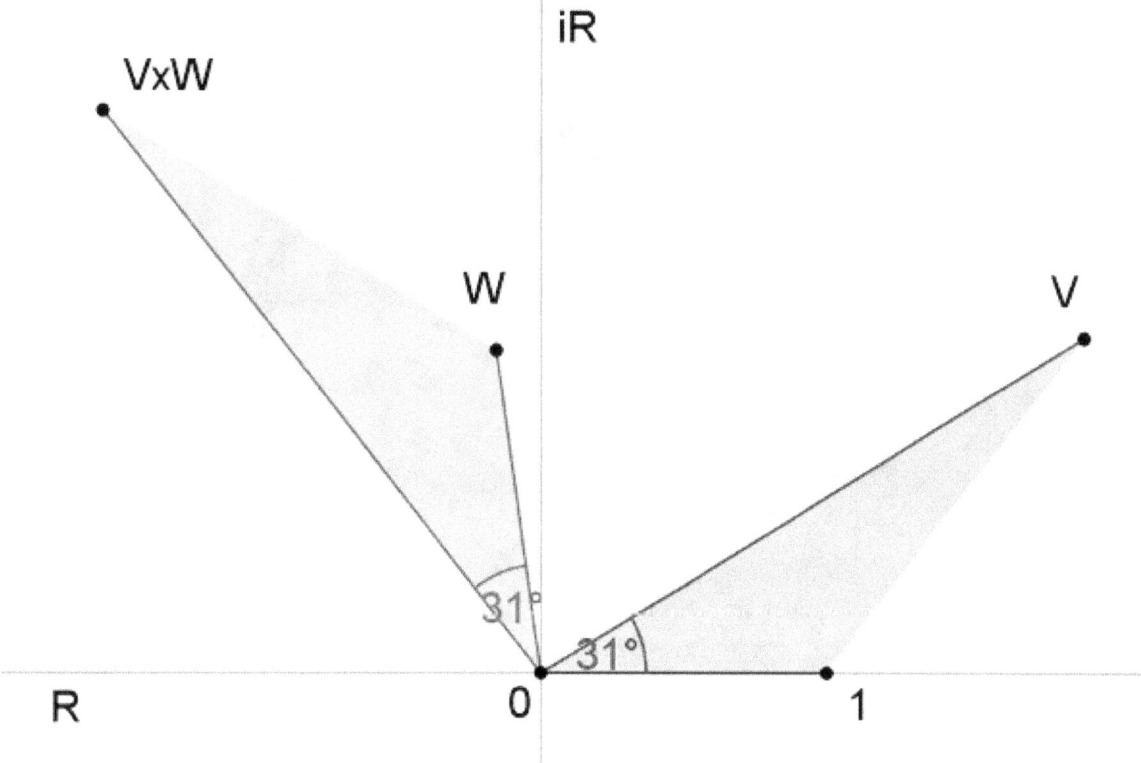

Multiplication of probability amplitudes as complex numbers

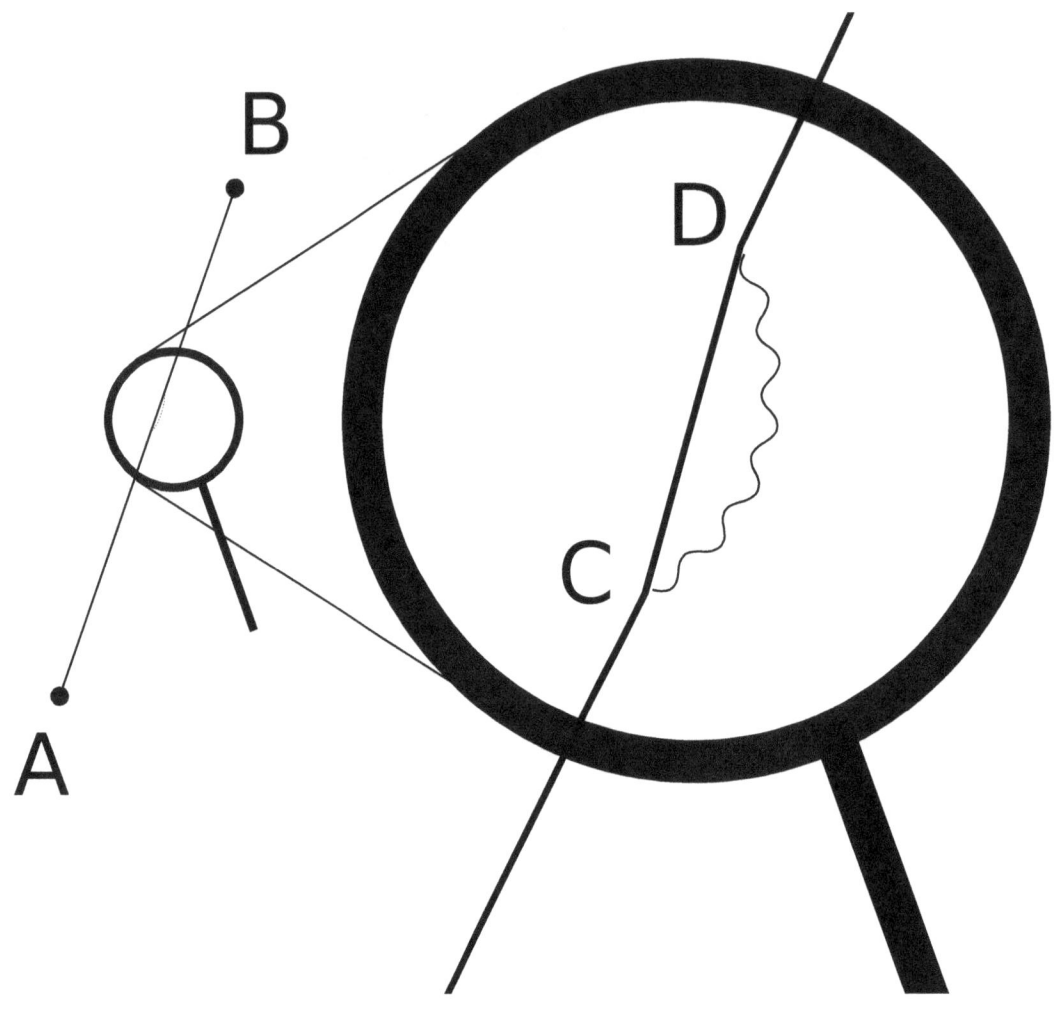

Electron self-energy loop

Chapter 37

Principle of least action

This article discusses the history of the principle of least action. For the application, please refer to action (physics).

In non-relativistic physics, the **principle of least action** – or, more accurately, the **principle of stationary action** – is a variational principle that, when applied to the action of a mechanical system, can be used to obtain the equations of motion for that system by stating a system follows the path where the average difference between the kinetic energy and potential energy is minimized or maximized over any time period. It is called stable if minimized. In relativity, a different average must be minimized or maximized. The principle can be used to derive Newtonian, Lagrangian, and Hamiltonian equations of motion. It was historically called "least" because its solution requires finding the path that has the least change from nearby paths.[1] Its classical mechanics and electromagnetic expressions are a consequence of quantum mechanics, but the stationary action method helped in the development of quantum mechanics. [2]

The principle remains central in modern physics and mathematics, being applied in the theory of relativity, quantum mechanics and quantum field theory, and a focus of modern mathematical investigation in Morse theory. Maupertuis' principle and Hamilton's principle exemplify the principle of stationary action.

The action principle is preceded by earlier ideas in surveying and optics. Rope stretchers in ancient Egypt stretched corded ropes to measure the distance between two points. Ptolemy, in his *Geography* (Bk 1, Ch 2), emphasized that one must correct for "deviations from a straight course". In ancient Greece, Euclid wrote in his *Catoptrica* that, for the path of light reflecting from a mirror, the angle of incidence equals the angle of reflection. Hero of Alexandria later showed that this path was the shortest length and least time.[3]

Scholars often credit Pierre Louis Maupertuis for formulating the principle of least action because he wrote about it in 1744[4] and 1746.[5] However, Leonhard Euler discussed the principle in 1744,[6] and evidence shows that Gottfried Leibniz preceded both by 39 years.[7][8][9]

In 1932, Paul Dirac discerned the quantum mechanical underpinning of the principle in the quantum interference of amplitudes:For macroscopic systems, the dominant contribution to the apparent path is the classical path (the stationary, action-extremizing one), while any other path is possible in the quantum realm.

37.1 General statement

The starting point is the *action*, denoted $\mathcal{S}$ (calligraphic S), of a physical system. It is defined as the integral of the Lagrangian L between two instants of time t_1 and t_2 - technically a functional of the N generalized coordinates $\mathbf{q} = (q_1, q_2 \ldots qN)$ which define the configuration of the system:

$$\mathcal{S}[\mathbf{q}(t)] = \int_{t_1}^{t_2} L(\mathbf{q}(t), \mathbf{q}(t), t) dt$$

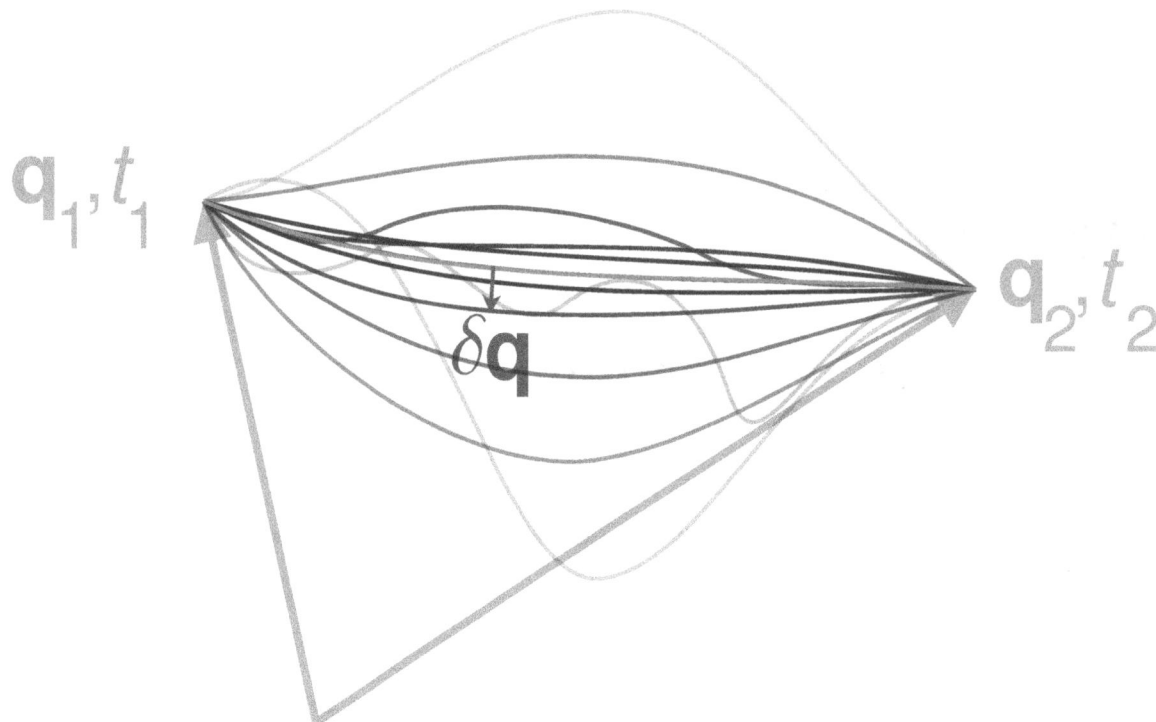

As the system evolves, $\mathbf{q}$ traces a path through configuration space (only some are shown). The path taken by the system (red) has a stationary action ($\delta S = 0$) under small changes in the configuration of the system ($\delta \mathbf{q}$).[10]

where the dot denotes the time derivative, and t is time.

Mathematically the principle is[11][12][13]

$$\delta \mathcal{S} = 0$$

where δ (Greek lowercase delta) means a *small* change. In words this reads:[10]

> The path taken by the system between times t_1 and t_2 is the one for which the **action** is **stationary (no change) to *first order***.

In applications the statement and definition of action are taken together:[14]

$$\delta \int_{t_1}^{t_2} L(\mathbf{q}, \mathbf{q}, t)\, dt = 0$$

The action and Lagrangian both contain the dynamics of the system for all times. The term "path" simply refers to a curve traced out by the system in terms of the coordinates in the configuration space, i.e. the curve $\mathbf{q}(t)$, parameterized by time (see also parametric equation for this concept).

37.2 Origins, statements, and controversy

37.2.1 Fermat

Main article: Fermat's principle

In the 1600s, Pierre de Fermat postulated that "*light travels between two given points along the path of shortest time,*" which is known as the **principle of least time** or **Fermat's principle**.[13]

37.2.2   Maupertuis

Main article: Maupertuis principle

Credit for the formulation of the **principle of least action** is commonly given to Pierre Louis Maupertuis, who felt that "Nature is thrifty in all its actions", and applied the principle broadly:

> The laws of movement and of rest deduced from this principle being precisely the same as those observed in nature, we can admire the application of it to all phenomena. The movement of animals, the vegetative growth of plants ... are only its consequences; and the spectacle of the universe becomes so much the grander, so much more beautiful, the worthier of its Author, when one knows that a small number of laws, most wisely established, suffice for all movements.
> —Pierre Louis Maupertuis[15]

This notion of Maupertuis, although somewhat deterministic today, does capture much of the essence of mechanics.

In application to physics, Maupertuis suggested that the quantity to be minimized was the product of the duration (time) of movement within a system by the "vis viva",

which is the integral of twice what we now call the kinetic energy T of the system.

37.2.3   Euler

Leonhard Euler gave a formulation of the action principle in 1744, in very recognizable terms, in the *Additamentum 2* to his *Methodus Inveniendi Lineas Curvas Maximi Minive Proprietate Gaudentes*. Beginning with the second paragraph:

As Euler states, $\int Mv ds$ is the integral of the momentum over distance travelled, which, in modern notation, equals the reduced action

Thus, Euler made an equivalent and (apparently) independent statement of the variational principle in the same year as Maupertuis, albeit slightly later. Curiously, Euler did not claim any priority, as the following episode shows.

37.2.4   Disputed priority

Maupertuis' priority was disputed in 1751 by the mathematician Samuel König, who claimed that it had been invented by Gottfried Leibniz in 1707. Although similar to many of Leibniz's arguments, the principle itself has not been documented in Leibniz's works. König himself showed a *copy* of a 1707 letter from Leibniz to Jacob Hermann with the principle, but the *original* letter has been lost. In contentious proceedings, König was accused of forgery,[7] and even the King of Prussia entered the debate, defending Maupertuis (the head of his Academy), while Voltaire defended König.

Euler, rather than claiming priority, was a staunch defender of Maupertuis, and Euler himself prosecuted König for forgery before the Berlin Academy on 13 April 1752.[7] The claims of forgery were re-examined 150 years later, and archival work by C.I. Gerhardt in 1898[8] and W. Kabitz in 1913[9] uncovered other copies of the letter, and three others cited by König, in the Bernoulli archives.

37.3 Further development

Euler continued to write on the topic; in his *Reflexions sur quelques loix generales de la nature* (1748), he called the quantity "effort". His expression corresponds to what we would now call potential energy, so that his statement of least action in statics is equivalent to the principle that a system of bodies at rest will adopt a configuration that minimizes total potential energy.

37.3.1 Lagrange and Hamilton

Main article: Hamilton's principle

Much of the calculus of variations was stated by Joseph-Louis Lagrange in 1760[17][18] and he proceeded to apply this to problems in dynamics. In *Méchanique Analytique* (1788) Lagrange derived the general equations of motion of a mechanical body.[19] William Rowan Hamilton in 1834 and 1835[20] applied the variational principle to the classical Lagrangian function

$$L = T - V$$

to obtain the Euler–Lagrange equations in their present form.

37.3.2 Jacobi and Morse

In 1842, Carl Gustav Jacobi tackled the problem of whether the variational principle always found minima as opposed to other stationary points (maxima or stationary saddle points); most of his work focused on geodesics on two-dimensional surfaces.[21] The first clear general statements were given by Marston Morse in the 1920s and 1930s,[22] leading to what is now known as Morse theory. For example, Morse showed that the number of conjugate points in a trajectory equalled the number of negative eigenvalues in the second variation of the Lagrangian.

37.3.3 Gauss and Hertz

Other extremal principles of classical mechanics have been formulated, such as Gauss's principle of least constraint and its corollary, Hertz's principle of least curvature.

37.4 Apparent teleology

The mathematical equivalence of the differential equations of motion and their integral counterpart has important philosophical implications. The differential equations are statements about quantities localized to a single point in space or single moment of time. For example, Newton's second law

$$\mathbf{F} = m\mathbf{a}$$

states that the *instantaneous* force $\mathbf{F}$ applied to a mass m produces an acceleration $\mathbf{a}$ at the same *instant*. By contrast, the action principle is not localized to a point; rather, it involves integrals over an interval of time and (for fields) an extended region of space. Moreover, in the usual formulation of classical action principles, the initial and final states of the system are fixed, e.g.,

> Given that the particle begins at position x_1 at time t_1 and ends at position x_2 at time t_2, the physical trajectory that connects these two endpoints is an extremum of the action integral.

In particular, the fixing of the *final* state appears to give the action principle a teleological character which has been controversial historically.[23] However, some critics maintain this apparent teleology occurs because of the way in which the question was asked. By specifying some but not all aspects of both the initial and final conditions (the positions but not the velocities) we are making some inferences about the initial conditions from the final conditions, and it is this "backward" inference that can be seen as a teleological explanation. Teleology can also be overcome if we consider the classical description as a limiting case of the quantum formalism of path integration, in which stationary paths are obtained as a result of interference of amplitudes along all possible paths.

The short story *Story of Your Life* by the speculative fiction writer Ted Chiang contains visual depictions of Fermat's Principle along with a discussion of its teleological dimension. Keith Devlin's *The Math Instinct* contains a chapter, "Elvis the Welsh Corgi Who Can Do Calculus" that discusses the calculus "embedded" in some animals as they solve the "least time" problem in actual situations.

37.5   More Fundamental Than Newton's 2nd Law

The principle of least action is mathematically more specific than Newton's 2nd law and more fundamental in theoretical physics because it explains a wider range of physical law. You can derive Newton's 2nd law from least action, but the converse is not true without also applying Newton's 1st and 3rd laws and disallowing non-conservative forces like friction. By being more specific and thereby explaining only conservative forces, the principle of least action is able to solve problems Newton's 2nd law can't, but the converse is not true. The principle of least action can be used to derive the conservation of momentum and energy if its symmetry in space and time are assumed.[24] It correctly does not allow non-conservative potential fields, but Newton's 2nd law allows for them by allowing for non-conservative momentums and forces (such as friction) which are not fundamental forces.[25] The mathematical basis for the difference is that Newton's 2nd law (correctly stated as F=dp/dt instead of F=ma) allows for momentums $p(t)=q(t)+C$ where $q(t)$ are conserved momentums allowed by least action and C is a constant that can be non-zero in Newton's 2nd law but not in least action. The constant allows for non-conservative momentums and therefore non-conservative forces and potentials in Newton's 2nd law. Newton's 2nd law explains conservation of energy and momentum and can be used to show equivalency with least action when forces are properly conserved, e.g. when forces are summed to zero in accordance with Newton's 1st and 3rd laws and when accounting for heat generated by friction. Derivations of Lagrangian and Hamiltonian methods do not begin with Newton's 2nd law, but with a more modern mathematical formulation of it that requires forces to be conservative.

37.6   See also

- Action (physics)
- Analytical mechanics
- Calculus of variations
- Hamiltonian mechanics
- Lagrangian mechanics
- Occam's razor
- Path of least resistance

37.7   Notes and references

[1] Chapter 19 of Volume II, Feynman R, Leighton R, and Sands M. *The Feynman Lectures on Physics* . 3 volumes 1964, 1966. Library of Congress Catalog Card No. 63-20717. ISBN 0-201-02115-3 (1970 paperback three-volume set); ISBN 0-201-50064-7 (1989 commemorative hardcover three-volume set); ISBN 0-8053-9045-6 (2006 the definitive edition (2nd printing); hardcover)

[2] "The Character of Physical Law" Richard Feynman

[3] Kline, Morris (1972). *Mathematical Thought from Ancient to Modern Times*. New York: Oxford University Press. pp. 167–68. ISBN 0-19-501496-0.

[4] P.L.M. de Maupertuis, *Accord de différentes lois de la nature qui avaient jusqu'ici paru incompatibles.* (1744) Mém. As. Sc. Paris p. 417. (English translation)

[5] P.L.M. de Maupertuis, *Le lois de mouvement et du repos, déduites d'un principe de métaphysique.* (1746) Mém. Ac. Berlin, p. 267.(English translation)

[6] Leonhard Euler, *Methodus Inveniendi Lineas Curvas Maximi Minive Proprietate Gaudentes.* (1744) Bousquet, Lausanne & Geneva. 320 pages. Reprinted in *Leonhardi Euleri Opera Omnia: Series I vol 24.* (1952) C. Cartheodory (ed.) Orell Fuessli, Zurich. scanned copy of complete text at *The Euler Archive*, Dartmouth.

[7] J J O'Connor and E F Robertson, "The Berlin Academy and forgery", (2003), at *The MacTutor History of Mathematics archive.*

[8] Gerhardt CI. (1898) "Über die vier Briefe von Leibniz, die Samuel König in dem Appel au public, Leide MDCCLIII, veröffentlicht hat", *Sitzungsberichte der Königlich Preussischen Akademie der Wissenschaften*, **I**, 419-427.

[9] Kabitz W. (1913) "Über eine in Gotha aufgefundene Abschrift des von S. König in seinem Streite mit Maupertuis und der Akademie veröffentlichten, seinerzeit für unecht erklärten Leibnizbriefes", *Sitzungsberichte der Königlich Preussischen Akademie der Wissenschaften*, **II**, 632-638.

[10] R. Penrose (2007). *The Road to Reality*. Vintage books. p. 474. ISBN 0-679-77631-1.

[11] Encyclopaedia of Physics (2nd Edition), R.G. Lerner, G.L. Trigg, VHC publishers, 1991, ISBN (Verlagsgesellschaft) 3-527-26954-1, ISBN (VHC Inc.) 0-89573-752-3

[12] McGraw Hill Encyclopaedia of Physics (2nd Edition), C.B. Parker, 1994, ISBN 0-07-051400-3

[13] Analytical Mechanics, L.N. Hand, J.D. Finch, Cambridge University Press, 2008, ISBN 978-0-521-57572-0

[14] Classical Mechanics, T.W.B. Kibble, European Physics Series, McGraw-Hill (UK), 1973, ISBN 0-07-084018-0

[15] Chris Davis. *Idle theory* (1998)

[16] Euler, Additamentum II (external link), ibid. (English translation)

[17] D. J. Struik, ed. (1969). *A Source Book in Mathematics, 1200-1800*. Cambridge, Mass: MIT Press. pp. 406-413

[18] Kline, Morris (1972). *Mathematical Thought from Ancient to Modern Times*. New York: Oxford University Press. ISBN 0-19-501496-0. pp. 582-589

[19] Lagrange, Joseph-Louis (1788). *Mécanique Analytique*. p. 226

[20] W. R. Hamilton, "On a General Method in Dynamics", *Philosophical Transaction of the Royal Society* Part I (1834) p.247-308; Part II (1835) p. 95-144. (*From the collection Sir William Rowan Hamilton (1805-1865): Mathematical Papers edited by David R. Wilkins, School of Mathematics, Trinity College, Dublin 2, Ireland. (2000); also reviewed as On a General Method in Dynamics*)

[21] G.C.J. Jacobi, *Vorlesungen über Dynamik, gehalten an der Universität Königsberg im Wintersemester 1842-1843*. A. Clebsch (ed.) (1866); Reimer; Berlin. 290 pages, available online Œuvres complètes volume **8** at Gallica-Math from the Gallica Bibliothèque nationale de France.

[22] Marston Morse (1934). "The Calculus of Variations in the Large", *American Mathematical Society Colloquium Publication* **18**; New York.

[23] Stöltzner, Michael (1994). *Inside Versus Outside: Action Principles and Teleology*. Springer. pp. 33–62. ISBN 978-3-642-48649-4.

[24] "The Character of Physical Law" Richard Feynman

[25] "The Principle of Least Action" Richard Feynman

37.8   External links

- Interactive explanation of the principle of least action

- Interactive applet to construct trajectories using principle of least action

- Georgi Yordanov Georgiev 2012 , A quantitative measure, mechanism and attractor for self-organization in networked complex systems, in Lecture Notes in Computer Science (LNCS 7166), F.A. Kuipers and P.E. Heegaard (Eds.): IFIP International Federation for Information Processing, Proceedings of the Sixth International Workshop on Self-Organizing Systems (IWSOS 2012), pp. 90–95, Springer-Verlag (2012).

- Georgi Yordanov Georgiev and Iskren Yordanov Georgiev 2002 , The least action and the metric of an organized system, in Open Systems and Information Dynamics, 9(4), p. 371-380 (2002)

Chapter 38

Fiber bundle

Not to be confused with an Optical fiber bundle.

In mathematics, and particularly topology, a **fiber bundle** (or, in British English, **fibre bundle**) is a space that is *locally*

A cylindrical hairbrush showing the intuition behind the term "fiber bundle". This hairbrush is like a fiber bundle in which the base space is a cylinder and the fibers (bristles) are line segments. The mapping π:E→B would take a point on any bristle and map it to its root on the cylinder.

a product space, but *globally* may have a different topological structure. Specifically, the similarity between a space E and a product space $B \times F$ is defined using a continuous surjective map

$$\pi \colon E \to B$$

that in small regions of E behaves just like a projection from corresponding regions of $B \times F$ to B. The map π, called the **projection** or **submersion** of the bundle, is regarded as part of the structure of the bundle. The space E is known as the **total space** of the fiber bundle, B as the **base space**, and F the **fiber**.

In the *trivial* case, E is just $B \times F$, and the map π is just the projection from the product space to the first factor. This is called a **trivial bundle**. Examples of non-trivial fiber bundles include the Möbius strip and Klein bottle, as well as nontrivial covering spaces. Fiber bundles such as the tangent bundle of a manifold and more general vector bundles play an important role in differential geometry and differential topology, as do principal bundles.

Mappings between total spaces of fiber bundles that "commute" with the projection maps are known as bundle maps, and the class of fiber bundles forms a category with respect to such mappings. A bundle map from the base space itself (with the identity mapping as projection) to E is called a section of E. Fiber bundles can be specialized in a number of ways, the most common of which is requiring that the transitions between the local trivial patches lie in a certain topological group, known as the **structure group**, acting on the fiber F.

38.1   Formal definition

A fiber bundle is a structure (E, B, π, F), where $E, B,$ and F are topological spaces and $\pi : E \to B$ is a continuous surjection satisfying a *local triviality* condition outlined below. The space B is called the **base space** of the bundle, E the **total space**, and F the **fiber**. The map π is called the **projection map** (or bundle projection). We shall assume in what follows that the base space B is connected.

We require that for every x in E, there is an open neighborhood $U \subset B$ of $\pi(x)$ (which will be called a trivializing neighborhood) such that there is a homeomorphism $\varphi \colon \pi^{-1}(U) \to U \times F$ (where $U \times F$ is the product space) in such a way that π agrees with the projection onto the first factor. That is, the following diagram should commute:

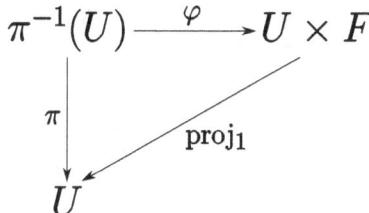

where $\text{proj}_1 : U \times F \to U$ is the natural projection and $\varphi : \pi^{-1}(U) \to U \times F$ is a homeomorphism. The set of all $\{(U_i, \varphi_i)\}$ is called a **local trivialization** of the bundle.

Thus for any p in B, the preimage $\pi^{-1}(\{p\})$ is homeomorphic to F (since $\text{proj}_1^{-1}(\{p\})$ clearly is) and is called the **fiber over p**. Every fiber bundle $\pi : E \to B$ is an open map, since projections of products are open maps. Therefore B carries the quotient topology determined by the map π.

A fiber bundle (E, B, π, F) is often denoted

$$F \longrightarrow E \xrightarrow{\ \pi\ } B$$

that, in analogy with a short exact sequence, indicates which space is the fiber, total space and base space, as well as the map from total to base space.

A **smooth fiber bundle** is a fiber bundle in the category of smooth manifolds. That is, $E, B,$ and F are required to be smooth manifolds and all the functions above are required to be smooth maps.

38.2   Examples

38.2.1 Trivial bundle

Let $E = B \times F$ and let $\pi : E \to B$ be the projection onto the first factor. Then E is a fiber bundle (of F) over B. Here E is not just locally a product but *globally* one. Any such fiber bundle is called a **trivial bundle**. Any fiber bundle over a contractible CW-complex is trivial.

38.2.2 Möbius strip

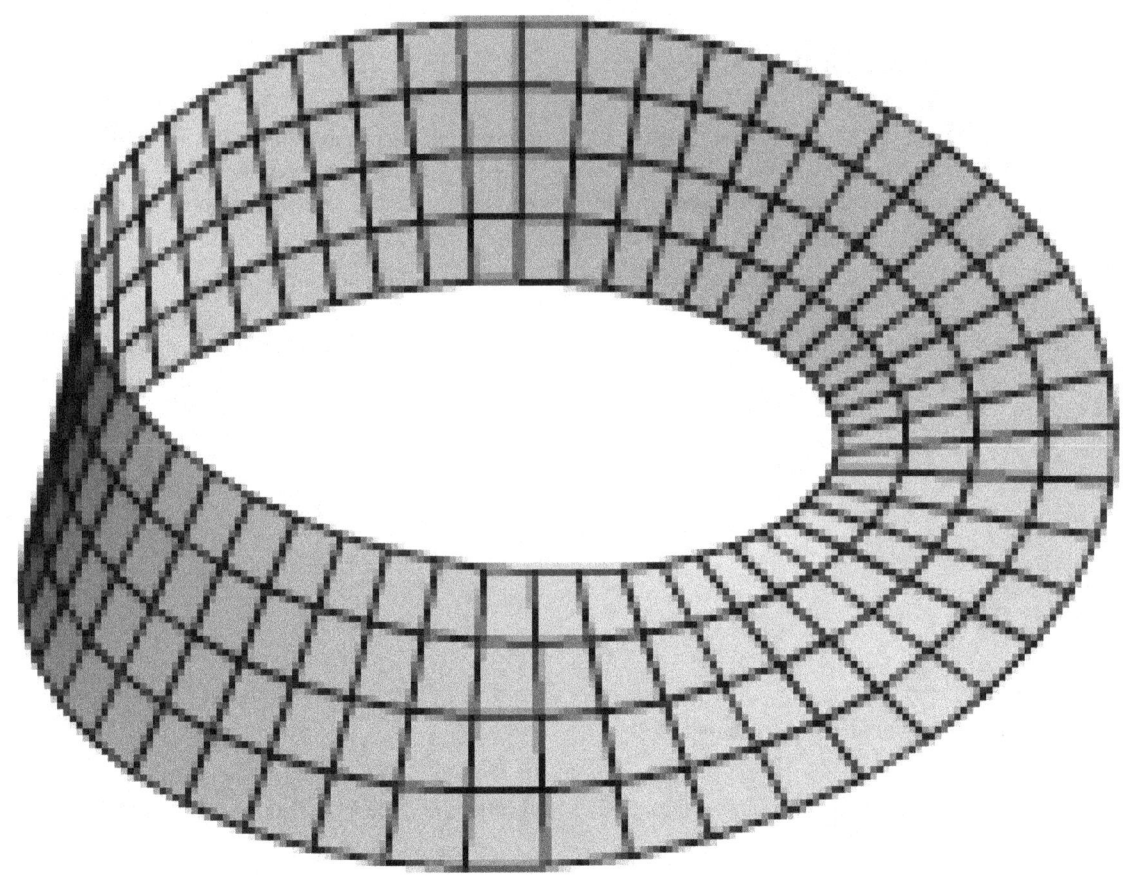

The Möbius strip is a nontrivial bundle over the circle.

Perhaps the simplest example of a nontrivial bundle E is the Möbius strip. It has the circle that runs lengthwise along the center of the strip as a base B and a line segment for the fiber F, so the Möbius strip is a bundle of the line segment over the circle. A neighborhood U of a point $x \in B$ is an arc; in the picture, this is the length of one of the squares. The preimage $\pi^{-1}(U)$ in the picture is a (somewhat twisted) slice of the strip four squares wide and one long. The homeomorphism φ maps the preimage of U to a slice of a cylinder: curved, but not twisted.

The corresponding trivial bundle $B \times F$ would be a cylinder, but the Möbius strip has an overall "twist". Note that this twist is visible only globally; locally the Möbius strip and the cylinder are identical (making a single vertical cut in either gives the same space).

38.2.3 Klein bottle

A similar nontrivial bundle is the Klein bottle which can be viewed as a "twisted" circle bundle over another circle. The corresponding non-twisted (trivial) bundle is the 2-torus, $S^1 \times S^1$.

38.2.4 Covering map

A **covering space** is a fiber bundle such that the bundle projection is a local homeomorphism. It follows that the fiber is a discrete space.

38.2.5 Vector and principal bundles

A special class of fiber bundles, called **vector bundles**, are those whose fibers are vector spaces (to qualify as a vector bundle the structure group of the bundle — see below — must be a linear group). Important examples of vector bundles include the tangent bundle and cotangent bundle of a smooth manifold. From any vector bundle, one can construct the frame bundle of bases which is a principal bundle (see below).

Another special class of fiber bundles, called **principal bundles**, are bundles on whose fibers a free and transitive action by a group G is given, so that each fiber is a principal homogeneous space. The bundle is often specified along with the group by referring to it as a principal G-bundle. The group G is also the structure group of the bundle. Given a representation ρ of G on a vector space V, a vector bundle with $\rho(G) \subseteq \text{Aut}(V)$ as a structure group may be constructed, known as the associated bundle.

38.2.6 Sphere bundles

A **sphere bundle** is a fiber bundle whose fiber is an n-sphere. Given a vector bundle E with a metric (such as the tangent bundle to a Riemannian manifold) one can construct the associated **unit sphere bundle**, for which the fiber over a point x is the set of all unit vectors in Ex. When the vector bundle in question is the tangent bundle T(M), the unit sphere bundle is known as the **unit tangent bundle**, and is denoted UT(M).

A sphere bundle is partially characterized by its Euler class, which is a degree n+1 cohomology class in the total space of the bundle. In the case n=1 the sphere bundle is called a circle bundle and the Euler class is equal to the first Chern class, which characterizes the topology of the bundle completely. For any n, given the Euler class of a bundle, one can calculate its cohomology using a long exact sequence called the Gysin sequence.

See also: Wang sequence

38.2.7 Mapping tori

If X is a topological space and $f:X \to X$ is a homeomorphism then the mapping torus Mf has a natural structure of a fiber bundle over the circle with fiber X. Mapping tori of homeomorphisms of surfaces are of particular importance in 3-manifold topology.

38.2.8 Quotient spaces

If G is a topological group and H is a closed subgroup, then under some circumstances, the quotient space G/H together with the quotient map $\pi : G \to G/H$ is a fiber bundle, whose fiber is the topological space H. A necessary and sufficient condition for $(G,G/H,\pi,H)$ to form a fiber bundle is that the mapping π admit local cross-sections (Steenrod 1951, §7).

The most general conditions under which the quotient map will admit local cross-sections are not known, although if G is a Lie group and H a closed subgroup (and thus a Lie subgroup by Cartan's theorem), then the quotient map is a fiber bundle. One example of this is the Hopf fibration, $S^3 \to S^2$ which is a fiber bundle over the sphere S^2 whose total space is S^3. From the perspective of Lie groups, S^3 can be identified with the special unitary group SU(2). The abelian subgroup of diagonal matrices is isomorphic to the circle group U(1), and the quotient SU(2)/U(1) is diffeomorphic to the sphere.

More generally, if G is any topological group and H a closed subgroup which also happens to be a Lie group, then $G \to G/H$ is a fiber bundle.

38.3 Sections

Main article: Section (fiber bundle)

A **section** (or **cross section**) of a fiber bundle π is a continuous map $f : B \to E$ such that $\pi(f(x))=x$ for all x in B. Since bundles do not in general have globally defined sections, one of the purposes of the theory is to account for their existence. The obstruction to the existence of a section can often be measured by a cohomology class, which leads to the theory of characteristic classes in algebraic topology.

The most well-known example is the hairy ball theorem, where the Euler class is the obstruction to the tangent bundle of the 2-sphere having a nowhere vanishing section.

Often one would like to define sections only locally (especially when global sections do not exist). A **local section** of a fiber bundle is a continuous map $f : U \to E$ where U is an open set in B and $\pi(f(x))=x$ for all x in U. If (U, φ) is a local trivialization chart then local sections always exist over U. Such sections are in 1-1 correspondence with continuous maps $U \to F$. Sections form a sheaf.

38.4 Structure groups and transition functions

Fiber bundles often come with a group of symmetries which describe the matching conditions between overlapping local trivialization charts. Specifically, let G be a topological group which acts continuously on the fiber space F on the left. We lose nothing if we require G to act effectively on F so that it may be thought of as a group of homeomorphisms of F. A **G-atlas** for the bundle (E, B, π, F) is a local trivialization such that for any two overlapping charts $(Ui, \varphi i)$ and $(Uj, \varphi j)$ the function

$$\varphi_i \varphi_j^{-1} : (U_i \cap U_j) \times F \to (U_i \cap U_j) \times F$$

is given by

$$\varphi_i \varphi_j^{-1}(x, \xi) = (x, t_{ij}(x)\xi)$$

where $tij : Ui \cap Uj \to G$ is a continuous map called a **transition function**. Two G-atlases are equivalent if their union is also a G-atlas. A **G-bundle** is a fiber bundle with an equivalence class of G-atlases. The group G is called the **structure group** of the bundle; the analogous term in physics is gauge group.

In the smooth category, a G-bundle is a smooth fiber bundle where G is a Lie group and the corresponding action on F is smooth and the transition functions are all smooth maps.

The transition functions tij satisfy the following conditions

1. $t_{ii}(x) = 1$
2. $t_{ij}(x) = t_{ji}(x)^{-1}$
3. $t_{ik}(x) = t_{ij}(x)t_{jk}(x)$.

The third condition applies on triple overlaps $Ui \cap Uj \cap Uk$ and is called the **cocycle condition** (see Čech cohomology). The importance of this is that the transition functions determine the fiber bundle (if one assumes the Čech cocycle condition).

A principal G-bundle is a G-bundle where the fiber F is a principal homogeneous space for the left action of G itself (equivalently, one can specify that the action of G on the fiber F is free and transitive). In this case, it is often a matter of convenience to identify F with G and so obtain a (right) action of G on the principal bundle.

38.5 Bundle maps

Main article: Bundle map

It is useful to have notions of a mapping between two fiber bundles. Suppose that M and N are base spaces, and $\pi E : E \to M$ and $\pi F : F \to N$ are fiber bundles over M and N, respectively. A bundle map (or **bundle morphism**) consists of a pair of continuous[1] functions

$$\varphi : E \to F, \quad f : M \to N$$

such that $\pi_F \circ \varphi = f \circ \pi_E$. That is, the following diagram commutes:

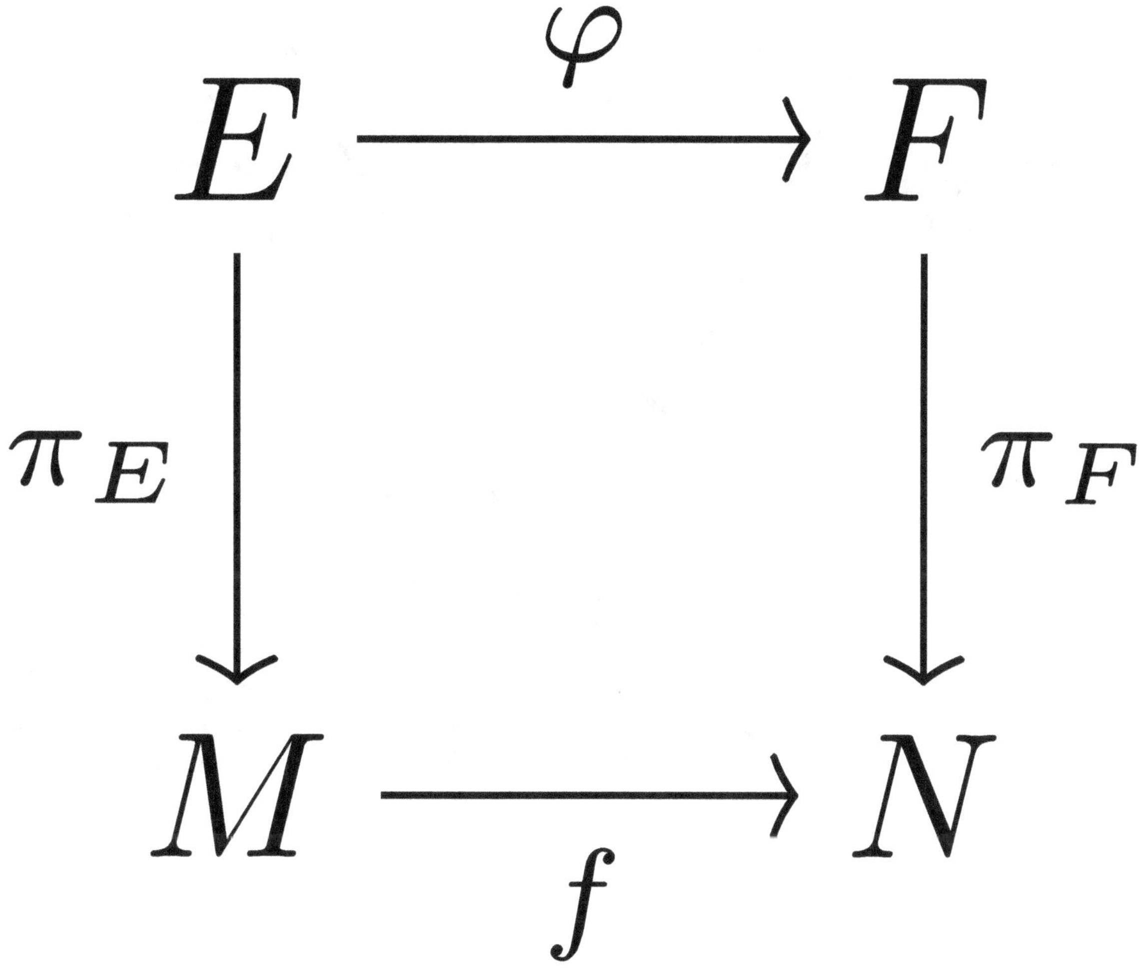

For fiber bundles with structure group G and whose total spaces are (right) G-spaces (such as a principal bundle), bundle morphisms are also required to be *G*-equivariant on the fibers. This means that $\varphi: E \to F$ is also *G*-morphism from one *G*-space to another, i.e., $\varphi(xs) = \varphi(x)s$ for all $x \in E$ and $s \in G$.

In case the base spaces *M* and *N* coincide, then a bundle morphism over *M* from the fiber bundle $\pi E: E \to M$ to $\pi F: F \to M$ is a map $\varphi: E \to F$ such that $\pi_E = \pi_F \circ \varphi$. This means that the bundle map $\varphi: E \to F$ covers the identity of *M*. That is, $f \equiv \mathrm{id}_M$ and the diagram commutes

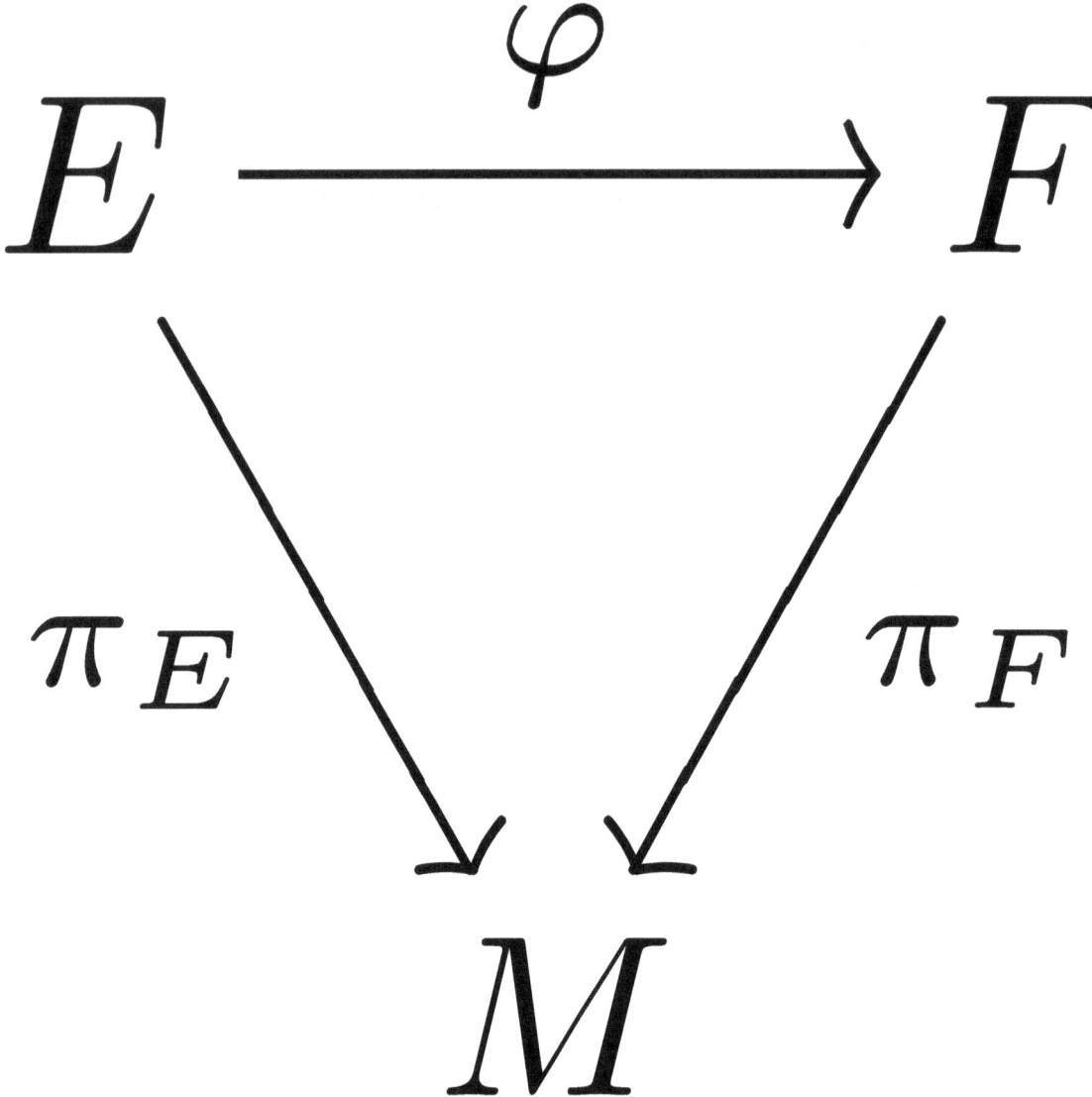

Assume that both $\pi E: E \to M$ and $\pi F: F \to M$ are defined over the same base space *M*. A bundle isomorphism is a bundle map (φ, f) between $\pi E: E \to M$ and $\pi F: F \to M$ such that $f \equiv \mathrm{id}_M$ and such that φ is also a homeomorphism.[2]

38.6 Differentiable fiber bundles

In the category of differentiable manifolds, fiber bundles arise naturally as submersions of one manifold to another. Not every (differentiable) submersion $f: M \to N$ from a differentiable manifold *M* to another differentiable manifold *N* gives rise to a differentiable fiber bundle. For one thing, the map must be surjective, and (M, N, f) is called a fibered manifold. However, this necessary condition is not quite sufficient, and there are a variety of sufficient conditions in common use.

If M and N are compact and connected, then any submersion $f : M \to N$ gives rise to a fiber bundle in the sense that there is a fiber space F diffeomorphic to each of the fibers such that $(E,B,\pi,F) = (M,N,f,F)$ is a fiber bundle. (Surjectivity of f follows by the assumptions already given in this case.) More generally, the assumption of compactness can be relaxed if the submersion $f : M \to N$ is assumed to be a surjective proper map, meaning that $f^{-1}(K)$ is compact for every compact subset K of N. Another sufficient condition, due to Ehresmann (1951), is that if $f : M \to N$ is a surjective submersion with M and N differentiable manifolds such that the preimage $f^{-1}\{x\}$ is compact and connected for all $x \in N$, then f admits a compatible fiber bundle structure (Michor 2008, §17).

38.7   Generalizations

- The notion of a bundle applies to many more categories in mathematics, at the expense of appropriately modifying the local triviality condition; cf. principal homogeneous space and torsor (algebraic geometry).

- In topology, a fibration is a mapping $\pi : E \to B$ which has certain homotopy-theoretic properties in common with fiber bundles. Specifically, under mild technical assumptions a fiber bundle always has the homotopy lifting property or homotopy covering property (see Steenrod (1951, 11.7) for details). This is the defining property of a fibration.

38.8   See also

38.9   Notes

[1] Depending on the category of spaces involved, the functions may be assumed to have properties other than continuity. For instance, in the category of differentiable manifolds, the functions are assumed to be smooth. In the category of algebraic varieties, they are regular morphisms.

[2] Or is, at least, invertible in the appropriate category; e.g., a diffeomorphism.

38.10   References

- Steenrod, Norman (1951), *The Topology of Fibre Bundles*, Princeton University Press, ISBN 0-691-08055-0

- Bleecker, David (1981), *Gauge Theory and Variational Principles*, Reading, Mass: Addison-Wesley publishing, ISBN 0-201-10096-7

- Ehresmann, C. "Les connexions infinitésimales dans un espace fibré différentiable". *Colloque de Topologie (Espaces fibrés), Bruxelles, 1950*. Georges Thone, Liège; Masson et Cie., Paris, 1951. pp. 29–55.

- Husemöller, Dale (1994), *Fibre Bundles*, Springer Verlag, ISBN 0-387-94087-1

- Michor, Peter W. (2008), *Topics in Differential Geometry*, Graduate Studies in Mathematics, Vol. 93, Providence: American Mathematical Society (*to appear*).

- Voitsekhovskii, M.I. (2001), "Fibre space", in Hazewinkel, Michiel, *Encyclopedia of Mathematics*, Springer, ISBN 978-1-55608-010-4

38.11   External links

- Fiber Bundle, PlanetMath

- Rowland, Todd, "Fiber Bundle", *MathWorld*.

- Making John Robinson's Symbolic Sculpture `Eternity'

- Sardanashvily, G., Fibre bundles, jet manifolds and Lagrangian theory. Lectures for theoreticians,arXiv: 0908.1886

38.12 Text and image sources, contributors, and licenses

38.12.1 Text

- **Gauge theory** *Source:* https://en.wikipedia.org/wiki/Gauge_theory?oldid=663144153 *Contributors:* The Anome, Michael Hardy, Tobias Bergemann, Ancheta Wis, TedPavlic, Xezbeth, MuDavid, Bender235, Pt, Phils, BD2412, Rjwilmsi, JocK, Modify, Teply, SmackBot, RDBury, Henning Makholm, Byelf2007, Michael C Price, Biblbroks, Headbomb, Nick Number, Fashionslide, VectorPosse, Magioladitis, Bakken, Email4mobile, JaGa, Policron, Squids and Chips, Cuzkatzimhut, VolkovBot, Red Act, Michael H 34, Setreset, Jwpitts, Tcamps42, Moonriddengirl, ClueBot, Mastertek, TimothyRias, XLinkBot, Addbot, Mortense, Eric Drexler, Bte99, Zorrobot, Luckas-bot, AnomieBOT, Christopher.Gordon3, Citation bot, Northryde, Xqbot, Pra1998, Gsard, A. di M., Erik9bot, FrescoBot, Fortdj33, Citation bot 1, Ganondolf, RedBot, RobinK, Mary at CERN, EmausBot, Brent Perreault, Slawekb, Cogiati, Maschen, Isocliff, ClueBot NG, Helpful Pixie Bot, Bibcode Bot, Dzustin, Brendan.Oz, ChrisGualtieri, SD5bot, Dexbot, Enyokoyama, Joeinwiki, Dath Thou Even Lift, Dhm4444, Dbw1976, KasparBot and Anonymous: 40

- **Introduction to gauge theory** *Source:* https://en.wikipedia.org/wiki/Introduction_to_gauge_theory?oldid=667466051 *Contributors:* Zundark, The Anome, Andre Engels, Boud, Michael Hardy, Isomorphic, Dante Alighieri, TakuyaMurata, Karada, SebastianHelm, CesarB, Ahoerstemeier, AugPi, Charles Matthews, Dysprosia, Zoicon5, Phys, Bevo, Robbot, Sverdrup, Ancheta Wis, Giftlite, Harp, BenFrantzDale, Lethe, Anville, Waltpohl, CryptoDerk, Lockeownzj00, AmarChandra, Lumidek, Mormegil, Pak21, Gianluigi, Mal~enwiki, MuDavid, Paul August, Gauge, Pt, El C, Lycurgus, Laurascudder, CDN99, Reinyday, Helix84, Pearle, Dachannien, Axl, Mac Davis, Count Iblis, Linas, Mindmatrix, Archie Paulson, Jfrancis, Mpatel, Isnow, Marudubshinki, Mandarax, Jmhodges, Pdelong, MarSch, R.e.b., Mathbot, Alfred Centauri, BjKa, Srleffler, Borgx, Bambaiah, RussBot, Bergsten, SpuriousQ, Archelon, Salsb, Rick Norwood, Buster79, Kymara, Light current, Nikkimaria, Netrapt, SmackBot, Tom Lougheed, Slashme, Melchoir, Bluebot, Complexica, Nbarth, Acipsen, QFT, Serenity-Fr, Mihovil, Akriasas, SS2005, Terry Bollinger, WhiteHatLurker, Mathsci, Igoldste, Vrkaul, Menswear, Pathosbot, Chetvorno, Harold f, Domanix, Vyznev Xnebara, Equendil, Cydebot, Xxanthippe, Michael C Price, Quibik, Mbell, 271828182, Headbomb, Escarbot, WinBot, Fashionslide, Shlomi Hillel, JAnDbot, 100110100, Arvindra, Smartcat, Bakken, Jpod2, Michael K. Edwards, Arturj, HEL, J.delanoy, Gotyear, TomyDuby, Tarotcards, STBotD, Useight, Lseixas, Sheliak, The Duke of Waltham, Butwhatdoiknow, HowardFrampton, Impunv, Mannafredo, Geometry guy, Lejarrag, Yartsa, YonaBot, Gerakibot, Enrico Poli, Wing gundam, Paolo.dL, Thehotelambush, Hamiltondaniel, StewartMH, ClueBot, General Epitaph, Rockfang, Masterpiece2000, DragonBot, CohesionBot, JavierReynaldo, Brews ohare, Hanzo 88, Dekisugi, Limitcycle, Qwfp, TimothyRias, YouRang?, Cmfuen, Mjamja, TStein, CosmiCarl, Lightbot, Cesiumfrog, מלמד כץ, Yobot, JohnHarold, Ht686rg90, Jskline, Matthil, AnomieBOT, Palpher, Citation bot, Bci2, Pra1998, Omnipaedista, Point-set topologist, Ace Antoni, A. di M., AllCluesKey, Green0eggs, Sławomir Biały, Steve Quinn, Tehminkeh, Amyamj, Rausch, Jordgette, GoingBatty, Addihockey10 (automated), Shedoblyde, Enyokoyama, Andyhowlett, Rudrene and Anonymous: 111

- **Quantum gauge theory** *Source:* https://en.wikipedia.org/wiki/Quantum_gauge_theory?oldid=666286220 *Contributors:* Karada, Charles Matthews, Grendelkhan, Phys, MuDavid, Pt, Jag123, David Haslam, Mpatel, BD2412, Conscious, SmackBot, Jeepday, Sheliak, Fuddle, Omnipaedista, Erik9bot, Hep thinker, Yellowweasel, Cogiati, SJ Defender and Anonymous: 3

- **Quantum field theory** *Source:* https://en.wikipedia.org/wiki/Quantum_field_theory?oldid=671846854 *Contributors:* AxelBoldt, CYD, Mav, The Anome, XJaM, Roadrunner, Stevertigo, Michael Hardy, Tim Starling, IZAK, TakuyaMurata, SebastianHelm, Looxix~enwiki, Ahoerstemeier, Cyp, Glenn, Rotem Dan, Stupidmoron, Charles Matthews, Timwi, Jitse Niesen, Kbk, Rudminjd, Wik, Phys, Bevo, BenRG, Northgrove, Robbot, Bkalafut, Gandalf61, Rursus, Fuelbottle, Tobias Bergemann, Ancheta Wis, Giftlite, Lethe, Dratman, Alison, St3vo, Mboverload, DefLog~enwiki, ConradPino, Amarvc, Pcarbonn, Karol Langner, APH, AmarChandra, D6, CALR, Urvabara, Discospinster, Guanabot, Igorivanov~enwiki, Masudr, Pjacobi, Vsmith, Nvj, MuDavid, Bender235, Pt, El C, Shanes, Sietse Snel, Physicistjedi, KarlHallowell, PWilkinson, Helix84, Thialfi, Varuna, Gcbirzan, Docboat, Count Iblis, Egg, Mpatel, Marudubshinki, Graham87, Opie, Vanderdecken, Rjwilmsi, MarSch, Earin, R.e.b., RE, Strobilomyces, Arnero, Itinerant1, Alfred Centauri, Srleffler, Chobot, UkPaolo, Wavelength, Bambaiah, Hairy Dude, RussBot, TimNelson, Archelon, CambridgeBayWeather, SCZenz, Odddmonster, E2mb0t~enwiki, Semperf, Tetracube, Garion96, Erik J, Robert L, Banus, RG2, SmackBot, Stephan Schneider, Tom Lougheed, Melchoir, KocjoBot~enwiki, Mcld, Dauto, Chris the speller, Complexica, Threepounds, RuudVisser, QFT, Jmnbatista, Cybercobra, Rebooted, Victor Eremita, DJIndica, Lambiam, Mgiganteus1, Zarniwoot, Jim.belk, Stwalkerster, SirFozzie, Hu12, Dan Gluck, Iridescent, Joseph Solis in Australia, Albertod4, Van helsing, BeenAroundAWhile, Witten Is God, Cydebot, Jamie Lokier, Meno25, Michael C Price, The 80s chick, Mendicus~enwiki, AstroPig7, Msebast~enwiki, Mbell, Headbomb, Nick Number, Mentifisto, AntiVandalBot, Bt414, Bananan~enwiki, Martin Kostner, Moltrix, Kasimann, Kromatol, Puksik, Lerman, LLHolm, RogueNinja, Tlabshier, JEH, Nikolas Karalis, Storkk, JAnDbot, Igodard, Four Dog Night, N shaji, Bongwarrior, Andrea Allais, Soulbot, Etale, Maliz, Custos0, HEL, J.delanoy, Acalamari, Jeepday, Policron, Blckavnger, Juliancolton, Skou, Telecomtom, GrahamHardy, Sheliak, Cuzkatzimhut, VolkovBot, Bktennis2006, Marksr, HowardFrampton, The Original Wildbear, Dj thegreat, Markisgreen, TBond, Lejarrag, Moose-32, Raphtee, Sue Rangell, Neparis, Drschawrz, YohanN7, SieBot, TCO, Yintan, Likebox, Paolo.dL, Tugjob, Henry Delforn (old), Jecht (Final Fantasy X), OKBot, StewartMH, ClueBot, EoGuy, Wwheaton, The Wild West guy, Shvav~enwiki, Bob108, Brews ohare, Thingg, Count Truthstein, XLinkBot, PSimeon, SilvonenBot, Truthnlove, HexaChord, Addbot, ConCompS, Pinkgoanna, Leapold~enwiki, Dmhowarth26, Glane23, Hanish.polavarapu, Lightbot, Scientryst, R.ductor, Ettrig, Yndurain, Legobot, Luckas-bot, Yobot, Ht686rg90, Niout, Tamtamar, AnomieBOT, Ciphers, Palpher, IRP, Gjsreejith, Materialscientist, Citation bot, Bci2, ArthurBot, Northryde, LilHelpa, Caracolillo, Amareto2, MIRROR, Professor J Lawrence, Plasmon1248, Omnipaedista, RibotBOT, Spellage, JayJay, FrescoBot, Kenneth Dawson, D'ohBot, Knowandgive, N4tur4le, Hyqeom, Newt Winkler, Hickorybark, Lotje, Dinamik-bot, LilyKitty, Fortesque666, Reaper Eternal, Minimac, Marie Poise, Yaush, Dylan1946, EmausBot, Racerx11, GoingBatty, Carbosi, Thecheesykid, ZéroBot, Cogiati, Jjspinorfield1, Suslindisambiguator, Quondum, Maschen, Zueignung, Davidaedwards, Lom Konkreta, ClueBot NG, Gilderien, Iloveandrea, Vacation9, Heyheyheyhohoho, Fortune432, The ubik, Zak.estrada, Widr, Helpful Pixie Bot, Evanescent7, Ykentluo, Martin.uecker, Walterpfeifer, Pfeiferwalter, Klilidiplomus, W.D., CarrieVS, Khazar2, Momo1381, Dexbot, Cerabot~enwiki, Garuda0001, AHusain314, Thepalerider2012, A.entropy, Mark viking, Faizan, Aj7s6, संजीव कुमार, Lemnaminor, BerFinelli, Axel.P.Hedstrom, Kclongstocking, Mutley1989, I art a troler, Liquidityinsta, Prokaryotes, DemonThuum, Dingdong2680, Asherkirschbaum, Monkbot, Gjbayes, Thedarkcheese, BradNorton1979, UareNumber6, Teelaskeletor, YeOldeGentleman, Mret81, KasparBot and Anonymous: 293

- **Field (physics)** *Source:* https://en.wikipedia.org/wiki/Field_(physics)?oldid=672213730 *Contributors:* Patrick, Michael Hardy, Dcljr, Angela, Andres, Wooster, Charles Matthews, Dino, Reddi, Phys, Cncs wikipedia, Robbot, Ojigiri~enwiki, Wikibot, Fuelbottle, Ancheta Wis, Giftlite,

Waltpohl, LucasVB, Antandrus, Karol Langner, AmarChandra, Mschlindwein, Starfoxy, Lucidish, CALR, Laoma, Masudr, YUL89YYZ, Kbh3rd, El C, Laurascudder, Army1987, Guiltyspark, Varuna, Bobrg~enwiki, MIT Trekkie, Linas, Natcase, Polyparadigm, Dodiad, Mpatel, Ketiltrout, Tbone, Nihiltres, Borgx, RobotE, Bambaiah, Stephenb, NawlinWiki, Albedo, Beanyk, Epipelagic, Arthur Rubin, Sbyrnes321, SmackBot, TheLeopard, Sbharris, Colonies Chris, Hongooi, Sergio.ballestrero, Vegard, John, JHunterJ, Mets501, Treyp, Trevor.tombe, Chmee2, Cydebot, WISo, Soetermans, Skittleys, Thijs!bot, Epbr123, Headbomb, Dalahäst, MichaelMaggs, JBouwman, JAnDbot, Husond, Jpod2, Ed!, Hdt83, R'n'B, Fconaway, HEL, Kimse, Metamusing, Maurice Carbonaro, Lamp90, BernardZ, Squids and Chips, VolkovBot, TXiKiBoT, Rei-bot, Thomas.schick, SieBot, OKBot, Laurentseries, Anchor Link Bot, ClueBot, Mild Bill Hiccup, Djr32, Brews ohare, Schreiber-Bike, PCHS-NJROTC, TimothyRias, Heinsaar, Mhsb, Truthnlove, Addbot, Fgnievinski, Bte99, Xgambler, Luckas-bot, Yobot, AnomieBOT, Palpher, Ulric1313, Citation bot, ArthurBot, Xqbot, Calcio33, J04n, Topherwhelan, Gsard, WaysToEscape, LucienBOT, Citation bot 1, Rapsar, Pinethicket, Loudubewe, Tom.Reding, TobeBot, Jfmantis, RjwilmsiBot, Ripchip Bot, EmausBot, Jjspinorfield1, Chrisman62, Maschen, RockMagnetist, ClueBot NG, Gilderien, Admock, Helpful Pixie Bot, Shivsagardharam, BG19bot, Jcdericco, Uioplk, Ema--or, Hmainsbot1, Katterjohn, DavidLeighEllis, Noyster, BHBrunt, Peterfreed, Mpcalkins, Isambard Kingdom, KasparBot, Crosleybendix and Anonymous: 84

- **Vector field** *Source:* https://en.wikipedia.org/wiki/Vector_field?oldid=672100944 *Contributors:* AxelBoldt, Chato, Patrick, Chas zzz brown, Michael Hardy, Tim Starling, Wshun, TakuyaMurata, Cyp, Stevenj, Andres, Charles Matthews, Reddi, Sbwoodside, Dysprosia, Jitse Niesen, Maximus Rex, Fibonacci, Phys, Jaredwf, MathMartin, Idoneus~enwiki, Tosha, Giftlite, BenFrantzDale, Ævar Arnfjörð Bjarmason, DefLog~enwiki, LiDaobing, MFNickster, Hellisp, JohnArmagh, Zowie, Klaas van Aarsen, Rich Farmbrough, ReiVaX, Mdd, Wendell, Oleg Alexandrov, Woohookitty, Linas, Jacobolus, Rjwilmsi, MarSch, HannsEwald, Salix alba, Dergrosse, Mo-Al, FlaBot, Margosbot~enwiki, Alfred Centauri, Chobot, 121a0012, WriterHound, YurikBot, Archelon, Buster79, Mgnbar, Darrel francis, Sbyrnes321, SmackBot, RDBury, Rex the first, Pokipsy76, Silly rabbit, Nbarth, DHN-bot~enwiki, Regford, Daqu, Pen of bushido, Andrei Stroe, Cronholm144, Jim.belk, Dwmalone, FelisSchrödingeris, Thijs!bot, KlausN~enwiki, JAnDbot, Rivertorch, Catgut, Sullivan.t.j, User A1, Martynas Patasius, JaGa, Rickard Vogelberg, R'n'B, TomyDuby, Policron, HyDeckar, Jaimeastorga2000, VolkovBot, LokiClock, Julian I Do Stuff, TXiKiBoT, A4bot, Anonymous Dissident, Michael H 34, Geometry guy, Antixt, SieBot, Soler97, Paolo.dL, JackSchmidt, OKBot, 7&6=thirteen, Wikidsp, Addbot, Fgnievinski, Topology Expert, EconoPhysicist, AndersBot, Mattmatt79, Jasper Deng, Zorrobot, Jarble, Snaily, Legobot, Luckas-bot, Naudefjbot~enwiki, AnomieBOT, Ciphers, ArthurBot, Titi2~enwiki, Point-set topologist, FrescoBot, Lookang, Sławomir Biały, Lost-n-translation, Foobarnix, Tcnuk, Rausch, EmausBot, Fly by Night, Slawekb, Hhhippo, ZéroBot, Qniemiec, Wikfr, Glosser.ca, Sp4cetiger, Helpsome, Wcherowi, Clearlyfakeusername, Snotbot, Mesoderm, Vinícius Machado Vogt, Helpful Pixie Bot, Shivsagardharam, Nawk, MusikAnimal, Cispyre, F=q(E+v^B), ChrisGualtieri, Oxherdn, Creepsevry1out and Anonymous: 76

- **Generating set of a group** *Source:* https://en.wikipedia.org/wiki/Generating_set_of_a_group?oldid=663569582 *Contributors:* AxelBoldt, Zundark, Tomo, Chas zzz brown, Michael Hardy, Chinju, Emperorbma, Charles Matthews, Dcoetzee, Dysprosia, Fibonacci, Romanm, Tobias Bergemann, Giftlite, Dbenbenn, Herbee, Tomruen, ArnoldReinhold, BD2412, Mathbot, YurikBot, Michael Slone, Lenthe, Eyal0, Eskimbot, Mhss, J. Finkelstein, Vp loreta, CRGreathouse, Dr.enh, RobHar, DorganBot, TXiKiBoT, Optimisteo, Flyer22, JackSchmidt, ClueBot, Watchduck, Bender2k14, MystBot, Addbot, VladimirReshetnikov, Artem M. Pelenitsyn, HRoestBot, EmausBot, ZéroBot, Wcherowi and Anonymous: 20

- **Quantum** *Source:* https://en.wikipedia.org/wiki/Quantum?oldid=667837208 *Contributors:* The Anome, Stevertigo, Ahoerstemeier, Smack, RodC, Charles Matthews, Jitse Niesen, Topbanana, Robbot, Sverdrup, Academic Challenger, Pengo, Graeme Bartlett, Lethe, Alison, Bensaccount, Finn-Zoltan, Jaan513, Smartcowboy, Raylu, Tsemii, Andreas Kaufmann, Freakofnurture, EugeneZelenko, Masudr, Vsmith, Too Old, El C, Laurascudder, RoyBoy, Omoo, Truthflux, Marco Polo, John Vandenberg, Viriditas, Dungodung, Hujaza, Jag123, Sam Korn, Nsaa, Anthony Appleyard, Atlant, Andrewpmk, Ricky81682, BryanD, Spangineer, Wtmitchell, Dalillama, Deathphoenix, Ceyockey, JHolman, Yougotavirus, Ashmoo, Bilbo1507, Qwertyus, Dennis Estenson II, Platypus222, Scorpionman, FlaBot, Nihiltres, Jeff02, Whodunit, Fresheneesz, WikiWikiPhil, Salvatore Ingala, DaGizza, Roboto de Ajvol, YurikBot, Mhocker, 4C~enwiki, Qwertzy2, Wimt, NawlinWiki, DragonHawk, Vanished user 1029384756, Mikeblas, Hosterweis, Orthografer, Geoffrey.landis, Garion96, Mebden, RG2, SmackBot, Melchoir, Vald, Bluebot, Complexica, Bbq332, Jfsamper, Scwlong, Stevenmitchell, Gragox, LoveEncounterFlow, Sadi Carnot, Fjjf, Mental Blank, Chymicus, SashatoBot, JorisvS, Bjankuloski06en~enwiki, Andypandy.UK, TastyPoutine, Laplace's Demon, Tawkerbot2, Filelakeshoe, Bupper, Wafulz, Mariodivece, GavinMorley, Dragon's Blood, Bicala, Xxanthippe, Thijs!bot, Epbr123, Davidhorman, Nick Number, Oreo Priest, AntiVandalBot, Drakonicon, Fdmt, JAnDbot, Db099221, Burakburak, Bongwarrior, VoABot II, Kinston eagle, Soulbot, Zamb, Crunchy Numbers, MartinBot, AlexiusHoratius, Juventas, C. Trifle, Maurice Carbonaro, NerdyNSK, Acalamari, Openforbusiness, Stewartrfc, Ontarioboy, Pundit, Pdcook, Leebo, DavidBrahm, TXiKiBoT, Djkrajnik, Kilmer-san, Yungjui, Synthebot, Lova Falk, Falcon8765, Grinq, HiDrNick, SieBot, Dannyeder, BotMultichill, Pengyanan, JerrySteal, Tiptoety, Oxymoron83, OKBot, Ngexpert5, Francvs, Seanruiz, FlamingSilmaril, Gratedparmesan, ClueBot, Abhinav, Alexbot, KnowledgeBased, Quantumpundit, SilvonenBot, NHJG, Truthnlove, NCDane, Addbot, War sharks, Favonian, Tide rolls, Lightbot, Deasmumhain, Zorrobot, Luckas-bot, Yobot, Nutfortuna, Julia W, Amirobot, Vortico, AnomieBOT, DemocraticLuntz, Greenbreen, AdjustShift, Jalexsmith1991, Materialscientist, Citation bot, ArthurBot, Parkyere, Capricorn42, Programming gecko, GrouchoBot, ProtectionTaggingBot, LtBert44, Nacefe, FrescoBot, Citation bot 1, Guruspiritual, Jsjunkie, Robo Cop, Lightlowemon, FoxBot, ಚೇ೦, Vrenator, Darsie42, Raidon Kane, Orphan Wiki, Booknotes, Wikipelli, K.zaman1710, Loggin12354, Grapeguy7, H3llBot, Makecat, Wayne Slam, Flightx52, Harishng, Kurt hueston, DASHBotAV, ClueBot NG, Dr Miles Long, Joefromrandb, Gexmeansgecko, Calabe1992, Bibcode Bot, Iisthphir, J991, Ajmah 200, Achowat, Vishakh24, BrightStarSky, Vogone, Reatlas, Ginsuloft, Monkbot, Jrafner, Farhan babra, Lllll2009, QuantaRaj39, Shubhamlanje1, KasparBot and Anonymous: 199

- **Lagrangian** *Source:* https://en.wikipedia.org/wiki/Lagrangian?oldid=672181273 *Contributors:* CYD, Zundark, The Anome, Tarquin, Awaterl, Andre Engels, Roadrunner, Peterlin~enwiki, Stevertigo, Michael Hardy, Anders Feder, AugPi, Andres, Schneelocke, Charles Matthews, Dysprosia, Rednblu, Patrick0Moran, Cameronc, Phys, Bevo, Robbot, Pps, Aetheling, Rho~enwiki, Jheise, Ahadley, Giftlite, BenFrantzDale, Tom harrison, Art Carlson, Wwoods, Dratman, FrYGuY, Jason Quinn, DefLog~enwiki, Zeimusu, Karol Langner, Balcer, AmarChandra, Zowie, Jim Fraser, Masudr, Mal~enwiki, Bender235, ChristophDemmer, Laurascudder, Army1987, Wisdom89, Matt McIrvin, Physicistjedi, Ardric47, Jérôme, Diego Moya, Wtmitchell, Tbsmith, Linas, Ae-a, Mpatel, SeventyThree, Salix alba, Mathbot, Srleffler, Chobot, ChrisChiasson, YurikBot, RussBot, RL0919, Voidxor, E2mb0t~enwiki, Larsobrien, Arthur Rubin, Reyk, RG2, Zvika, Eigenlambda, SmackBot, Melchoir, Unyoyega, Papa November, Complexica, Hongooi, Berland, Twalton, Rajkishan211990, Jgates, DJIndica, Lambiam, WhiteHatLurker, Makyen, Dr.K., JRSpriggs, NormHardy, Cydebot, Xxanthippe, Dr.enh, Dchristle, Hugozam, Mbell, LeBofSportif, JAnDbot, MER-C, Magioladitis, WolfmanSF, Jpod2, User A1, Sprevrha, R'n'B, Lilac Soul, DrKiernan, Maurice Carbonaro, KIAaze, Tarotcards, Plasticup, Lseixas, Red Act, Michael H 34, Geometry guy, Lejarrag, Ketyner, Antixt, Austinstkong, YohanN7, SieBot, Gerakibot, MikeGogulski, Tugjob, Correogsk,

MenoBot, Copyeditor42, SchreiberBike, 1ForTheMoney, Crowsnest, Addbot, CUSENZA Mario, Erik Streb, Luckas-bot, Yobot, AnomieBOT, Xqbot, Gsard, FrescoBot, Craig Pemberton, Biniamin, Kiefer.Wolfowitz, 124Nick, Puzl bustr, Fermat618, Amirhdrz 91, EmausBot, Rami radwan, GoingBatty, HolyCookie, Quondum, Zueignung, Davidaedwards, ChuispastonBot, Ebehn, Gilderien, Diogenes2000, Helpful Pixie Bot, CedricMC, Bart van den Broek, Odysei, F=q(E+v^B), Pratyush Sarkar, Andyhowlett, Saloon.cat, Pdecalculus, Msurajit, MutluMan, Lý Minh Nhật, Ndeine, Susilehtola, Cap'n Squid, Anthul, Differential 0celo7, TCAllen07 and Anonymous: 125

- **Invariant (physics)** *Source:* https://en.wikipedia.org/wiki/Invariant_(physics)?oldid=665432316 *Contributors:* Charles Matthews, Ancheta Wis, Karol Langner, Sam Hocevar, Rgdboer, Keenan Pepper, Dirkb, Loxley~enwiki, LOL, BD2412, Pnrj, SamuelRiv, Sbharris, Javalenok, Timdream, Generalcp702, Btate, Cydebot, Headbomb, Lantonov, Tbhartman, Thurth, StewartMH, Djr32, Sergey kudryavtsev, Sk8ter395425, Crowsnest, Quidproquo2004, D.M. from Ukraine, Addbot, Vasil', Luckas-bot, Yobot, AnomieBOT, Palpher, FoxBot, Nilock, EmausBot, ZéroBot, RockMagnetist, Helpful Pixie Bot, Faus, Pdecalculus, Prokaryotes, Jwratner1, Fench and Anonymous: 13

- **Lie algebra** *Source:* https://en.wikipedia.org/wiki/Lie_algebra?oldid=668578572 *Contributors:* AxelBoldt, Zundark, Miguel~enwiki, Michael Hardy, Wshun, Joel Koerwer, TakuyaMurata, Suisui, Kragen, Rossami, Iorsh, Loren Rosen, Charles Matthews, Dysprosia, Michael Larsen, Grendelkhan, Phys, Tobias Bergemann, David Gerard, Weialawaga~enwiki, Tosha, Giftlite, BenFrantzDale, Lethe, Fropuff, Curps, Jeremy Henty, Jason Quinn, Python eggs, Chameleon, DefLog~enwiki, CryptoDerk, CSTAR, Pyrop, Guanabot, Pj.de.bruin, Vsmith, Gauge, Pt, Kwamikagami, Wood Thrush, Reinyday, Foobaz, Msh210, Arthena, Spangineer, Dirac1933, Drbreznjev, Oleg Alexandrov, Linas, Isnow, BD2412, NatusRoma, MarSch, Mathbot, Margosbot~enwiki, RexNL, Masnevets, YurikBot, Wavelength, Hairy Dude, Michael Slone, Lenthe, Stephenb, Grubber, Trovatore, Asimy, Crasshopper, Curpsbot-unicodify, Sbyrnes321, SmackBot, Incnis Mrsi, Grokmoo, Kmarinas86, Bluebot, Silly rabbit, Nbarth, Thomas Bliem, Chlewbot, BlackFingolfin, Noegenesis, Rschwieb, AlainD, Harold f, CmdrObot, Shirulashem, Headbomb, Second Quantization, Dachande, RobHar, B-80, Jrw@pobox.com, Deflective, Englebert, Vanish2, R'n'B, Bogey97, Maurice Carbonaro, Supermanifold, Policron, Fylwind, Cuzkatzimhut, VolkovBot, JohnBlackburne, LokiClock, Ndbrian1, Hesam7, Geometry guy, Drorata, Arcfrk, StevenJohnston, YohanN7, SieBot, Stca74, Jenny Lam, Paolo.dL, JackSchmidt, Mr. Stradivarius, Fatchat, Veromies, JP.Martin-Flatin, Count Truthstein, Addbot, Roentgenium111, Lightbot, Legobot, Luckas-bot, Yobot, Niout, Jason Recliner, Esq., Delilahblue, AnomieBOT, Twri, SassoBot, Kaoru Itou, D'ohBot, Darij, Juniuswikiae, Prtmrz, Rausch, Jkock, Adam cohenus, TobeBot, Lotje, Doctor Zook, Slawekb, Quondum, Mikhail Ryazanov, ClueBot NG, Dd314, Teika kazura, Walterpfeifer, Pfeiferwalter, IkamusumeFan, Flbsimas, Deltahedron, Saung Tadashi, Mark L MacDonald, Danielbrice, Enyokoyama, CsDix, 314Username, Forgetfulfunctor00, CaptainLama, KasparBot, Texnico and Anonymous: 91

- **Electromagnetic four-potential** *Source:* https://en.wikipedia.org/wiki/Electromagnetic_four-potential?oldid=662366388 *Contributors:* The Anome, Stevan White, Charles Matthews, Ancheta Wis, Giftlite, Elroch, Icairns, Markalex, L-H, NeilTarrant, Laurascudder, Linuxlad, Mpatel, Sbyrnes321, That Guy, From That Show!, SmackBot, Stepa, Tpellman, Dauto, Ligulembot, Dicklyon, Comech, JRSpriggs, Pph~enwiki, LAUBO, Cydebot, Dr.enh, Headbomb, Escarbot, Pervect, Dekimasu, Lseixas, Antixt, Enthusiastic Student, BOTarate, MystBot, Addbot, Мыша, TStein, Luckas-bot, Citation bot, GrouchoBot, FoxBot, Jordgette, EmausBot, ZéroBot, Quondum, Maschen, Helpful Pixie Bot, Steve86au, F=q(E+v^B), AHusain314, SpecialPiggy and Anonymous: 17

- **Gauge boson** *Source:* https://en.wikipedia.org/wiki/Gauge_boson?oldid=662605501 *Contributors:* Bryan Derksen, Andre Engels, Michael Hardy, Ahoerstemeier, Bueller 007, LouI, Phys, Robbot, Gwrede, Rholton, Rursus, Davidl9999, Giftlite, Xerxes314, Alison, JeffBobFrank, Chinasaur, Andris, Garth 187, Beland, Setokaiba, Icairns, AmarChandra, Lumidek, Vsmith, Roybb95~enwiki, Mal~enwiki, La goutte de pluie, Nk, Kusma, Ringbang, Mpatel, Nakos2208~enwiki, Tevatron~enwiki, Kbdank71, Chobot, Roboto de Ajvol, Hairy Dude, Salsb, StuRat, ArielGold, RG2, InverseHypercube, Niels Olson, Sadi Carnot, TriTertButoxy, Ekjon Lok, Bjankuloski06en~enwiki, Phatom87, Headbomb, Tyco.skinner, Knotwork, Swpb, Maurice Carbonaro, Gombang, TXiKiBoT, Odellus, Antixt, AlleborgoBot, SieBot, Jim E. Black, Homonihilis, BOTarate, DumZiBoT, SilvonenBot, Addbot, Bertman600, NjardarBot, Numbo3-bot, Lightbot, Zorrobot, Luckas-bot, Yobot, Citation bot, ArthurBot, A. di M., Rameshngbot, RedBot, RobinK, Mary at CERN, TjBot, EmausBot, ZéroBot, StringTheory11, Mentibot, Dsperlich, CeraBot, Galactic Messiah, DerekWinters, Fisherv, KasparBot and Anonymous: 41

- **Elementary particle** *Source:* https://en.wikipedia.org/wiki/Elementary_particle?oldid=671875382 *Contributors:* CYD, Mav, Bryan Derksen, XJaM, Heron, Stevertigo, Patrick, Fbjon, Looxix~enwiki, Александър, Julesd, Glenn, AugPi, Mxn, Timwi, Reddi, Tpbradbury, Furrykef, Bevo, Donarreiskoffer, Robbot, Craig Stuntz, Nurg, Papadopc, Wikibot, Jimduck, Anthony, Ancheta Wis, Giftlite, DavidCary, Mikez, Haselhurst, Monedula, Xerxes314, Alison, Guanaco, Greydream, Anythingyouwant, Bodnotbod, Kate, Brianjd, Mormegil, Urvabara, Rich Farmbrough, Guanabot, Qutezuce, Hidaspal, Dmr2, Goplat, RJHall, RoyBoy, Robotje, Neonumbers, ליאור, Dirac1933, DV8 2XL, Azmaverick623, Blaxthos, Kay Dekker, Joriki, Simetrical, TomTheHand, Mpatel, Isnow, Ggonnell, Palica, Strait, Miserlou, Ligulem, Naraht, DannyWilde, Lmatt, Srleffler, Chobot, Cactus.man, Roboto de Ajvol, YurikBot, Hairy Dude, NTBot~enwiki, Ohwilleke, Albert Einsteins pipe, Stephenb, Chaos, Vibritannia, SCZenz, Edwardlalone, Larsobrien, Bota47, BraneJ, Dna-webmaster, Arthur Rubin, Oyvind, GrinBot~enwiki, SmackBot, Mrcoolbp, Bomac, GrGBL~enwiki, Chris the speller, MalafayaBot, George Rodney Maruri Game, Silly rabbit, Complexica, MovGP0, Fmalan, Scwlong, Amazins490, Cybercobra, EPM, Garry Denke, Drphilharmonic, Sadi Carnot, ArglebargleIV, Tktktk, NongBot~enwiki, WhiteHatLurker, Jonhall, Dekaels~enwiki, Jynus, Newone, Courcelles, Laplace's Demon, SchmittM, J Milburn, Fordmadoxfraud, Cydebot, Bvcrist, Kozuch, Thijs!bot, Lord Hawk, Headbomb, MichaelMaggs, Escarbot, Ssr, JAnDbot, Eurobas, Acroterion, VoABot II, Appraiser, R'n'B, Sgreddin, MikeBaharmast, Lk69, Acalamari, DraakUSA, TomasBat, Joshua Issac, Kenneth M Burke, Ken g6, Idioma-bot, VolkovBot, SarahLawrence Scott, Nxavar, JhsBot, Abdullais4u, Lejarrag, Antixt, PGWG, SieBot, Timb66, Sonicology, PlanetStar, Bamkin, Dhatfield, Byrialbot, Svick, Perfectapproach, Thorncrag, Big55e, ClueBot, Jmorris84, Maxtitan, Alexbot, Dekisugi, Paradoxalterist, Saintlucifer2008, Cockshut12345, Rreagan007, RP459, Truthnlove, Addbot, Yakiv Gluck, Draco 2k, Mac Dreamstate, Funky Fantom, CarsracBot, HerculeBot, Legobot, Blah28948, Luckas-bot, Zhitelew, KamikazeBot, Kulmalukko, Orion11M87, AnomieBOT, Girl Scout cookie, Templatehater, Icalanise, Citation bot, Onesius, Vuerqex, Bci2, ArthurBot, Rightly, Xqbot, Phazvmk, Kirin13, FrescoBot, Delphinus1997, Steve Quinn, Robo37, SuperJew, HRoestBot, Sthyne, Hellknowz, Yahia.barie, Skyerise, Tobi - Tobsen, FoxBot, Physics therapist, Think!97, Bj norge, RjwilmsiBot, Beyond My Ken, EmausBot, John of Reading, Mnkyman, GoingBatty, Mthorndill, ZéroBot, Bollyjeff, StringTheory11, Markinvancouver, Quantumor, RolteVolte, Negovori, NTox, I hate whitespace, ClueBot NG, CocuBot, Widr, Micah.yannatos1, Helpful Pixie Bot, Guzman.c, Bibcode Bot, BG19bot, Spaceawesome, Rainbot, Leaverward, Let'sBuildTheFuture, Eduardofeld, Sha-256, Dr.RobertTweed, ZX95, Joeinwiki, Mark viking, Cephas Atheos, Yo butt, Snakeboy666, Psyruby42, Haminoon, Sardeth42, TaiSakuma, LadyCailin, Morph dtlr, Delbert7, Karam adel, GottaGoFast, KasparBot and Anonymous: 183

ing, RetiredUser2, Icairns, Mike Rosoft, Vsmith, Gianluigi, Kjoonlee, Drhex, Obradovic Goran, Jérôme, Fkbreitl, Cameron.simpson, Gene Nygaard, Linas, LoopZilla, Graham87, Kbdank71, Rjwilmsi, Strait, Mike Peel, Lmatt, Goudzovski, Chobot, FrankTobia, Roboto de Ajvol, Ugha, Mushin, Bambaiah, Wester, Hairy Dude, Hellbus, Salsb, Seb35, Długosz, Turbolinux999, Ravedave, Scottfisher, Dna-webmaster, Modify, Argo Navis, Teply, Sbyrnes321, SmackBot, Tom Lougheed, Jagged 85, ZerodEgo, Dauto, Bluebot, Shaggorama, Sbharris, Niels Olson, Radagast83, Acdx, John, Lottamiata, Happy-melon, Tubezone, MightyWarrior, Joelholdsworth, Tangobot, Michael C Price, Quibik, Dchristle, Realjanuary, Headbomb, Davidhorman, Nosirrom, Certain, Gökhan, JAnDbot, Tigga, Omeganian, Brimofinsanity, TheEditrix2, Trapezoidal, Magioladitis, ThoHug, Leyo, Lilac Soul, HEL, Rod57, Y2H, HiEv, Adam Zivner, Madblueplanet, Sheliak, Dextrose, Anonymous Dissident, Synthebot, Antixt, Coronellian~enwiki, SieBot, STANMAR725, Jim E. Black, Gerakibot, Martin Kealey, CutOffTies, Fratrep, ClueBot, Mild Bill Hiccup, Alexbot, Carsrac, SkyLined, Dieppu, Stephen Poppitt, Addbot, Eric Drexler, Toyokuni3, Mjamja, Ronkonkaman, Download, CarsracBot, ChenzwBot, Lightbot, M sotirov, Luckas-bot, Yobot, Jim1138, MehrdadAfshari, ArthurBot, Ernsts, A. di M., Howard McCay, FrescoBot, Paine Ellsworth, D'ohBot, Citation bot 1, Gil987, Tom.Reding, Swallerick, FoxBot, Earthandmoon, Tm1729, TjBot, Антон Гліністы, Newty23125, EmausBot, Mnkyman, StringTheory11, Quondum, MisterDub, WaterCrane, Whoop whoop pull up, ClueBot NG, Helpful Pixie Bot, Bibcode Bot, BG19bot, Bakkedal, JYBot, Mamaphyskerin, Anrnusna, MartinNicklin and Anonymous: 137

- **Gluon** *Source:* https://en.wikipedia.org/wiki/Gluon?oldid=672019780 *Contributors:* AxelBoldt, CYD, Bryan Derksen, Gdarin, TakuyaMurata, Card~enwiki, Looxix~enwiki, Ellywa, Ahoerstemeier, Med, Schneelocke, Phys, Phil Boswell, Donarreiskoffer, Fredrik, Merovingian, Hadal, Giftlite, Herbee, Xerxes314, Eequor, Darrien, Keith Edkins, RetiredUser2, Icairns, Mike Rosoft, AlexChurchill, HedgeHog, Kenny TM~~enwiki, David Schaich, Ioliver, Mashford, El C, Kwamikagami, Ardric47, Obradovic Goran, Alansohn, Guy Harris, Dachannien, Ricky81682, Batmanand, Velella, Kazvorpal, April Arcus, Forteblast, Mpatel, Palica, BD2412, Kbdank71, Rjwilmsi, Macumba, Strait, Mike Peel, Bubba73, Klortho, FlaBot, Srleffler, Chobot, YurikBot, Wavelength, Bambaiah, Hairy Dude, Jimp, JabberWok, Zelmerszoetrop, Salsb, SCZenz, Randolf Richardson, Ravedave, Danlaycock, Bota47, LeonardoRob0t, Anclation~enwiki, Physicsdavid, Erudy, GrinBot~enwiki, Kgf0, SmackBot, Melchoir, Cessator, Benjaminevans82, Abtal, MK8, Colonies Chris, Can't sleep, clown will eat me, Decltype, Qcdmaestro, Edconrad, Darkpoison99, FredrickS, Omsharan, Pegasusbot, Gregbard, ProfessorPaul, Thijs!bot, Headbomb, Rriegs, Oreo Priest, AntiVandalBot, Shambolic Entity, Deflective, Mujokan, Yill577, Happycool, Mother.earth, Martynas Patasius, WiiWillieWiki, HEL, Hans Dunkelberg, Gombang, Inwind, Sheliak, Jonthaler, VolkovBot, TXiKiBoT, Davehi1, Kriak, Anonymous Dissident, Imasleepviking, AlleborgoBot, EJF, SieBot, Steven Zhang, OKBot, ClueBot, Wwheaton, Qsaw, Nucularphysicist, Ottava Rima, Gordon Ecker, Rhododendrites, Brews ohare, Cacadril, RexxS, JKeck, Against the current, SkyLined, Addbot, DOI bot, Lightbot, Skippy le Grand Gourou, Luckas-bot, Planlips, AnomieBOT, Jim1138, JackieBot, Citation bot, Bci2, ArthurBot, Xqbot, Neil95, Triclops200, Omnipaedista, TorKr, ⁇⁇, Paine Ellsworth, Ivoras, Citation bot 1, Pekayer11, Rameshngbot, PNG, RjwilmsiBot, TjBot, Lilcal89012, EmausBot, Socob, JSquish, StringTheory11, Quondum, TyA, Maschen, RolteVolte, ClueBot NG, Timothy jordan, Maplelanefarm, Bibcode Bot, Gravitoweak, Cadiomals, Tropcho, Fraulein451, DrHjmHam, Rhlozier, D.shinkaruk, Yaara dildaara, BronzeRatio, KasparBot and Anonymous: 138

- **Unitary group** *Source:* https://en.wikipedia.org/wiki/Unitary_group?oldid=662230305 *Contributors:* AxelBoldt, Zundark, The Anome, Michael Hardy, Looxix~enwiki, Charles Matthews, Topbanana, Giftlite, Fropuff, Vivacissamamente, Oleg Alexandrov, Linas, Ruud Koot, JATerg, BD2412, MarSch, HappyCamper, R.e.b., LeonardoRob0t, KnightRider~enwiki, Jjalexand, Silly rabbit, Nbarth, Aghitza, JarahE, Simon Brady, Keyi, Headbomb, RobHar, Dispenser, Fylwind, VolkovBot, LokiClock, Conformancenut347, Yartsa, Drschawrz, JackSchmidt, Mr. Stradivarius, Winston365, Count Truthstein, DumZiBoT, Addbot, Roentgenium111, LaaknorBot, Dr Zimbu, Legobot, Luckas-bot, Yobot, Niout, Amirobot, AnomieBOT, CXCV, KonradVoelkel, ZéroBot, Meng6, Quondum, Rcsprinter123, Alexjbest, Enyokoyama, CsDix, GodMadeTheIntegers and Anonymous: 25

- **Special unitary group** *Source:* https://en.wikipedia.org/wiki/Special_unitary_group?oldid=672311996 *Contributors:* AxelBoldt, Taral, Stevertigo, Michael Hardy, Looxix~enwiki, Charles Matthews, Dysprosia, Rudminjd, Phys, Robbot, Robinh, Tobias Bergemann, Giftlite, BenFrantzDale, Lethe, Fropuff, Jason Quinn, Eequor, Lumidek, Vivacissamamente, 4pq1injbok, Xezbeth, MuDavid, Paul August, Spoon!, Giraffedata, Eric Kvaalen, Fourthords, Joriki, Simetrical, JATerg, GregorB, BD2412, Rjwilmsi, HappyCamper, Marozols, Nowhither, Itinerant1, Roboto de Ajvol, YurikBot, StuffOfInterest, RussBot, JabberWok, Archelon, Crasshopper, Tetracube, Reyk, Banus, KnightRider~enwiki, SmackBot, Incnis Mrsi, Tom Lougheed, Movementarian, Silly rabbit, Nbarth, Tamfang, Chlewbot, Speedplane, Harryboyles, Ryulong, Vaughan Pratt, Vyznev Xnebara, MatthewMain, WISo, Dr.enh, Michael C Price, Quibik, Thijs!bot, Koeplinger, Headbomb, Savant13, Magioladitis, VoABot II, Jlenthe, Etale, David Eppstein, Haseldon, Cuzkatzimhut, JohnBlackburne, LokiClock, Anonymous Dissident, StevenJohnston, Yartsa, Drschawrz, YohanN7, Jasondet, JackSchmidt, OKBot, Mr. Stradivarius, Ideal gas equation, Gigacephalus, Cacadril, Count Truthstein, Addbot, Eric Drexler, Morriswa, Shender, Luckas-bot, Yobot, Niout, Dickdock, AnomieBOT, Collieuk, Citation bot, Waltruda, Tkuvho, LittleWink, Jonesey95, Tim1357, EmausBot, Groemaer, NN22, Zueignung, Bomazi, David C Bailey, ⁇⁇⁇, Lemingue, BG19bot, Trodemaster, Hillbillyholiday, CsDix, Zimboras, Impsswoon, MarkovianStumble and Anonymous: 95

- **Lanczos tensor** *Source:* https://en.wikipedia.org/wiki/Lanczos_tensor?oldid=671494971 *Contributors:* SebastianHelm, Bender235, Oleg Alexandrov, Mpatel, Markdroberts, BradBeattie, Hillman, Salsb, Teply, SmackBot, Lantonov, ChrisHodgesUK, Pqnelson, Point-set topologist, Molitorppd22, Rausch, Quondum, Brandmeister and Anonymous: 7

- **General covariance** *Source:* https://en.wikipedia.org/wiki/General_covariance?oldid=653903947 *Contributors:* The Anome, Charles Matthews, Aenar, Sdedeo, Isopropyl, Giftlite, Anythingyouwant, Lumidek, PhotoBox, Pearle, Oleg Alexandrov, Cleonis, Mpatel, MarSch, Ligulem, Ems57fcva, Reedbeta, Mathbot, Hillman, Salsb, Petri Krohn, SmackBot, Bytesmythe, Polonium, Ligulembot, JRSpriggs, Storm63640, Michael C Price, Thijs!bot, Headbomb, DAGwyn, BatteryIncluded, J Hill, JCarlos, Kevin aylward, Lantonov, Tarotcards, Henry Delforn (old), Michel421, Brews ohare, MystBot, Addbot, Ozob, AnomieBOT, Citation bot, MauritsBot, Xqbot, Point-set topologist, Gsard, Anterior1, HRoestBot, RobinK, Rausch, Gildorien, Helpful Pixie Bot, I3roly and Anonymous: 19

- **Graviton** *Source:* https://en.wikipedia.org/wiki/Graviton?oldid=671813129 *Contributors:* CYD, Bryan Derksen, Timo Honkasalo, XJaM, Fubar Obfusco, Maury Markowitz, Kaczor~enwiki, Jketola, TakuyaMurata, Eric119, Looxix~enwiki, Glenn, Cyan, Wooster, Charles Matthews, Timwi, Wik, BenRG, Donarreiskoffer, Scott McNay, Stephan Schulz, Arkuat, Chris Roy, Merovingian, David9999, Giftlite, Xerxes314, Jason Quinn, Matt Crypto, CryptoDerk, RetiredUser2, Icairns, Zfr, Lumidek, Ukexpat, Urvabara, Discospinster, Pjacobi, Vapour, Brian0918, El C, Joanjoc~enwiki, Dalf, Army1987, Mpvdm, La goutte de pluie, Physicistjedi, Daniel Arteaga~enwiki, Zenosparadox, Dethtron5000, Keenan Pepper, Viridian, Falcorian, Skeejay, Simetrical, Dr Archeville, Mpatel, Kyleca, Tmassey, Christopher Thomas, Tevatron~enwiki, Kbdank71, Nightscream, Koavf, Mike Peel, Ems57fcva, FlaBot, RexNL, Chobot, DVdm, Roboto de Ajvol, Spacepotato, Anonymous editor, SnoopY~enwiki, Salsb, Bachrach44, Hyperbrand, NickBush24, Pnrj, RL0919, EEMIV, IslandGyrl, Bota47, C h fleming, Petri Krohn, Mario23,

Alias Flood, Tim314, Teply, GrinBot~enwiki, SmackBot, Amcbride, Melchoir, Eskimbot, Gilliam, Skizzik, Timneu22, Complexica, Villarinho, Colonies Chris, Vladis1av, Chlewbot, Xyzzyplugh, Jmnbatista, Fuhghettaboutit, Sadi Carnot, Yevgeny Kats, TenPoundHammer, Lambiam, Zaphraud, JorisvS, Mr Stephen, Ramuman, Quasar Jarosz, Lottamiata, Firewall62, Kurtan~enwiki, CmdrObot, BeenAroundAWhile, WeggeBot, Shultz IV, UncleBubba, Michael C Price, Anthmoo, Thijs!bot, Epbr123, Headbomb, KevinS06, Opelio, Spartaz, JAnDbot, Xoneca, SHCarter, Pikazilla, Robin S, STBot, Kostisl, J.delanoy, Tarotcards, Coppertwig, Wesino, Sava ankit2006, Tygrrr, Idioma-bot, Sheliak, JoAnneThrax, TXiKiBoT, WilliamSommerwerck, Hqb, Anonymous Dissident, Antixt, SieBot, Flyer22, Henry Delforn (old), ClueBot, Ergn, Darkicebot, DenverRedhead, Addbot, Eric Drexler, Uruk2008, DOI bot, BrianBop, PJonDevelopment, F Notebook, Legobot, Picturesofnothing, Dov Henis, Alfredschrader, Eric-Wester, AnomieBOT, VanishedUser sdu9aya9fasdsopa, Jim1138, Materialscientist, Citation bot, Tomflaherty, ProtectionTaggingBot, Waleswatcher, FrescoBot, Juto20, LucienBOT, Paine Ellsworth, I dream of horses, Tom.Reding, RedBot, Omar.tigereyes, IVAN3MAN, Ashish.kotwal, Michael9422, D0wnfalle, EmausBot, Octaazacubane, 8digits, Slightsmile, K6ka, Thecheesykid, User10 5, Rcsprinter123, Orbjeeples, Puffin, Herk1955, ClueBot NG, Raidr, Helpful Pixie Bot, Bibcode Bot, BG19bot, Shapoopy178, ServiceAT, PhnomPencil, Trevayne08, Brainssturm, Tjamcclain2, ChrisGualtieri, Ariscod, TheUyulala, LightandDark2000, Jessybun, Makecatbot, Kryomaxim, JRYon, Andyhowlett, Mark viking, Yorsh07, CensoredScribe, WPratiwi, Monkbot, Bryan Paul Senior, Dr.Begich, Nompynuthead, Jacobflarsen and Anonymous: 196

- **Diffeomorphism** *Source:* https://en.wikipedia.org/wiki/Diffeomorphism?oldid=665107482 *Contributors:* JeLuF, Maury Markowitz, Michael Hardy, TakuyaMurata, AugPi, Poor Yorick, Med, Charles Matthews, Dysprosia, Kuszi, MathMartin, Pascalromon, Tosha, Connelly, Giftlite, Lethe, Fropuff, CryptoDerk, Paul August, Rgdboer, Physistjedi, Msh210, BRW, Oleg Alexandrov, R.e.b., Mathbot, Lmatt, Bgwhite, Mhwu, YurikBot, Wavelength, RussBot, Woseph, Gaius Cornelius, SmackBot, Silly rabbit, Nakon, Dreadstar, Mathsci, Myasuda, Dharma6662000, Thijs!bot, Headbomb, LachlanA, Nosirrom, Ensign beedrill, Policron, LokiClock, Geometry guy, Rybu, BotMultichill, JerroldPease-Atlanta, He7d3r, Topology Expert, PV=nRT, Legobot, Yobot, AnomieBOT, Point-set topologist, RibotBOT, Sławomir Biały, Citation bot 1, Åkebråke, Redrose64, MondalorBot, Fly by Night, Slaweko, ZéroBot, Chester Markel, Uni.Liu, Helpful Pixie Bot, BG19bot, Muses' house, Herve.lombaert, Jeremy112233, Hillbillyholiday, CsDix, Pwm86 and Anonymous: 44

- **Gauge theory gravity** *Source:* https://en.wikipedia.org/wiki/Gauge_theory_gravity?oldid=639808633 *Contributors:* Michael Hardy, Giftlite, Chris Howard, JHCaufield, Teply, Tom.Reding, Quondum, Bibcode Bot, BG19bot, Anrnusna and Anonymous: 2

- **Gauge gravitation theory** *Source:* https://en.wikipedia.org/wiki/Gauge_gravitation_theory?oldid=662966519 *Contributors:* Michael Hardy, Gabbe, Drernie, MBisanz, Crasshopper, Teply, Stifle, Silly rabbit, CmdrObot, Adavidb, Moonriddengirl, TrulyBlue, Addbot, LaaknorBot, Xqbot, Gsard, Tom.Reding, Pmokeefe, ChrisGualtieri, Garuda0001 and Anonymous: 3

- **Quantum gravity** *Source:* https://en.wikipedia.org/wiki/Quantum_gravity?oldid=667958630 *Contributors:* AstroNomer~enwiki, Matusz, Miguel~enwiki, Roadrunner, Stevertigo, Ubiquity, Bobby D. Bryant, Mcarling, NuclearWinner, Anders Feder, Susurrus, Coren, Charles Matthews, Timwi, Reddi, Tpbradbury, Phys, Bevo, Raul654, BenRG, Frazzydee, Jeffq, Sdedeo, Rholton, Wereon, Ilya (usurped), Seth Ilys, Ancheta Wis, Giftlite, Herbee, Fropuff, Endlessnameless, Malyctenar, Jason Quinn, Finn-Zoltan, YapaTi~enwiki, Lumidek, Marcus2, Joyous!, TJSwoboda, Vitaleyes, Davidclifford, JimJast, Guanabot, FT2, Masudr, Pjacobi, Pie4all88, David Schaich, Bender235, Clement Cherlin, El C, PhilHibbs, Army1987, Apyule, VBGFscJUn3, PWilkinson, Daniel Arteaga~enwiki, Keenan Pepper, Cjthellama, DonJStevens, Velella, Dabbler, Tycho, Cal 1234, RJFJR, Count Iblis, ThomasWinwood, Anarchimede, Scarykitty, Woohookitty, Igny, ToddFincannon, Mpatel, GregorB, Joke137, Christopher Thomas, Marudubshinki, Graham87, Yurik, Kroggz, Rjwilmsi, Eoghanacht, Jrasowsky, JHMM13, Smithfarm, Ems57fcva, FayssalF, Itinerant1, Lmatt, Chobot, Hmonroe, YurikBot, Hillman, ErkDemon, JocK, SCZenz, Roy Brumback, Bota47, Zunaid, JonathanD, 2over0, Arthur Rubin, Modify, LeonardoRob0t, Caco de vidro, RG2, KasugaHuang, Resolute, SmackBot, Samdutton, Vald, Eskimbot, Hbackman, Onebravemonkey, Chris the speller, Ben.c.roberts, Cthuljew, Silly rabbit, Complexica, Colonies Chris, QFT, Soosed, Theanphibian, Shushruth, Ck lostsword, Yevgeny Kats, DJIndica, Lambiam, Vampus, Vincenzo.romano, Jaganath, JorisvS, RoboDick~enwiki, IronGargoyle, Dicklyon, SirFozzie, Treyp, Twunchy, Piccor, Kurtan~enwiki, Harold f, CalebNoble, Duduong, Paulmlieberman, TVC 15, UncleBubba, TAz69x, Sam Staton, ST47, B, Patrick O'Leary, Epbr123, Koeplinger, Klasovsky, Markus Pössel, Keraunos, Headbomb, Marek69, MichaelMaggs, Tim Shuba, MER-C, ParadiZio, Clementvidal, Perlygatekeeper, VoABot II, Alvatros~enwiki, Bdalevin, SHCarter, Jpod2, DAGwyn, Nucleophilic, LorenzoB, Rickard Vogelberg, DancingPenguin, Rettetast, Victor Blacus, AstroHurricane001, Yonidebot, Acalamari, Mstuomel, Fullmetal2887, NewEnglandYankee, DorganBot, CardinalDan, Idioma-bot, Sheliak, VolkovBot, Pleasantville, Seattle Skier, AlnoktaBOT, TXiKiBoT, Dllahr, Rdekleer, Saibod, Cyberchip, Wikiwikimoore, Carlorovelli, LoreMiles, StevenJohnston, SieBot, LeadSongDog, Bentogoa, Coldcreation, ReluctantPhilosopher, StaticG, GarbagEcol, ClueBot, The Thing That Should Not Be, EoGuy, Polyamorph, Andwor9, Notburnt, Tms9, Alexbot, Resoru, Eeekster, Tamaratrouts, Brews ohare, SchreiberBike, Askahrc, BOTarate, Lambtron, DumZiBoT, XLinkBot, Rror, Facts707, SilvonenBot, Theonlydavewilliams, Mhsb, Truthnlove, Ttimespan, Trifonov~enwiki, Addbot, Mortense, Grayfell, Eric Drexler, Gravitophoton, DOI bot, AkhtaBot, CanadianLinuxUser, Frosty726, LaaknorBot, Delaszk, Tassedethe, Tide rolls, Taketa, Titan1129, Krano, Luckas-bot, Yobot, WikiDan61, Pigetrational, Wireader, Allowgolf~enwiki, Wiki Roxor, Jim1138, IRP, Sz-iwbot, Quantity, Materialscientist, Citation bot, ArthurBot, LilHelpa, Amareto2, Ekwos, KrisBogdanov, Rolfguthmann, StealthCopyEditor, 配配, Dan6hell66, Rabsmith, Hep thinker, Paine Ellsworth, DrArthurRubinPHD, Lagelspeil, Nunc aut numquam, Vacuunaut, Van Speijk, Knowandgive, Craig Pemberton, Udifuchs, Citation bot 2, Citation bot 1, Citation bot 4, Jonesey95, Hirvenkürpa, Tom.Reding, Pmokeefe, Casimir9999, Dac04, Dude1818, Valeriy Pischenko, Follyland, TrueTeargem, N0814444, Earthandmoon, Korepin, DARTH SIDIOUS 2, Musictime4me, RjwilmsiBot, EmausBot, Francophile124, Octaazacubane, Fotoni, Slightsmile, Garfield Salazar, Hhhippo, JSquish, John Cline, Fæ, Brazmyth, Throwmeaway, Arbnos, Ebrambot, Kusername, DanielBurnstein, TonyMath, L Kensington, Maschen, Donner60, Parusaro, Apratim07, Terra Novus, Isocliff, Googledin!, ClueBot NG, SpikeTorontoRCP, Science writer, Preon, Raidr, Jhmmok, 336, Widr, Helpful Pixie Bot, Bibcode Bot, Bardsley Rides a Segway, Apelikedawg, FiveColourMap, Trevayne08, Mr.viktor.stepanov, Brainssturm, BattyBot, Jimw338, Ryanr666, Kryomaxim, Garuda0001, Saehry, Sanathdevalapurkar, Andyhowlett, Gabelglesia, Sanathlab, Roiwallace, Spencer.mccormick, Spencerfjase, MrShlongNo1, Marc D. Garrett, D00d00ballz, Gigantmozg, Polytope24, Frinthruit, Anrnusna, Dfyytj, Monkbot, Umut Alihan Dikel, Amortias, Klj1234, Pfpguy, KasparBot and Anonymous: 290

- **Circle group** *Source:* https://en.wikipedia.org/wiki/Circle_group?oldid=648705186 *Contributors:* Zundark, Michael Hardy, TakuyaMurata, Karada, Eric119, Revolver, Charles Matthews, Giftlite, Fropuff, HorsePunchKid, Eep², ZeroOne, Rgdboer, Keenan Pepper, Oleg Alexandrov, Linas, Juan Marquez, Mathbot, Elpaw, Dmharvey, Archelon, Netrapt, SmackBot, Incnis Mrsi, Melchoir, Bluebot, Richard L. Peterson, Jim.belk, JoeBot, Freelance Intellectual, Tac-Tics, WISo, Kilva, RobHar, Dogru144, LokiClock, Hesam7, Arcfrk, JackSchmidt, Mr. Stradivarius, Addbot, Fgnievinski, CanadianLinuxUser, HerculeBot, Yobot, Jgmoxness, AnomieBOT, Ciphers, Sławomir Biały, Chricho, QuantumSquirrel, Bulldog73, The Anonymouse, CsDix, Blackbombchu and Anonymous: 15

- **Spinor** *Source:* https://en.wikipedia.org/wiki/Spinor?oldid=671915285 *Contributors:* AxelBoldt, CYD, The Anome, Jeronimo, XJaM, Youandme, Gabbe, AugPi, Rossami, Charles Matthews, David Newton, Dcoetzee, Rudminjd, Fibonacci, Phys, Phil Boswell, Rorro, Aetheling, Giftlite, BenFrantzDale, Fropuff, Jason Quinn, Jorge Stolfi, Dan Gardner, Lumidek, Chris Howard, Pjacobi, Francis Davey, Dbachmann, Gauge, Pearle, Fkbreitl, Jheald, RJFJR, Killing Vector, Oleg Alexandrov, Linas, Mpatel, Tabletop, Rjwilmsi, MarSch, Dennis Estenson II, Natkuhn, Brad-Beattie, Chobot, Roboto de Ajvol, Wavelength, Presscorr, Crasshopper, BOT-Superzerocool, Larsobrien, Orthografer, SmackBot, Incnis Mrsi, Eskimbot, Hmains, Bluebot, Silly rabbit, Nbarth, Mungbean, QFT, Wiki me, Daqu, Martijn Hoekstra, Vina-iwbot~enwiki, JorisvS, JarahE, Dan Gluck, JoeBot, Mattbr, JasonHise, Myasuda, Michael C Price, Difty, Headbomb, RobHar, CarlAB, Antic-Hay, .anacondabot, Kborland, RogierBrussee, Vinograd19, Maurice Carbonaro, Peskydan, Jqar, Policron, JohnBlackburne, CarlBrannen, Michael H 34, Geometry guy, Shadoweye, Neparis, YohanN7, Liszter, KoenDelaere, Sfan00 IMG, ArdClose, Ericlord, SchreiberBike, Count Truthstein, Chadoh, TimothyRias, SkyLined, Addbot, Mathieu Perrin, Download, Lightbot, Yobot, Bjarnec~enwiki, ^musaz, Galoubet, NickK, Citation bot, ArthurBot, LilHelpa, RibotBOT, FrescoBot, Almuhammedi, Sławomir Biały, Sae1962, Citation bot 1, Casimir9999, Rimpulili, EmausBot, Slawekb, Midas02, Quondum, Maschen, Anagogist, Helpful Pixie Bot, Bibcode Bot, Krastanov, Jimw338, A.entropy, AptitudeDesign, Impsswoon, SkateTier, Omikr0n.poland and Anonymous: 62

- **Fundamental interaction** *Source:* https://en.wikipedia.org/wiki/Fundamental_interaction?oldid=671928308 *Contributors:* AxelBoldt, Zundark, The Anome, Tarquin, AstroNomer~enwiki, William Avery, Roadrunner, Ellmist, Robert Foley, Heron, Isis~enwiki, Stevertigo, Patrick, Michael Hardy, Gdarin, CesarB, Looxix~enwiki, Cyp, William M. Connolley, Theresa knott, Mxn, Bemoeial, Reddi, Zoicon5, Finlay McWalter, Robbot, Lowellian, Brjaga, Roscoe x, Seth Ilys, Ancheta Wis, Giftlite, Christopher Parham, Herbee, Monedula, Xerxes314, Alison, Pcarbonn, Beland, Melikamp, Karol Langner, AmarChandra, Mike Rosoft, Jørgen Friis Bak, JimJast, Discospinster, Guanabot, FT2, Harriv, Quietly, GoldenRing, Clement Cherlin, El C, Lycurgus, Joanjoc~enwiki, Alereon, Euyyn, Kanzure, Army1987, Rbj, Haham hanuka, Nsaa, Jumbuck, Foant, Dachannien, Kdau, ReyBrujo, Reaverdrop, BDD, Someoneinmyheadbutit'snotme, DV8 2XL, Kazvorpal, Woohookitty, Linas, Mindmatrix, Sabejias, StradivariusTV, Mpatel, Miss Madeline, Isnow, Elvey, Chun-hian, Koavf, Strait, Jmcc150, RE, Gadha, FlaBot, DClement, ZoneSeek, Alfred Centauri, Lmatt, Rell Canis, Mstroeck, Chobot, Subtractive, Visor, GangofOne, Mysekurity, YurikBot, Ashleyisachild, Bambaiah, Lucinos~enwiki, Wavesmikey, Chaos, FFLaguna, Dbfirs, Trigger hippie77, Enormousdude, Shimei, RG2, Bweenie, Phr en, GrinBot~enwiki, SmackBot, Unyoyega, Andy M. Wang, Vvarkey, Jjalexand, Mithaca, Acipsen, DHN-bot~enwiki, Colonies Chris, Andy120290, Addshore, SundarBot, Jgwacker, LeoNomis, Sadi Carnot, TTE, SashatoBot, FrozenMan, Philosophus, A. Parrot, Fangfufu, GDallimore, Avanishsharma, CRGreathouse, Green caterpillar, McVities, MaxEnt, A. Exeunt, Scott.medling, LouisBB, Thijs!bot, Mojo Hand, Headbomb, Dfrg.msc, Dodecahedron~enwiki, JAnDbot, The penfool, Fordskydog, MER-C, TheEditrix2, Fabrictramp, Leyo, Trusilver, Joshuaali, Idioma-bot, VolkovBot, TXiKiBoT, Anonymous Dissident, MackSalmon, Praveen pillay, BotKung, Gnomon13, RMW42, EmxBot, Neparis, SieBot, WereSpielChequers, ToePeu.bot, Avargasm, RadicalOne, Dhatfield, SuperSpy00bob, Sbowers3, Beast of traal, Lightmouse, Nskillen, Sunrise, OKBot, Bpeps, C0nanPayne, StewartMH, Sfan00 IMG, ClueBot, MichaelVernonDavis, SuperHamster, Djr32, Sadiqsaleem09, PixelBot, Eeekster, Zamis45, Yonskii, 1ForTheMoney, Noctibus, Truthnlove, Addbot, Mabdul, LinkFA-Bot, F Notebook, Lightbot, Legobot, Clay Juicer, Luckas-bot, Yobot, II MusLiM HyBRiD II, Rifter0x0000, AnomieBOT, Glen Dillon, Girl Scout cookie, Cleroth, JackieBot, Piano non troppo, Flewis, AthenaO, Xqbot, Omnipaedista, RibotBOT, A. di M., Ironboy11, Goodbye Galaxy, Jmbenham, Unkownkid2400, Rameshngbot, Jschnur, RedBot, Σ, Frankjohnson123, IVAN3MAN, Right-wing genius, Lokentaren, Setsuna29, EngineerFromVega, RjwilmsiBot, Anuandraj, Beyond My Ken, Deadlyops, Carbo1200, Kbasford, ClueBot NG, Greedohun, MelbourneStar, Grannis3, Kasirbot, Kaos Magician, Einsteiner900, Widr, Shelbylv, CasualVisitor, Hz.tiang, JimmyMachineGunHand, Cengime, BattyBot, Dexbot, Mogism, Makecat-bot, Ryan.laff, Jamesx12345, Ttitts, CsDix, Kenanwang, GregRos, Prokaryotes, Robertpb97, Occurring, Basedrawnz, Learnerktm, Julietvbarbara, Barbarousbunch815, Brendapallister, Nicholaspurcellstudio, Pickleslover, Tetra quark, Claudio.nahmad.arcaraz, KasparBot, The oracle 2015, Mitzionne and Anonymous: 249

- **Weak interaction** *Source:* https://en.wikipedia.org/wiki/Weak_interaction?oldid=672253610 *Contributors:* AxelBoldt, Chenyu, Sodium, Bryan Derksen, Tarquin, AstroNomer~enwiki, Andre Engels, XJaM, Heron, JohnOwens, Gdarin, Delirium, Andrewa, Andres, Emperorbma, Timwi, Fibonacci, Phys, Phil Boswell, Lowellian, Mayooranathan, Tobias Bergemann, Giftlite, Sj, Herbee, Xerxes314, Jcobb, Mckaysalisbury, Munkee, Toby Woodwark, Bbbl67, Icairns, AmarChandra, Lumidek, Jørgen Friis Bak, Discospinster, Roybb95~enwiki, Gianluigi, Joanjoc~enwiki, Shanes, AJP, AtomicDragon, Danski14, Alansohn, Arthena, Axl, SidneySM, Hwefhasvs, DV8 2XL, Nightstallion, Kazvorpal, Linas, StradivariusTV, Benbest, Bbatsell, Palica, Tevatron~enwiki, Graham87, BD2412, Ketiltrout, Rjwilmsi, Strait, Erkcan, The wub, FlaBot, Naraht, Itinerant1, Srleffler, Chobot, Krishnavedala, YurikBot, Borgx, Bambaiah, Hairy Dude, Jimp, Sillybilly, Conscious, Epolk, JabberWok, Gaius Cornelius, Shaddack, SCZenz, Irishguy, Shimei, Willtron, RG2, Phr en, That Guy, From That Show!, Luk, SmackBot, David Kernow, Tom Lougheed, WookieInHeat, Dauto, Chris the speller, Philosopher, Moshe Constantine Hassan Al-Silverburg, Complexica, DHN-bot~enwiki, Zirconscot, Robma, "alyosha", Maxwahrhaftig, Akriasas, Vina-iwbot~enwiki, Bdushaw, TTE, SashatoBot, Fontenello, Herr apa, Condem, Tony Fox, MottyGlix, JRSpriggs, Heartofgoldfish, Calmargulis, Green caterpillar, Joelholdsworth, Cydebot, Michael C Price, Mtpaley, Thijs!bot, ChKa, Kichwa Tembo, Headbomb, Hcobb, Icep, Escarbot, AntiVandalBot, Jimeree, Steelpillow, JAnDbot, Magioladitis, Swpb, باسم, Wormcast, DAGwyn, Giggy, Khalid Mahmood, Gah4, Tarotcards, 2help, Lighted Match, DorganBot, Halmstad, Idioma-bot, VolkovBot, Jcuadros, Hilarious Bookbinder, TXiKiBoT, Rei-bot, CaptinJohn, Awl, Shenanegins, BotKung, Antixt, Xxxlilbritxxx, Ptrslv72, Monty845, AlleborgoBot, SieBot, Paolo.dL, Skyentist, Ptr123, ClueBot, Bondchic007, SuperHamster, Erudecorp, Rotational, Jackey0105, Alexbot, Cenarium, Zomno, Zahnrad, He6kd, TimothyRias, InternetMeme, Timo Metzemakers, Stephen Poppitt, Addbot, Some jerk on the Internet, Markdman, ChenzwBot, Ehrenkater, Tide rolls, Luckas-bot, Yobot, Les boys, Kilom691, THEN WHO WAS PHONE?, Rifter0x0000, Duping Man, Dickdock, Magog the Ogre, AnomieBOT, Materialscientist, Citation bot, Quebec99, Kreigiron, Xqbot, Drilnoth, BurntSynapse, GrouchoBot, Omnipaedista, RibotBOT, Workanode, Jaz1305, Mnmngb, Dave3457, FrescoBot, Charles.walker, LucienBOT, Ionutzmovie, Grandiose, Pinethicket, Boulaur, Rameshngbot, RedBot, 23790ΛD, Tea with toast, Jauhicnij, FoxBot, Earthandmoon, RjwilmsiBot, Itamarhason, Newty23125, EmausBot, WikitanvirBot, GA bot, GoingBatty, Splibubay, StringTheory11, Braswiki, Git2010, Wayne Slam, Jsayre64, ChuispastonBot, ClueBot NG, VinculumMan, Physics is all gnomes, Fjpyanez, Mouse20080706, Helpful Pixie Bot, Geo7777, Bibcode Bot, Junaid2754, Bolatbek, Phbarnacle, Neutral current, Glevum, Idenshi, Marioedesouza, Dexbot, Spray787, Reatlas, CsDix, Jamesmcmahon0, Ihatedirac2k13, Jwratner1, YimmyYohnson, Monkbot, BalderdashVonDrivel, ASCarretero, Malerisch, Lachlan Newland, Tetra quark, KasparBot and Anonymous: 153

- **Electromagnetism** *Source:* https://en.wikipedia.org/wiki/Electromagnetism?oldid=670451101 *Contributors:* AxelBoldt, Magnus Manske, Trelvis, Carey Evans, CYD, Mav, Bryan Derksen, Zundark, Szopen, The Anome, Malcolm Farmer, FreddyZ, Miguel~enwiki, William Avery, Maury Markowitz, Heron, Patrick, Tim Starling, Kku, Bcrowell, Delirium, Ahoerstemeier, William M. Connolley, Khorn, Babbo, Darkwind,

Kevin Baas, Julesd, Glenn, Sray, Poor Yorick, Mm, Rawr, Mxn, Hike395, Emperorbma, Wikiborg, Reddi, Phys, Lumos3, Phil Boswell, Robbot, Hankwang, Pigsonthewing, ZimZalaBim, Arkuat, Hemanshu, Texture, Roscoe x, Sunray, Wikibot, Fuelbottle, Lupo, Diberri, Wile E. Heresiarch, Tobias Bergemann, Ancheta Wis, Decumanus, Giftlite, DocWatson42, Andries, Wolfkeeper, Lethe, Koyn~enwiki, Everyking, Snowdog, Dratman, Curps, Ssd, Tom-, Jason Quinn, Brockert, Sohanley, Karol Langner, APH, Maximaximax, Bodnotbod, Icairns, Lumidek, Iantresman, Tsemii, Slipstream (usurped), Adashiel, Mike Rosoft, EugeneZelenko, Rich Farmbrough, Bedel23, Pjacobi, Vsmith, Theiloth, MuDavid, Paul August, Bender235, El C, Shanes, Mkosmul, RoyBoy, Femto, Matt McIrvin, Bert Hickman, Physicistjedi, Sam Korn, Signor Giuseppe, Mareino, Ranveig, Jumbuck, Red Winged Duck, Alansohn, Gary, Pinar, ChristopherWillis, Arthena, Atlant, Andrewpmk, Lectonar, Snowolf, Melaen, BRW, Wtshymanski, Evil Monkey, Bob1817, Mikeo, Gene Nygaard, Ttownfeen, Oleg Alexandrov, Cimex, TheNightFly, MONGO, Nakos2208~enwiki, Macaddct1984, Eras-mus, Gimboid13, Rnt20, Graham87, Qwertyus, Ando228, Rjwilmsi, Collins.mc, Tangotango, Ligulem, SeanMack, Bhadani, Yamamoto Ichiro, Cjpuffin, FlaBot, Nihiltres, Nivix, RexNL, Ewlyahoocom, Gurch, Otets, Fresheneesz, TeaDrinker, Srleffler, Physchim62, Chobot, Roboto de Ajvol, YurikBot, Wavelength, Spacepotato, JabberWok, CambridgeBayWeather, Salsb, Wimt, David R. Ingham, NawlinWiki, Bachrach44, Buster79, Tearlach, Anetode, Scottfisher, Figaro, Bota47, Nick123, FF2010, Light current, Orioane, Enormousdude, 21655, 2over0, JoanneB, Phil Holmes, Willtron, Sizarieldor, AGToth, Katieh5584, RG2, GrinBot~enwiki, Sbyrnes321, DVD R W, Luk, SmackBot, Manu0x0~enwiki, PEHowland, Prodego, KnowledgeOfSelf, McGeddon, Jagged 85, Jrockley, Swerdnaneb, Rpmorrow, Skizzik, JAn Dudík, LinguistAtLarge, Master of Puppets, Complexica, CMacMillan, TheGerm, Frap, Ioscius, Avoidance, SundarBot, Stevenmitchell, Cybercobra, MichaelBillington, Blake-, Akriasas, Illnab1024, Daniel.Cardenas, LeoNomis, Sadi Carnot, Apoorvchebolu, Skinnyweed, TTE, Sarfa, DJIndica, Nmnogueira, Ozhiker, Wvbailey, Finejon, UberCryxic, Philosophus, Cronholm144, Ckatz, El Dahveed, Grapetonix, Alatius, Sinistrum, Dicklyon, Jon186, Waggers, Ryulong, Describer, KJS77, Newone, Adambiswanger1, Courcelles, Ziusudra, Tawkerbot2, Yanah, Xcentaur, Mosaffa, JForget, KyraVixen, Baiji, Vyznev Xnebara, Fjomeli, MarsRover, Musicalantonio, Marly88, Peripitus, Fifo, Ssilvers, UberScienceNerd, Thijs!bot, VoABot, Jb.schneider-electric, Headbomb, Marek69, Gerry Ashton, Leon7, D.H, MichaelMaggs, AntiVandalBot, Majorly, Seaphoto, Quintote, Gnixon, Jnyanydts, Tyco.skinner, Eleos, Steelpillow, JAnDbot, Matthew Fennell, Mkch, Bongwarrior, VoABot II, Tails4, SHCarter, TxAlien, GBYork, WhatamIdoing, Vanished user ty12kl89jq10, Adrian J. Hunter, 28421u2232nfenfcenc, Coldwarrier, User A1, Khalid Mahmood, Ruhihumphries, PrattTA1, InvertRect, MartinBot, Jargon777, LedgendGamer, J.delanoy, Sasajid, Abecedare, Bogey97, JohnPritchard, Maurice Carbonaro, Extransit, Tarotcards, NewEnglandYankee, Wesino, Shoessss, Cometstyles, ACBest, DorganBot, Treisijs, Bently34, Lights, 28bytes, Part Deux, Thedjatclubrock, Constant314, Philip Trueman, TXiKiBoT, The Original Wildbear, Guillaume2303, Anonymous Dissident, Kevin Steinhardt, Monkey Bounce, CaptinJohn, Imasleepviking, Mbarrieau, DoktorDec, Atomicswoosh, TongueSpeaker, Andy Dingley, Dirkbb, Lova Falk, @pple, Sylent, Doc James, PGWG, NHRHS2010, SieBot, Coffee, Tresiden, Graham Beards, Work permit, Hertz1888, Avargasm, Winchelsea, Dawn Bard, Caltas, Yintan, Zoragotcha, Keilana, Flyer22, Qst, Csblack, Klmeze~enwiki, Joseph Banks, Oxymoron83, Fbarw, Maelgwnbot, StaticGull, Dolphin51, TreeSmiler, Kanonkas, Ainlina, ElectronicsEnthusiast, Llywelyn2000, WikipedianMarlith, Bschaeffer~enwiki, Atif.t2, Martarius, ClueBot, Stevekirst7, The Thing That Should Not Be, Jan1nad, Uncle Milty, Blanchardb, Twicemost, LizardJr8, Jackey0105, Electromagnetic, Otto Tanaka, Excirial, Kocher2006, BlueLikeYou, Jusdafax, Abrech, Plastic Fish, MacedonianBoy, PhySusie, Tseno Maximov, Thingg, Jesus.murphy55567, Versus22, InternetMeme, XLinkBot, Emmette Hernandez Coleman, Jovianeye, Rror, Boratlike, HarlandQPitt, Rogimoto, Deineka, Addbot, Dustbin123, Allfor12008, CanadianLinuxUser, Download, EconoPhysicist, Redheylin, WikiDegausser, K Eliza Coyne, LinkFA-Bot, Kisbesbot, AgadaUrbanit, VASANTH S.N., Tide rolls, Lightbot, Gail, Kurtis, Will.M.Thompson, Luckas-bot, Yobot, Ptbotgourou, Lichen from Hell, ScienceMind, Tempodivalse, Orion11M87, AnomieBOT, Jim1138, Piano non troppo, AdjustShift, Penguinatortoo, Materialscientist, Citation bot, Vuerqex, Xqbot, Konor org, Lolman33, Plumpurple, Melmann, DSisyphBot, GrouchoBot, Nayvik, RibotBOT, Jsleaby, Maplestory101, Slowart, A. di M., GliderMaven, Tobby72, Wikipe-tan, Sum33, Ian88800, Hippycaller, Steve Quinn, HamburgerRadio, Citation bot 1, Чаховіч Уладзіслаў, Pinethicket, Elockid, Hard Sin, LittleWink, PvsKllKsVp, Tinton5, Yahia.barie, Jschnur, Ezhuttukari, Corinne68, FoxBot, Рыцарь поля, Retired user 0001, SchreyP, Hickorybark, ItsZippy, Lotje, Cjlim, Antipastor, Reaper Eternal, DARTH SIDIOUS 2, Triden, RjwilmsiBot, Alph Bot, Agent Smith (The Matrix), WildBot, DASHBot, EmausBot, John of Reading, WikitanvirBot, Mnkyman, Never give in, ITshnik, IncognitoErgoSum, RA0808, Tommy2010, Wikipelli, Thecheesykid, JSquish, ZéroBot, Harddk, Leptonoggin, Fæ, Maypigeon of Liberty, WFarver, H3llBot, Quondum, Git2010, Makecat, Sonygal, Coasterlover1994, L Kensington, Rr2wiki, Maschen, Donner60, Rjowsey, Rock-Magnetist, Peter Karlsen, Sir sachin, Teapeat, Planetscared, Cgt, Xanchester, ClueBot NG, Jack Greenmaven, MelbourneStar, Hiperfelix, Hallaman3, Ant.acke, Enopet, Frietjes, Milikguay, Widr, Helpful Pixie Bot, Eemcginnis, Vagobot, Bolatbek, Manuelfeliz, MusikAnimal, Stephenwanjau, Rm1271, Thekillerpenguin, Cadiomals, Mariano Blasi, Whyamiadampandalol, Franz99, Physicsch, Brad7777, Glacialfox, Neumannjo, Acul132, ChrisGualtieri, Dexbot, Marlowfrontier, Lugia2453, Frosty, Hamerbro, SubratamindPal, Trillig, Abhaikumar10, Cs-Dix, Ihatedirac2k13, Hell to earth88, CROY123, Zenibus, Jwratner1, YiFeiBot, JosephSpiral, SpecialPiggy, Kdmeaney, Yashshroff97, Csutric, Mahusha, Venomous Cobra, Trackteur, Gronk Oz, Meski33, Hkeyser, Lolbob12345, Nanophysisct12345, Simonessnygg, Noobmagnet, Mediavalia, Slayeredwarrior, Tetra quark, Isambard Kingdom, SergioCruz2015, Aenfinger, KasparBot, Zeke Essiestudy, Kafishabbir, JJMC89, Cesarnajera56 and Anonymous: 698

- **Strong interaction** *Source:* https://en.wikipedia.org/wiki/Strong_interaction?oldid=672231991 *Contributors:* AxelBoldt, Sodium, Bryan Derksen, RK, Andre Engels, Danny, Peterlin~enwiki, Heron, Xavic69, Tim Starling, EddEdmondson, Gdarin, Ixfd64, Wintran, Salsa Shark, AugPi, Timwi, Bemoeial, Wikiborg, Fuzheado, ElusiveByte, Populus, Phys, Omegatron, Bevo, PuzzletChung, Robbot, Mayooranathan, Henrygb, Giftlite, DocWatson42, Sj, Harp, Monedula, Xerxes314, Remy B, Jason Quinn, Utcursch, Zfr, AmarChandra, Sam Hocevar, Nicobn~enwiki, Jørgen Friis Bak, Discospinster, FT2, Vsmith, ArnoldReinhold, Trekie8472, Roybb95~enwiki, David Schaich, Gianluigi, Marknewlyn~enwiki, Drhex, CDN99, Army1987, Matt McIrvin, Danski14, VivaEmilyDavies, TenOfAllTrades, Vuo, DV8 2XL, Kazvorpal, Oleg Alexandrov, Superstring, Spettro9, BillC, Eras-mus, Tevatron~enwiki, Mendaliv, Zbxgscqf, Strait, Loudenvier, Donotresus, Yamamoto Ichiro, Siv0r, Lmatt, Srleffler, Chobot, DVdm, Wavelength, Borgx, Bambaiah, Phmer, Jimp, Supasheep, Limulus, Salsb, RazorICE, Expensivehat, Dhollm, Lobwedge, Superiority, Tetracube, Willtron, GrinBot~enwiki, Asterion, Finell, AndrewWTaylor, SmackBot, RDBury, Tom Lougheed, Trojo~enwiki, Jrockley, Dauto, Chris the speller, Jjalexand, Complexica, Sbharris, Colonies Chris, Richard001, TTE, Spiritia, Titus III, FrozenMan, Newone, Happy-melon, Conrad.Irwin, Rowellcf, Chrisahn, Cydebot, Danny Bierek, Mtpaley, Wannabe Runny, Irigi, Headbomb, Niduzzi, KP Botany, Tlabshier, Hanzoro5, Dougher, Steelpillow, JAnDbot, Supertheman, Roleplayer, Kopovoi, Vssun, Hoverfish, Khalid Mahmood, Pan Dan, Robin S, Vortimer, Sujaybhu, Natsirtguy, Peter Chastain, Cpiral, Tygrrr, Treisijs, Alpvax, VolkovBot, JohnBlackburne, TXiKiBoT, Marskuzz, Muro de Aguas, Qxz, Sintaku, SieBot, Gerakibot, Escape Artist Swyer, Proton666, ObfuscatePenguin, ClueBot, GorillaWarfare, Jackey0105, DragonBot, Jefflayman, Nownownow, Cenarium, Razorflame, Zahnrad, Silvercromagnon, InternetMeme, Rreagan007, WikiHead, Drogs630, Addbot, Guoguo12, Omega Squad, ThisIsMyWikipediaName, Seratna, CarsracBot, Purple Emu, CosmiCarl, AgadaUrbanit, Tide rolls, Lightbot, Luckas-bot, Timeroot, Donthedev, Rifter0x0000, Umnum, AnomieBOT, VanishedUser sdu9aya9fasdsopa, Orange Knight of Passion, Piano non troppo, Citation bot, Obersachsebot, Xqbot, DSisyphBot, Barelistido, Almabot, GrouchoBot, RibotBOT, SassoBot, Mnmngb, CES1596, Gummer85, Citation bot 1, Boulaur, RedBot, Jauhienij, Surf5270, ElPeste, Slon02, EmausBot, John of Reading, Mnkyman, JSquish, Cogiati, Bamyers99, Rexprimoris, Donner60, ChuispastonBot, ClueBot NG, Jj1236, Helpful Pixie Bot, Bolatbek, ElphiBot, J.wong.wiki, Glevum, Zedtwitz, Zedshort, Nishantkumar19, Kisokj, YFdyh-bot, Andyhowlett, Reatlas, CsDix, EvergreenFir, Aurelianjh, Jwratner1, Diggerh, Kshitizarora2993, Tetra quark, KasparBot and Anonymous: 169

- **Standard Model (mathematical formulation)** *Source:* https://en.wikipedia.org/wiki/Standard_Model_(mathematical_formulation)?oldid= 667464814 *Contributors:* The Anome, Phys, BenRG, Giftlite, Alison, Wmahan, Beland, Setokaiba, Kaldari, Masudr, Bender235, Mykhal, Pt, Cmdrjameson, Kuratowski's Ghost, Wdyoung, Mandarax, BD2412, Qwertyus, Mathbot, Tone, Whosasking, Bambaiah, Dna-webmaster, Closedmouth, Caco de vidro, SmackBot, Maksim-e~enwiki, Robotbeat, Incnis Mrsi, Baad, Dauto, Chris the speller, Acipsen, QFT, Grover cleveland, Akriasas, Yevgeny Kats, Xxanthippe, Headbomb, Marwie, Gnixon, Christopher Cooper, Magioladitis, DinoBot, HEL, Lseixas, Schucker, Themel, Flyte35, Ptrslv72, Wing gundam, Bamkin, General Epitaph, Kitsunegami, Brews ohare, NuclearWarfare, SkyLined, Truthnlove, Jeffrey.Yepez, Lightbot, Yobot, Mateusz.Kwasnicki, Bci2, Omnipaedista, John S. Peterson, Ernsts, A. di M., FrescoBot, Ganondolf, Wikipelli, Maschen, Zueignung, Isocliff, Clearlyfakeusername, Asadsphotogremlin, Dsperlich, MerllwBot, Qwerty12345, Randomnonsense, Cjean42, Epicgenius, Euan Richard, Dimension10, Impsswoon, Lathamboyle and Anonymous: 62

- **Quantum electrodynamics** *Source:* https://en.wikipedia.org/wiki/Quantum_electrodynamics?oldid=663169179 *Contributors:* Bryan Derksen, DWeir, Stevertigo, Spsprez, Tim Starling, TakuyaMurata, Looxix~enwiki, Ahoerstemeier, Wooster, Joshuabowman, Greenrd, Patrick0Moran, Phys, Ortonmc, Robbot, Sanders muc, Diberri, Ancheta Wis, Giftlite, Sj, Wolfkeeper, Lethe, Dratman, JeffBobFrank, Markus Kuhn, Waltpohl, Sukael, Jason Quinn, DefLog~enwiki, LucasVB, Beland, Icairns, AmarChandra, Sctfn, Gcanyon, Lucidish, Urvabara, Pjacobi, Dbachmann, Mani1, Tooto, Triona, Neilrieck, Harley peters, Viriditas, .:Ajvol:., Matt McIrvin, Photonique, Rje, Helix84, Keenan Pepper, Count Iblis, DV8 2XL, Daubigne, Linas, Mpatel, Joke137, Alan Canon, ObsidianOrder, Magister Mathematicae, Snafflekid, Rjwilmsi, Strait, Patrick Gill, Salix alba, Rangek, FlaBot, Weihao.chiu~enwiki, Jsldub, Margosbot~enwiki, Srleffler, Chobot, YurikBot, Ugha, RussBot, JabberWok, Salsb, Spike Wilbury, Janke, SCZenz, Larsobrien, Gadget850, Ott2, Light current, Tribaal, Enormousdude, 2over0, Smoggyrob, Aeosynth, Didymos~enwiki, GrinBot~enwiki, Sbyrnes321, Chipef, Artemisfowl3rd, SmackBot, F, Rex the first, Tom Lougheed, Rjtucke, Melchoir, Unyoyega, Hmains, Chris the speller, Complexica, Aolanonawanabe, QFT, Ignirtoq, Voyajer, Aolanaonwaswronglyaccused, Sadi Carnot, DJIndica, Lambiam, MobyDikc, EGGS, JorisvS, Zarniwoot, RafaelRGarcia, Phuzion, Eyebum, JMK, Achoo5000, Albertod4, Retune, CmdrObot, Van helsing, Runningonbrains, Cydebot, Thijs!bot, Headbomb, Speedyboy, Second Quantization, D.H, Nick Number, Escarbot, Stannered, AntiVandalBot, Stripsi, Krapsin, And4e, Blaine Steinert, Yill577, Hroðulf, Bongwarrior, VoABot II, Bakken, Wormcast, Hekerui, Helianthi, Maliz, Daniel james, Alro, R'n'B, CommonsDelinker, Andrej.westermann, Science5, HEL, Maurice Carbonaro, Cpiral, Speed8ump, Omeganumber, Max woolfson, Squids and Chips, Idioma-bot, Sheliak, VolkovBot, Antoni Barau, Rei-bot, Richwil, Moose-32, Ivanivanovich, Minestrone Soup, SieBot, Timb66, Graham Beards, Peeter.joot, LeadSongDog, Iameukarya, C0nanPayne, Niceguyedc, DnetSvg, Estirabot, Brews ohare, Kmellem, Lazyrussian, Ost316, WikHead, Truthnlove, Jabberwoch, RandomTool2, Addbot, Mathieu Perrin, Gravitophoton, LaaknorBot, CarsracBot, AgadaUrbanit, Tide rolls, Lightbot, Zorrobot, Legobot, Ptbotgourou, TaBOT-zerem, TestEditBot, AnomieBOT, Materialscientist, Citation bot, Northryde, Alexshag, Plumpurple, Br77rino, Pra1998, Salbers, Peter484, A. di M., 🔲🔲, FrescoBot, Paine Ellsworth, UCSD3.14159, Techibun, Jonesey95, Skyerise, LiborX, MondalorBot, RobinK, Merlion444, Jensen.andrew, Dylan1946, EmausBot, Quondum, TonyMath, Brandmeister, Lulzprotuns, Maschen, HCPotter, CocuBot, Deep Thought, Helpful Pixie Bot, Addihockey10 (automated), Bibcode Bot, Mark Arsten, F=q(E+v^B), TrollWizard, Makecat-bot, Paul adrien, Cltschirhart, Edward.hughes, Mahusha, Delbert7, Mywalnut, Ggf4t, KasparBot and Anonymous: 155

- **Principle of least action** *Source:* https://en.wikipedia.org/wiki/Principle_of_least_action?oldid=671863766 *Contributors:* The Anome, XJaM, Deb, Michael Hardy, DIG~enwiki, Kragen, Palfrey, Charles Matthews, Timwi, Dysprosia, Patrick0Moran, Phys, Nagelfar, Snobot, Ancheta Wis, Giftlite, Wolfkeeper, Jrdioko, DefLog~enwiki, Phe, Ganymead, SamuelScarano, L-H, Dbachmann, Jag123, Linas, StradivariusTV, Hfarmer, Btyner, Gerbrant, MZMcBride, Mathbot, Tone, Larsobrien, Socrates123b, Enormousdude, 2over0, Esprit15d, Finell, Bluebot, Silly rabbit, Complexica, Colonies Chris, Serenity-Fr, Stevenmitchell, Jbergquist, Sadi Carnot, Richard L. Peterson, Makyen, Norm mit, Hetar, JRSpriggs, Cydebot, WillowW, Dr.enh, Raoul NK, Headbomb, Lawilkin, First Harmonic, E104421, HEL, Elugelab, Reedy Bot, Salih, Cuzkatzimhut, Inflector, Idlex, Lamro, Tugjob, Pmigdal, Rumping, ArdClose, RODERICKMOLASAR, Drmies, Subversive.sound, Addbot, Fgnievinski, AsphyxiateDrake, Yobot, TaBOT-zerem, Ywaz, Nathanielvirgo, 🔲🔲, Machine Elf 1735, Jgalleg5, Codwiki, Jordgette, RjwilmsiBot, Solarra, Physicist-theorist, Maschen, ClueBot NG, Helpful Pixie Bot, Daniel McMahon, Djrfree, F=q(E+v^B), Hublolly, Mark viking and Anonymous: 36

- **Fiber bundle** *Source:* https://en.wikipedia.org/wiki/Fiber_bundle?oldid=671894163 *Contributors:* AxelBoldt, Zundark, Wshun, TakuyaMurata, Charles Matthews, Dfeuer, Terse, Dysprosia, Jitse Niesen, Phys, Robbot, Rvollmert, Tobias Bergemann, Tosha, Giftlite, Donvinzk, Paisa, BenFrantzDale, Lethe, Fropuff, Mcapdevila, Semorrison, Just Another Dan, D3, Xubia, DefLog~enwiki, Sam nead, Mat cross, ArnoldReinhold, Paul August, Gauge, Tompw, PhilHibbs, Teorth, Varuna, Reaverdrop, SteinbDJ, Oleg Alexandrov, JoeRiel, Linas, Mpatel, BD2412, MarSch, Juan Marquez, YurikBot, Ugha, KSmrq, Archelon, Gaius Cornelius, Tong~enwiki, Crasshopper, Zwobot, Cjfsyntropy, RonnieBrown, SmackBot, KelleyCook, Chris the speller, Jweimar, Silly rabbit, Nbarth, Akriasas, Jon Awbrey, Tesseran, FrozenMan, Almeo, JarahE, Yuide, RJChapman, CBM, 🔲🔲, Sgb235, Hans Lundmark, Sullivan.t.j, Leyo, Aretakis, Stootoon, LokiClock, Ambrose H. Field, Marcosaedro, Geometry guy, Mark zweers, Rybu, YohanN7, SieBot, JL-Bot, Nsk92, Cacadril, Alexey Muranov, Rswarbrick, Addbot, Topology Expert, Eat-windmills, Bongilles, Matěj Grabovský, Yobot, Compsonheir, Ciphers, Churchill17, Point-set topologist, Gsard, Some standardized rigour, Thehelpfulbot, Acannas, Sławomir Biały, Craig Pemberton, DreamingInRed~enwiki, Rausch, Jowa fan, Roman3, Mgvongoeden, IkamusumeFan, AHusain314, Boonoboo, K9re11, KasparBot and Anonymous: 49

38.12.2 Images

- **File:Acap.svg** *Source:* https://upload.wikimedia.org/wikipedia/commons/5/52/Acap.svg *License:* Public domain *Contributors:* Own work *Original artist:* F l a n k e r

- **File:AdditionComplexes.svg** *Source:* https://upload.wikimedia.org/wikipedia/commons/6/67/AdditionComplexes.svg *License:* CC BY-SA 3.0 *Contributors:* Own work *Original artist:* <a href='//commons.wikimedia.org/wiki/User:-Strogoff-' title='User:-Strogoff-'>*Frédéric MICHEL*</a>

38.12.3   Content license

- Creative Commons Attribution-Share Alike 3.0